Let Us Use
White Noise

Let Us Use White Noise

Editors

T Hida
Nagoya University, Japan & Meijo University, Japan

L Streit
University of Bielefeld, Germany & University of Madeira, Portugal

 **World Scientific**

NEW JERSEY · LONDON · SINGAPORE · BEIJING · SHANGHAI · HONG KONG · TAIPEI · CHENNAI · TOKYO

Published by

World Scientific Publishing Co. Pte. Ltd.

5 Toh Tuck Link, Singapore 596224

USA office: 27 Warren Street, Suite 401-402, Hackensack, NJ 07601

UK office: 57 Shelton Street, Covent Garden, London WC2H 9HE

Library of Congress Cataloging-in-Publication Data

Names: Hida, Takeyuki, 1927–   editor. | Streit, Ludwig, 1938–   editor.

Title: Let us use white noise / edited by T. Hida (Nagoya University, Japan &
    Meijo University, Japan), L. Streit (University of Bielefeld, Germany &
    University of Madeira, Portugal).

Description: New Jersey : World Scientific, 2017. |
    Includes bibliographical references and subject index.

Identifiers: LCCN 2017002271 | ISBN 9789813220935 (hardcover : alk. paper)

Subjects: LCSH: White noise theory. | Stochastic analysis. | Stationary processes.

Classification: LCC QA274.29 .A66 2017 | DDC 519.2/2--dc23

LC record available at https://lccn.loc.gov/2017002271

British Library Cataloguing-in-Publication Data

A catalogue record for this book is available from the British Library.

Printed in Singapore

Preface

Why Should We?
Some Personal Comments from One Happy User.

When I embarked into the world of mathematical physics, learning about "axiomatic" quantum field theory from H. Lehmann and W. Zimmermann, and reading Borchers, Symanzik, Haag, Streater, and Wightman, I was impressed with the beauty and clarity of the LSZ and Wightman frameworks, - and quite depressed afterwards. In his book on the general theory of quantized fields, the great Res Jost wrote at the time:[1] "We had very compelling reasons for not mentioning any models except free fields. No interesting models are known ...", bad news for a junior researcher who wondered: "Will there ever be any?" And if so: "How to construct them?" Of course there were attempts; the best of them were visionary - and bad mathematics. Let me single out Feynman's "sum over histories" and the observation of Coester and Haag[2] that quantum field theory dynamics is in fact encoded in the vacuum. We knew even then that the Feynman integral was not an integral, and that the manipulations of Coester and Haag could not be justified mathematically, but it also became quite clear that there was by far not enough of infinite dimensional analysis in the physicists' mathematical tool kit. Things did get better with the physics breakthrough that goes under the name of "constructive quantum field theory", and when, on the mathematical side, 40 years ago the foundations of white noise analysis were laid.

Of course white noise analysis does not claim a monopoly: Mallavin calculus is a close relative, much like in finite dimensional analysis where there are many different Gelfand triples, suited to address particular needs. As Paul André Meyer once said in a heated debate - don't argue about the advantages of one approach or the other, show what you can do with the one that you prefer. My good friend J-A. Yan, together with Z-Y. Huang, presents the two approaches side by side in his beautiful book.[3]

So why should we use white noise analysis? Well one reason is of course that it fills that earlier gap in the tool kit. As Hida would put it, white noise provides us with a useful set of *independent* coordinates, parametrized by "time". And there is a feature which makes white noise analysis extremely user-friendly. Typically the physicist — and not only he — sits there with some heuristic ansatz, like e.g. the famous Feynman "integral", wondering whether and how this might make sense mathematically. In many cases the characterization theorem of white noise analysis provides the user with a sweet and easy answer. Feynman's "integral" can now be understood, the ansatz of Haag and Coester is now making sense via Dirichlet forms, and so on in many fields of application. There is mathematical finance, there have been applications in biology, and engineering,[4] many more than we could collect in the present volume, for some of them see e.g. Bernido and Bernido.[5]

Finally, there is one extra benefit: when we internalize the structures of Gaussian white noise analysis we will be ready to meet another close relative — we will enjoy the important similarities and differences which we encounter in the Poisson case, championed in particular by Y. Kondratiev and his group, let us look forward to a companion volume on the uses of Poisson white noise.

The present volume is essentially a collection of autonomous contributions. Fortunately however, the introductory chapter on white noise analysis was made available to the other authors early on for reference and to facilitate their efforts towards conceptual and notational coherence.

At the end of such a preface one has the right of a note of gratitude to friends and teachers. Some of the latter I have already mentioned. Then there is the "white noise community", too big by now to list it here. But I guess I have made the acquaintance and won the friendship of almost all of the white noise mathematicians you find quoted in the present volume. I also thank all of them for what they taught me. I thank the authors of the different chapters, I thank S. C. Lim of World Scientific for his invitation, help, and great patience, and can now finally, with the contributions in hand, enjoy the encouragement I got for this undertaking. May the readers enjoy those contributions and may they feel encouraged to use white noise.

Ludwig Streit, August 2016

References

1. Res Jost: The General Theory of Quantized Fields. AMS, 1965.
2. Fritz Coester and Rudolf Haag: Representation of States in a Field Theory with Canonical Variables. Phys. Rev. 117, 1137 (1960).
3. Zhi-Yuan Huang, Jia-An Yan; Introduction to Infinite Dimensional Stochastic Analysis. Springer, 2000.
4. I am thinking particular of the work of Roger Ghanem who struggled, unfortunately in vain, to meet the deadlines of this book.
5. Christopher and Victoria Bernido: Methods And Applications Of White Noise Analysis In Interdisciplinary Sciences. World Scientific, 2015.

References

Contents

Contents

Chapter 1

White Noise Analysis: An Introduction

Maria João Oliveira

Universidade Aberta, P 1269-001 Lisbon, Portugal
CMAF-CIO, University of Lisbon, P 1749-016 Lisbon, Portugal
mjoliveira@ciencias.ulisboa.pt

The starting point of White Noise Analysis[11] and[2,14–16,20,21,34,39] is a real separable Hilbert space $\mathcal{H}$ with inner product $(\cdot, \cdot)$ and the corresponding norm $|\cdot|$, and a nuclear triple

$$\mathcal{N} \subset \mathcal{H} \subset \mathcal{N}',$$

where $\mathcal{N}$ is a nuclear space densely and continuously embedded in $\mathcal{H}$. Of course, in a general framework, *a priori* there are several different possible nuclear spaces. However, in concrete applications, the application will suggest the use of particular nuclear triples. For example, in the study of intersection local times of d-dimensional Brownian motions it is natural to consider the space $\mathcal{H} = L^2(\mathbb{R}, \mathbb{R}^d) =: L_d^2(\mathbb{R})$ of all vector valued square integrable functions with respect to the Lebesgue measure on $\mathbb{R}$ and the Schwartz space $\mathcal{N} = S(\mathbb{R}, \mathbb{R}^d) =: S_d(\mathbb{R})$ of vector valued test functions, while in the treatment of Feynman integrals the spaces $L^2(\mathbb{R}) := L^2(\mathbb{R}, \mathbb{R})$, $S(\mathbb{R}) := S(\mathbb{R}, \mathbb{R})$ are the natural ones.

Since nuclear triples are the basis of the whole White Noise Analysis, we start by briefly recalling the main background of the theory of nuclear spaces, due to A. Grothendieck.[7] For simplicity, instead of general nuclear spaces, cf. e.g.,[40,42,45,50] we just consider nuclear Fréchet spaces, which are the only ones needed in this book. For more details and the proofs see e.g.[2,3,9,14].

1. Nuclear Triples

As before, let $\mathcal{H}$ be a real separable Hilbert space. We consider a family of real separable Hilbert spaces $\mathcal{H}_p$, $p \in \mathbb{N}$, with Hilbertian norm $|\cdot|_p$ such that

$$\mathcal{H} \supset \mathcal{H}_1 \supset \ldots \supset \mathcal{H}_p \supset \ldots$$

so that the corresponding system of norms is ordered:

$$|\cdot| \leq |\cdot|_1 \leq \ldots \leq |\cdot|_p \leq \ldots.$$

In addition, we assume that the intersection of the Hilbert spaces $\mathcal{H}_p$, denoted by

$$\mathcal{N} := \bigcap_{p \in \mathbb{N}} \mathcal{H}_p, \tag{1}$$

is dense in each space $\mathcal{H}_p$, $p \in \mathbb{N}$.

Definition 1. The linear space $\mathcal{N}$ is called nuclear whenever for every $p \in \mathbb{N}$ there is a $q > p$ such that the canonical embedding $\mathcal{H}_q \hookrightarrow \mathcal{H}_p$ is a Hilbert-Schmidt operator.

From now on we shall assume that all spaces (1) are nuclear and fix on $\mathcal{N}$ the *projective limit topology*, that is, the coarsest topology on $\mathcal{N}$ with respect to which all canonical embeddings $\mathcal{N} \hookrightarrow \mathcal{H}_p$, $p \in \mathbb{N}$, are continuous. Or, in an equivalent way, a sequence $(\xi_n)_{n \in \mathbb{N}}$ of elements in $\mathcal{N}$ converges to $\xi \in \mathcal{N}$ if and only if $(\xi_n)_{n \in \mathbb{N}}$ converges to ξ in every Hilbert space $\mathcal{H}_p$, $p \in \mathbb{N}$. It turns out that a nuclear space $\mathcal{N}$ endowed with the projective limit topology is a complete metrizable locally convex space, meaning that it is a Fréchet space. In order to mention explicitly this topology fixed on $\mathcal{N}$, we shall use the notation

$$\mathcal{N} = \operatorname*{prlim}_{p \in \mathbb{N}} \mathcal{H}_p$$

and call such a topological space a *projective limit* or a *countable limit of the family* $(\mathcal{H}_p)_{p \in \mathbb{N}}$.

For each $p \in \mathbb{N}$, let now $\mathcal{H}_{-p}$ be the Hilbertian dual space of $\mathcal{H}_p$ with respect to $\mathcal{H}$ with the corresponding Hilbertian norm $|\cdot|_{-p}$. By the general duality theory cf. e.g.,[9] we have

$$\mathcal{N}' = \bigcup_{p \in \mathbb{N}} \mathcal{H}_{-p},$$

where $\mathcal{N}'$ is the dual space of $\mathcal{N}$ with respect to $\mathcal{H}$. Unless stated otherwise, we shall consider $\mathcal{N}'$ endowed with the inductive limit topology, that is, the finest topology on $\mathcal{N}'$ with respect to which all embeddings $\mathcal{H}_{-p} \hookrightarrow \mathcal{N}'$ are continuous. As a topological space, we shall denote it by

$$\mathcal{N}' = \operatorname*{indlim}_{p \in \mathbb{N}} \mathcal{H}_{-p}$$

and call it an *inductive limit* of the family $(\mathcal{H}_{-p})_{p \in \mathbb{N}}$.

In this way, using the Riesz representation theorem to identify $\mathcal{H}$ with its dual space $\mathcal{H}'$, we have defined a so-called *nuclear* or *Gelfand triple*:

$$\mathcal{N} \subset \mathcal{H} \subset \mathcal{N}'.$$

By construction, it turns out that the bilinear dual pairing $\langle \cdot, \cdot \rangle$ between $\mathcal{N}'$ and $\mathcal{N}$ is defined as an extension of the inner product on $\mathcal{H}$:

$$\langle h, \xi \rangle = (h, \xi), \quad h \in \mathcal{H}, \xi \in \mathcal{N}.$$

Example 1. (i) The Schwartz space $S(\mathbb{R})$ of rapidly decreasing C^∞-functions on $\mathbb{R}$ endowed with its usual topology given by the system of seminorms

$$\sup_{u \in \mathbb{R}} \left| u^n \frac{d^m \xi}{du^m}(u) \right|, \quad \xi \in S(\mathbb{R}), m, n \in \mathbb{N}_0 := \mathbb{N} \cup \{0\}$$

is a first example of a nuclear space. Indeed, given the Hamiltonian of the quantum harmonic oscillator, that is, the self-adjoint operator on $L^2(\mathbb{R})$ defined on $S(\mathbb{R})$ by

$$(H\xi)(u) := -\frac{d^2 \xi}{du^2}(u) + (u^2 + 1)\xi(u), \quad u \in \mathbb{R},$$

we can define a system of norms $| \cdot |_p$ by setting

$$|\xi|_p := |H^p \xi|, \quad \xi \in S(\mathbb{R}), p \in \mathbb{N},$$

where the last norm is the one on $L^2(\mathbb{R})$. It turns out (cf. e.g.,[12,43,47]) that this system of norms is equivalent to the initial system of seminorms, and thus both systems lead to equivalent topologies on $S(\mathbb{R})$. In addition, the completion of $S(\mathbb{R})$ with respect to each norm $| \cdot |_p$ yields a family of Hilbert spaces $\mathcal{H}_p$ and

$$S(\mathbb{R}) = \operatorname*{pr\,lim}_{p \in \mathbb{N}} \mathcal{H}_p,$$

see e.g.,[14]. Therefore, for the dual space $S'(\mathbb{R})$ of $S(\mathbb{R})$ (with respect to $L^2(\mathbb{R})$) of Schwartz tempered distributions we have

$$S'(\mathbb{R}) = \operatorname*{ind\,lim}_{p \in \mathbb{N}} \mathcal{H}_{-p}.$$

(ii) The previous example extends to the space $S_d(\mathbb{R})$ of vector valued Schwartz test functions for the operator H defined on $S_d(\mathbb{R})$ by

$$(H\boldsymbol{\xi})(u) := ((H\boldsymbol{\xi})_1(u), \ldots, (H\boldsymbol{\xi})_d(u)), \quad \boldsymbol{\xi} = (\xi_1, \ldots, \xi_d), \xi_i \in S(\mathbb{R}) \quad (2)$$

with

$$(H\boldsymbol{\xi})_i(u) := -\frac{d^2\xi_i}{du^2}(u) + (u^2+1)\xi_i(u) = (H\xi_i)(u), \quad i = 1,\ldots,d, u \in \mathbb{R}.$$

This leads to the following system of increasing Hilbertian norms $|\cdot|_p$, $p \in \mathbb{N}$,

$$|\boldsymbol{\xi}|_p^2 := \sum_{i=1}^{d} |\xi_i|_p^2 = \sum_{i=1}^{d} |H^p\xi_i|^2, \quad \boldsymbol{\xi} = (\xi_1,\ldots,\xi_d), \xi_i \in S(\mathbb{R}), i = 1,\ldots,d, \quad (3)$$

where the last sum in (3) is the square of the $L_d^2(\mathbb{R})$-norm of (2), and to the corresponding Hilbert spaces $\mathcal{H}_p$ defined by completion of $S_d(\mathbb{R})$ with respect to the norms (3). As in (i), we have

$$S_d(\mathbb{R}) = \operatorname*{prlim}_{p\in\mathbb{N}}\mathcal{H}_p, \quad S_d'(\mathbb{R}) = \operatorname*{indlim}_{p\in\mathbb{N}}\mathcal{H}_{-p},$$

being $S_d'(\mathbb{R})$ the space of vector valued Schwartz tempered distributions.

(iii) Example (i) also extends to the Schwartz space $S(\mathbb{R}^d, \mathbb{R})$ of smooth functions on $\mathbb{R}^d$, $d \geq 2$, of rapid decrease (shortly $S(\mathbb{R}^d)$) and to its dual space $S'(\mathbb{R}^d)$ of Schwartz tempered distributions. In this case, the usual topology on $S(\mathbb{R}^d)$ is given by the family of seminorms indexed by multi-indices $(\alpha_1,\ldots,\alpha_d)$, $(\beta_1,\ldots,\beta_d)$ in $\mathbb{N}_0^d$,

$$\sup_{\mathbf{u}=(u_1,\ldots,u_d)\in\mathbb{R}^d} \left| u_1^{\alpha_1}\ldots u_d^{\alpha_d} \left(\partial_1^{\beta_1}\ldots\partial_d^{\beta_d}\xi\right)(\mathbf{u}) \right|, \quad \xi \in S(\mathbb{R}^d),$$

where ∂_i, $i = 1,\ldots,d$, is the partial derivative on $\mathbb{R}^d$ with respect to the i-th coordinate. Given the Hamiltonian of the quantum harmonic oscillator, that is, the self-adjoint operator on $L^2(\mathbb{R}^d, \mathbb{R}) =: L^2(\mathbb{R}^d)$ defined on $S(\mathbb{R}^d)$ by

$$(H\xi)(\mathbf{u}) := -(\Delta\xi)(\mathbf{u}) + (|\mathbf{u}|^2+1)\xi(\mathbf{u}), \quad \mathbf{u} \in \mathbb{R}^d,$$

being Δ the Laplacian on $\mathbb{R}^d$, we define a system of norms $|\cdot|_p$ on $S(\mathbb{R}^d)$ by

$$|\xi|_p := |H^p\xi|, \quad \xi \in S(\mathbb{R}^d), p \in \mathbb{N},$$

where the last norm is the one on $L^2(\mathbb{R}^d)$. As in Example (i), it turns out cf. e.g.,[12,43,47] that such a system is equivalent to the above system of seminorms, leading then to equivalent topologies on $S(\mathbb{R}^d)$. In addition, cf. e.g.,[14] we have

$$S(\mathbb{R}^d) = \operatorname*{prlim}_{p\in\mathbb{N}}\mathcal{H}_p,$$

where each $\mathcal{H}_p$, $p \in \mathbb{N}$, is the Hilbert space obtained by completion of $S(\mathbb{R}^d)$ with respect to the norm $|\cdot|_p$. Thus

$$S'(\mathbb{R}^d) = \operatorname*{indlim}_{p\in\mathbb{N}}\mathcal{H}_{-p}.$$

2. Gaussian Space

Given a nuclear triple $\mathcal{N} \subset \mathcal{H} \subset \mathcal{N}'$, let $\mathcal{C}_\sigma(\mathcal{N}')$ be the σ-algebra on $\mathcal{N}'$ generated by the *cylinder sets*

$$\{x \in \mathcal{N}' : (\langle x, \varphi_1\rangle, \ldots, \langle x, \varphi_n\rangle) \in B, \varphi_1, \ldots, \varphi_n \in \mathcal{N}, B \in \mathcal{B}(\mathbb{R}^n), n \in \mathbb{N}\},$$

where $\mathcal{B}(\mathbb{R}^n)$, $n \in \mathbb{N}$, is the Borel σ-algebra on $\mathbb{R}^n$.

Theorem 1. *(The Minlos Theorem[37]) Let C be a complex-valued function on $\mathcal{N}$ fulfilling the following three properties:*

(i) $C(0) = 1$,
(ii) C *is continuous on* $\mathcal{N}$,
(iii) C *is positive definite, i.e.,*

$$\sum_{i,j=1}^{n} C(\xi_i - \xi_j) z_i \overline{z_j} \geq 0, \quad \xi_1, \ldots, \xi_n \in \mathcal{N}, z_1, \ldots, z_n \in \mathbb{C}, n \in \mathbb{N}.$$

Then, there is a unique probability measure μ_C on $(\mathcal{N}', \mathcal{C}_\sigma(\mathcal{N}'))$ which characteristic function is equal to C, that is, for all $\xi \in \mathcal{N}$

$$\int_{\mathcal{N}'} \exp\left(i\langle x, \xi\rangle\right) \, d\mu_C(x) = C(\xi). \tag{4}$$

For a presentation of the Minlos theorem, including support properties of the probability measure given by this theorem see.[10]

Remark 1. The analogous statement of the Minlos theorem for the nuclear space $\mathcal{N}$ replaced by the finite dimensional space $\mathbb{R}^d$ is the well-known Bochner theorem. Because of this, in the literature Theorem 1 is quite often called the Bochner-Minlos theorem as well.

Consider now the following particular positive definite continuous function defined on $\mathcal{N}$ by

$$C(\xi) = \exp\left(-\frac{1}{2}|\xi|^2\right), \quad \xi \in \mathcal{N}. \tag{5}$$

Then, by the Minlos theorem, we are given a (Gaussian) measure μ on $(\mathcal{N}', \mathcal{C}_\sigma(\mathcal{N}'))$ defined by (4) and (5).

Definition 2. We call the probability space $(\mathcal{N}', \mathcal{C}_\sigma(\mathcal{N}'), \mu)$ the Gaussian space associated with $\mathcal{N}$ and $\mathcal{H}$.

In particular, if $\mathcal{N} = S(\mathbb{R}^d)$ with the topology described in Example 1, the space $(S'(\mathbb{R}^d), \mathcal{C}_\sigma(S'(\mathbb{R}^d)), \mu)$ is called white noise with d-dimensional time parameter. If $d = 1$, we simply call it white noise.

Definition 3. For short we set

$$(L^2) := L^2(\mathcal{N}', \mathcal{C}_\sigma(\mathcal{N}'), \mu)$$

for the complex L^2 space.

In applications of White Noise Analysis, the space (L^2) plays an essential role. In order to distinguish clearly the inner product $(\cdot, \cdot)$ and the Hilbertian norm $|\cdot|$ on the real space $\mathcal{H}$ from those defined on the complex space (L^2), we shall denote the inner product on (L^2) by $((\cdot, \cdot))$ and the corresponding norm by $\|\cdot\|$. Furthermore, we shall assume that $((\cdot, \cdot))$ is linear in the first factor and antilinear in the second one, that is,

$$((F_1, F_2)) := \int_{\mathcal{N}'} F_1(x) \bar{F}_2(x) \, d\mu(x), \quad F_1, F_2 \in (L^2),$$

where $\bar{F}_2$ is the complex conjugate function of F_2.

From the definition of the Gaussian measure μ given by (4) and (5), it follows straightforwardly that for every $\xi \in \mathcal{N}$, $\langle \cdot, \xi \rangle$ is a normally distributed random variable with variance $|\xi|^2$. Thus, for all $\xi \in \mathcal{N}$, $\xi \neq 0$,

$$\|\langle \cdot, \xi \rangle\|^2 = \int_{\mathcal{N}'} \langle x, \xi \rangle^2 \, d\mu(x) = \frac{1}{\sqrt{2\pi|\xi|^2}} \int_{-\infty}^{+\infty} u^2 \exp\left(-\frac{u^2}{2|\xi|^2}\right) du = |\xi|^2.$$

Moreover, again by (4) and (5), the real process X defined on $\mathcal{N}' \times \mathcal{N}$ by $X_\xi(x) = \langle x, \xi \rangle$ is centered Gaussian with covariance

$$\int_{\mathcal{N}'} \langle x, \xi_1 \rangle \langle x, \xi_2 \rangle \, d\mu(x) = \frac{1}{2} \left(\|\langle \cdot, \xi_1 + \xi_2 \rangle\|^2 - \|\langle \cdot, \xi_1 \rangle\|^2 - \|\langle \cdot, \xi_2 \rangle\|^2 \right) = (\xi_1, \xi_2).$$

As we have mentioned above, in this book we shall mostly choose $\mathcal{N}$ to be the Schwartz space $S(\mathbb{R}^d)$, $S_d(\mathbb{R})$, or $S(\mathbb{R})$ of test functions and $\mathcal{H}$ to be $L^2(\mathbb{R}^d)$, $L_d^2(\mathbb{R})$, or $L^2(\mathbb{R})$, respectively. In all these cases, $\mathcal{N}$ is dense in $\mathcal{H}$. This is an assumption fixed on general $\mathcal{N}$ and $\mathcal{H}$ from the very beginning. Therefore, the above considerations allow an extension of the mapping

$$\mathcal{N} \ni \xi \mapsto \langle \cdot, \xi \rangle \in (L^2)$$

to a bounded linear operator

$$\mathcal{H} \ni f \mapsto \langle \cdot, f \rangle \in (L^2)$$

defined at each $f \in \mathcal{H}$ by

$$\langle \cdot, f \rangle := (L^2) - \lim_n \langle \cdot, \xi_n \rangle,$$

where $(\xi_n)_{n \in \mathbb{N}}$ is any sequence in $\mathcal{N}$ converging to f in $\mathcal{H}$. Moreover, $\|\langle \cdot, f \rangle\| = |f|$ for all $f \in \mathcal{H}$.

Proposition 1 (^{14}). *The process X defined on $\mathcal{N}' \times \mathcal{H}$ by $X_f(x) = \langle x, f \rangle$ is centered Gaussian with covariance*

$$(\!(\langle \cdot, f \rangle, \langle \cdot, g \rangle)\!) = \int_{\mathcal{N}'} \langle x, f \rangle \langle x, g \rangle \, d\mu(x) = (f, g), \quad f, g \in \mathcal{H}.$$

In particular, for every $f \in \mathcal{H}$, $\langle \cdot, f \rangle$ is normally distributed with variance $|f|^2$. Thus, from its characteristic function we have

$$\int_{\mathcal{N}'} \exp\left(i \langle x, f \rangle \right) d\mu(x) = \exp\left(-\frac{1}{2} |f|^2 \right), \tag{6}$$

which extends (4) and (5) to $f \in \mathcal{H}$.

More generally, for every $n \in \mathbb{N}_0$ and every $f \in \mathcal{H}$, $f \neq 0$, we can derive from the characteristic function (6),

$$\int_{\mathcal{N}'} \langle x, f \rangle^{2n} \, d\mu(x) = \frac{1}{\sqrt{2\pi |f|^2}} \int_{-\infty}^{+\infty} u^{2n} \exp\left(-\frac{u^2}{2|f|^2} \right) du = \frac{(2n)!}{n! 2^n} |f|^{2n}$$

$$\int_{\mathcal{N}'} \langle x, f \rangle^{2n+1} \, d\mu(x) = 0$$

and, by the polarization identity,

$$\int_{\mathcal{N}'} \langle x, f_1 \rangle \ldots \langle x, f_n \rangle \, d\mu(x)$$

$$= \frac{1}{n!} \sum_{k=1}^{n} (-1)^{n-k} \sum_{i_1 < \ldots < i_k} \int_{\mathcal{N}'} \langle x, f_{i_1} + \ldots + f_{i_k} \rangle^n \, d\mu(x),$$

for every $f_1, \ldots, f_n \in \mathcal{H}$, $n \in \mathbb{N}$.

Example 2. Coming back to the white noise space $(S'(\mathbb{R}), \mathcal{C}_\sigma(S'(\mathbb{R})), \mu)$, the previous proposition allows us to consider the Gaussian centered process X with independent increments,

$$X_{\mathbb{1}_{[0,t)}}(x) = \langle x, \mathbb{1}_{[0,t)} \rangle, \quad t \geq 0,$$

being $\mathbb{1}_B$ the *indicator function* of a Borel set $B \subseteq \mathbb{R}$. This process has covariance

$$(\!(\langle \cdot, \mathbb{1}_{[0,t)} \rangle, \langle \cdot, \mathbb{1}_{[0,s)} \rangle)\!) = (\mathbb{1}_{[0,t)}, \mathbb{1}_{[0,s)}) = s \wedge t,$$

and thus X is a one-dimensional Brownian motion starting at the origin at time zero. We shall denote this Brownian motion by B and $X_{\mathbb{1}_{[0,t)}}$ by B_t or $B(t, \cdot)$. Informally, note that

$$B_t(x) = \langle x, \mathbb{1}_{[0,t)} \rangle = \int_0^t x(s) \, ds,$$

which suggests considering $x(t)$ as the time derivative of the Brownian motion. Of course, this time derivative does not exist in a pointwise sense. However, it exists as a distribution. From now on, we shall denote $x(t)$ by ω_t or $\omega(t)$ and call it **white noise**. As an aside, let us mention that this example is the connecting point for another direction inside infinite dimensional analysis, the well-known Malliavin Calculus.[36] For a clear explanation about the relation between both infinite dimensional analyses see e.g.[16,38].

Within the more general setting of the Gaussian space

$$(S'_d(\mathbb{R}), \mathcal{C}_\sigma(S'_d(\mathbb{R})), \mu), \ d > 1,$$

we can then introduce a d-dimensional Brownian motion $\mathbf{B}$ starting at the origin at time zero by

$$\mathbf{B}_t(\omega_1, \ldots, \omega_d) := \left(\langle \omega_1, \mathbb{1}_{[0,t)} \rangle, \ldots, \langle \omega_d, \mathbb{1}_{[0,t)} \rangle \right), \ (\omega_1, \ldots, \omega_d) \in S'_d(\mathbb{R}), t \geq 0.$$

3. Itô-Segal-Wiener Isomorphism

We verify from equalities above Example 2 that the important monomials of the type

$$\langle \cdot, f \rangle^n = \langle \cdot^{\otimes n}, f^{\otimes n} \rangle,$$

$$\langle \cdot, f_1 \rangle \cdots \langle \cdot, f_n \rangle = \langle \cdot^{\otimes n}, f_1 \otimes \ldots \otimes f_n \rangle = \langle \cdot^{\otimes n}, f_1 \widehat{\otimes} \ldots \widehat{\otimes} f_n \rangle,$$

do not verify an orthogonal relation. This fact is a reason for introducing the orthogonalized so-called Wick-ordered polynomials, a class of functions closely related to the orthogonal Hermite polynomials.

For each $x \in \mathcal{N}'$, let $: x^{\otimes n} :\in \mathcal{N}'^{\widehat{\otimes} n}$, $n \in \mathbb{N}_0$ (Appendix A.1.3) be the so-called *Wick power of order* n, inductively defined by

$$: x^0 := 1,$$

$$: x^1 := x,$$

$$: x^{\otimes n} :=: x^{\otimes(n-1)} : \widehat{\otimes} x - (n-1) : x^{\otimes(n-2)} : \widehat{\otimes} \mathrm{Tr}, \quad n \geq 2,$$

where $\mathrm{Tr} \in \mathcal{N}'^{\widehat{\otimes}2}$ is given by

$$\langle \mathrm{Tr}, \xi_1 \otimes \xi_2 \rangle = \langle \xi_1, \xi_2 \rangle, \quad \xi_1, \xi_2 \in \mathcal{N}.$$

Thus, by induction, for all $x \in \mathcal{N}'$ and all $\xi \in \mathcal{N}$ we have

$$\langle : x^{\otimes n} :, \xi^{\otimes n} \rangle = \sum_{k=0}^{[\frac{n}{2}]} \binom{n}{2k} \frac{(2k)!}{k! 2^k} (-\langle \xi, \xi \rangle)^k \langle x, \xi \rangle^{n-2k}, \tag{7}$$

where the right-hand side is the so-called Hermite polynomial in $\langle x, \xi \rangle$ of order n and parameter $\sqrt{\langle \xi, \xi \rangle} = |\xi|$. We recall that given a constant $\sigma > 0$, the *n-th Hermite polynomial in $u \in \mathbb{R}$ with parameter σ is defined by*

$$: u^n :_{\sigma^2} := (-\sigma)^n \exp\left(\frac{u^2}{2\sigma^2}\right) \frac{d^n}{du^n} \exp\left(-\frac{u^2}{2\sigma^2}\right)$$

$$= \left(\frac{\sigma}{\sqrt{2}}\right)^n H_n\left(\frac{u}{\sqrt{2}\sigma}\right),$$

being H_n the *Hermite polynomial of order* n,

$$H_n(u) := (-1)^n \exp\left(u^2\right) \frac{d^n}{du^n} \exp\left(-u^2\right) = 2^n : u^n :_{\frac{1}{2}}, \quad u \in \mathbb{R}, n \in \mathbb{N}_0.$$

That is,

$$H_n(u) = 2^n \sum_{k=0}^{\left[\frac{n}{2}\right]} \binom{n}{2k} \frac{(2k)!}{k! 2^k} \left(-\frac{1}{2}\right)^k u^{n-2k}, \quad u \in \mathbb{R}, n \in \mathbb{N}_0.$$

Hence, for each $n \in \mathbb{N}_0$ and every $\xi \in \mathcal{N}$, $\xi \neq 0$, we have

$$\langle : x^{\otimes n} :, \xi^{\otimes n} \rangle =: \langle x, \xi \rangle^n :_{\langle \xi, \xi \rangle} = \left(\frac{|\xi|}{\sqrt{2}}\right)^n H_n\left(\frac{\langle x, \xi \rangle}{\sqrt{2}|\xi|}\right),$$

in accordance with (7). Of course, by the polarization identity, (7) also holds for $\xi \in \mathcal{N}_{\mathbb{C}} := \{\xi_1 + i\xi_2 : \xi_1, \xi_2 \in \mathcal{N}\}$ with

$$\langle x, \xi_1 + i\xi_2 \rangle := \langle x, \xi_1 \rangle + i\langle x, \xi_2 \rangle, \quad x \in \mathcal{N}', \xi_1, \xi_2 \in \mathcal{N},$$

meaning that for $f \in \mathcal{H}$ or, more generally, for $f \in \mathcal{H}_{\mathbb{C}}$,

$$\langle f, \xi_1 + i\xi_2 \rangle = (f, \xi_1) + i(f, \xi_2), \quad \xi_1, \xi_2 \in \mathcal{N}.$$

Proposition 2. *For all $\varphi^{(n)} \in \mathcal{N}_{\mathbb{C}}^{\widehat{\otimes} n}$ and all $\phi^{(m)} \in \mathcal{N}_{\mathbb{C}}^{\widehat{\otimes} m}$ the following orthogonal relation holds:*

$$((\langle : x^{\otimes n} :, \varphi^{(n)} \rangle, \langle : x^{\otimes m} :, \phi^{(m)} \rangle)) = \delta_{n,m} n! (\varphi^{(n)}, \phi^{(n)}). \tag{8}$$

Proof. (Sketch) Since elements in $\mathcal{N}^{\widehat{\otimes} n}$, $n \in \mathbb{N}_0$, are linear combinations of elements of the form $\xi^{\otimes n}$ with $\xi \in \mathcal{N}$, it is sufficient to prove (8) for $\varphi^{(n)}$, $\phi^{(m)}$ of the form $\varphi^{(n)} = \xi_1^{\otimes n}$, $\phi^{(m)} = \xi_2^{\otimes m}$, $\xi_1, \xi_2 \in \mathcal{N}$. In this case, the proof follows from the orthogonality relation between Hermite polynomials,

$$\int_{-\infty}^{+\infty} H_n(u) H_m(u) \exp\left(-u^2\right) du = \delta_{n,m} \sqrt{\pi} 2^n n!,$$

cf. e.g.,[2,14,39]. As before, the general case can then be derived from the real case by means of polarization identity. $\square$

Conversely, since each monomial $u \mapsto u^n$, $n \in \mathbb{N}_0$ can be written as linear combination of Hermite polynomials in $u \in \mathbb{R}$ with any given parameter $\sigma > 0$,

$$u^n = \sum_{k=0}^{[\frac{n}{2}]} \binom{n}{2k} \frac{(2k)!}{k!2^k} \sigma^{2k} : u^{n-2k} :_{\sigma^2}, \quad u \in \mathbb{R},$$

then, by the polarization identity, each monomial $\langle \cdot^{\otimes n}, \xi^{\otimes n} \rangle$, $\xi \in \mathcal{N}_\mathbb{C}$, can be written as

$$\langle x^{\otimes n}, \xi^{\otimes n} \rangle = \langle x, \xi \rangle^n = \sum_{k=0}^{[\frac{n}{2}]} \binom{n}{2k} \frac{(2k)!}{k!2^k} \langle \xi, \xi \rangle^k : \langle x, \xi \rangle^{n-2k} :_{\langle \xi, \xi \rangle}$$

$$= \sum_{k=0}^{[\frac{n}{2}]} \binom{n}{2k} \frac{(2k)!}{k!2^k} \langle \xi, \xi \rangle^k \langle : x^{\otimes(n-2k)} :, \xi^{\otimes(n-2k)} \rangle, \quad x \in \mathcal{N}'.$$

Therefore, the linear space of the so-called *smooth Wick-ordered polynomials*,

$$\mathcal{P}(\mathcal{N}') := \left\{ \Phi : \Phi(x) = \sum_{n=0}^{N} \langle : x^{\otimes n} :, \varphi^{(n)} \rangle, \varphi^{(n)} \in \mathcal{N}_\mathbb{C}^{\widehat{\otimes} n}, x \in \mathcal{N}', N \in \mathbb{N}_0 \right\}$$

coincides with the linear space

$$\left\{ \Phi : \Phi(x) = \sum_{n=0}^{N} \langle x^{\otimes n}, \varphi^{(n)} \rangle, \varphi^{(n)} \in \mathcal{N}_\mathbb{C}^{\widehat{\otimes} n}, x \in \mathcal{N}', N \in \mathbb{N}_0 \right\}.$$

In terms of (L^2) properties, it turns out that $\mathcal{P}(\mathcal{N}')$ is dense in (L^2).[48] As a consequence, for any $F \in (L^2)$ there is a sequence $(f^{(n)})_{n \in \mathbb{N}_0}$ in the Fock space $\mathrm{Exp}(\mathcal{H}_\mathbb{C})$ (Appendix A.1.2) such that

$$F = \sum_{n=0}^{\infty} \langle : \cdot^{\otimes n} :, f^{(n)} \rangle \tag{9}$$

and, moreover, by the orthogonality property (Proposition 2),

$$\|F\|^2 = \sum_{n=0}^{\infty} n! \left| f^{(n)} \right|^2 = \left\| (f^{(n)})_{n \in \mathbb{N}_0} \right\|^2_{\mathrm{Exp}(\mathcal{H}_\mathbb{C})}.$$

And vice versa, any series of the form (9) with $(f^{(n)})_{n \in \mathbb{N}_0} \in \mathrm{Exp}(\mathcal{H}_\mathbb{C})$ defines a function in (L^2). In other words, the expansion (9) yields a unitary isomorphism between the space (L^2) and the symmetric Fock space $\mathrm{Exp}(\mathcal{H}_\mathbb{C})$.

Definition 4. We call this unitary isomorphism the Itô-Segal-Wiener isomorphism. The expansion (9) with $(f^{(n)})_{n \in \mathbb{N}_0} \in \mathrm{Exp}(\mathcal{H}_\mathbb{C})$ is called the Itô-Segal-Wiener chaos decomposition or simply the chaos decomposition of $F \in (L^2)$ and $f^{(n)}$, $n \in \mathbb{N}_0$, the kernels of F.

Remark 2. According to Section 2 and the considerations done just before Proposition 2, equality (7) can be extended to $f \in \mathcal{H}_{\mathbb{C}}$:

$$\langle : x^{\otimes n} :, f^{\otimes n} \rangle =: \langle x, f \rangle^n :_{\langle f,f \rangle} = \sum_{k=0}^{[\frac{n}{2}]} \binom{n}{2k} \frac{(2k)!}{k!2^k} (-\langle f, f \rangle)^k \langle x, f \rangle^{n-2k}, \quad x \in \mathcal{N}'.$$

This yields an alternative approach to introduce the Itô-Segal-Wiener isomorphism. Let I be the set of all sequences $\alpha := (\alpha_n)_{n \in \mathbb{N}}$ such that all terms vanish except finitely many ones. For each $\alpha \in I$ set

$$\alpha! = \prod_{n=1}^{\infty} \alpha_n!.$$

Given an orthonormal basis $\{e_n\}_{n \in \mathbb{N}}$ of $\mathcal{H}_{\mathbb{C}}$, it turns out that the family of functions H_α in (L^2) defined by

$$H_\alpha(x) := \prod_{n=1}^{\infty} : \langle x, e_n \rangle^{\alpha_n} :_{\langle e_n, e_n \rangle}, \quad x \in \mathcal{N}', \alpha \in I$$

is an orthogonal basis of (L^2) such that $((H_\alpha, H_\beta)) = \delta_{\alpha,\beta} \alpha!$ for all $\alpha, \beta \in I$. Moreover, the space spanned by this family is dense in (L^2), leading then to the Itô-Segal-Wiener isomorphism cf. e.g.[2,14].

4. *S*- and *T*-transform

Among the (L^2) functions, we now consider in particular the class of functions with chaos decomposition of the form

$$\sum_{n=0}^{\infty} \frac{1}{n!} \langle : .^{\otimes n} :, f^{\otimes n} \rangle, \quad f \in \mathcal{H}_{\mathbb{C}}. \tag{10}$$

Observe that their image under the Itô-Segal-Wiener isomorphism is equal to the exponential vectors $e(f) \in \mathrm{Exp}(\mathcal{H}_{\mathbb{C}})$ (Appendix A.1.2). Therefore,

$$\left\| \sum_{n=0}^{\infty} \frac{1}{n!} \langle : .^{\otimes n} :, f^{\otimes n} \rangle \right\| = \exp \left(\frac{|f|^2}{2} \right)$$

and, more generally,

$$\left(\left(\sum_{n=0}^{\infty} \frac{1}{n!} \langle : .^{\otimes n} :, f^{\otimes n} \rangle, \sum_{n=0}^{\infty} \frac{1}{n!} \langle : .^{\otimes n} :, g^{\otimes n} \rangle \right) \right) = e^{(f,g)}, \quad f, g \in \mathcal{H}_{\mathbb{C}}.$$

We shall call (10) the *Wick or normalized exponential corresponding to f* and denote it by $: e^{\langle \cdot, f \rangle} :$.

As a first step towards an explicit form for the Wick exponentials (10), we observe that from the definition of the Hermite polynomials $: u^n :_{\sigma^2}$, $u \in \mathbb{R}$, we have

$$: u^n :_{\sigma^2} = \frac{d^n}{d\lambda^n} \exp\left(\lambda u - \frac{1}{2}\sigma^2\lambda^2\right)\Bigg|_{\lambda=0}.$$

Thus, for all $\xi \in \mathcal{N}_\mathbb{C}$,

$$: e^{\langle x,\xi\rangle} := \sum_{n=0}^{\infty} \frac{1}{n!} : \langle x,\xi\rangle^n :_{\langle\xi,\xi\rangle} = \exp\left(\langle x,\xi\rangle - \frac{1}{2}\langle\xi,\xi\rangle\right), \quad x \in \mathcal{N}'. \quad (11)$$

Definition 5. The S-transform of $F \in (L^2)$ is the mapping defined on $\mathcal{N}_\mathbb{C}$ by

$$(SF)(\xi) = \int_{\mathcal{N}'} : e^{\langle x,\xi\rangle} : F(x)\,d\mu(x), \quad \xi \in \mathcal{N}_\mathbb{C}.$$

Since the S-transform of a function F in (L^2) is defined by

$$(SF)(\xi) = \int_{\mathcal{N}'} : e^{\langle x,\xi\rangle} : F(x)\,d\mu(x) = \left(\left(: e^{\langle \cdot,\xi\rangle} :, \bar{F}\right)\right), \quad \xi \in \mathcal{N}_\mathbb{C},$$

being $\bar{F}$ the complex conjugate function of F, then in terms of chaos decomposition

$$F = \sum_{n=0}^{\infty} \langle : \cdot^{\otimes n} :, f^{(n)}\rangle, \quad (12)$$

it follows from Proposition 2 that

$$(SF)(\xi) = \sum_{n=0}^{\infty} n! \left(\frac{\xi^{\otimes n}}{n!}, \overline{f^{(n)}}\right) = \sum_{n=0}^{\infty} \left(\xi^{\otimes n}, \overline{f^{(n)}}\right), \quad \xi \in \mathcal{N}_\mathbb{C}. \quad (13)$$

In particular, for $F =: e^{\langle \cdot,f\rangle} :$, $f \in \mathcal{H}_\mathbb{C}$,

$$\left(S : e^{\langle \cdot,f\rangle} :\right)(\xi) = e^{\langle\xi,\bar{f}\rangle}, \quad \xi \in \mathcal{N}_\mathbb{C}.$$

Remark 3. Equality (13) is of considerable practical importance. Whenever we can compute the S-transform of a $F \in (L^2)$, its expansion as in (13) immediately gives us the kernel functions of its Itô-Segal-Wiener decomposition (12).

The next result states another characterization of the S-transform, which is closely related to the Radon-Nikodym derivative of the translation of the Gaussian measure μ,

$$\frac{d\mu(\cdot - \xi)(x)}{d\mu(x)} =: e^{\langle x,\xi\rangle} :, \quad x \in \mathcal{N}', \xi \in \mathcal{N}.$$

See e.g.[2,12,39].

Proposition 3. *Let $F \in (L^2)$. Then, for all $\xi \in \mathcal{N}$,*

$$(SF)(\xi) = \int_{\mathcal{N}'} F(x + \xi)\, d\mu(x).$$

As a mapping, it is clear that the S-transform is linear on (L^2). Moreover, it is injective. In fact, since $\mathcal{N}$ is dense in $\mathcal{H}$, it follows from Proposition 5 in Appendix A.1.2 and the Itô-Segal-Wiener isomorphism that the space spanned by the set of Wick exponentials : $e^{\langle \cdot, \xi \rangle}$:, $\xi \in \mathcal{N}_{\mathbb{C}}$, is dense in (L^2). Therefore, if $SF = 0$, we have

$$0 = (SF)(\xi) = \left(\left(: e^{\langle \cdot, \xi \rangle} :, \bar{F} \right) \right), \quad \forall \xi \in \mathcal{N}_{\mathbb{C}},$$

which implies $F = 0$. As a particular application of the injective property of the S-transform we can now extend the explicit form (11) to $\mathcal{H}_{\mathbb{C}}$. For this purpose, we first observe that

$$\int_{\mathcal{N}'} \exp(\langle x, f \rangle)\, d\mu(x) = \exp\left(\frac{\langle f, f \rangle}{2} \right), \quad f \in \mathcal{H}_{\mathbb{C}}. \tag{14}$$

See e.g.[39] Thus $\exp(\langle \cdot, f \rangle - \frac{1}{2}\langle f, f \rangle) \in (L^2)$ and, yet by (14), for all $\xi \in \mathcal{N}_{\mathbb{C}}$ we have

$$S\left(\exp\left(\langle \cdot, f \rangle - \frac{1}{2}\langle f, f \rangle \right) \right)(\xi)$$

$$= \exp\left(-\frac{1}{2}(\langle \xi, \xi \rangle + \langle f, f \rangle) \right) \int_{\mathcal{N}'} \exp\left(\langle x, \xi + f \rangle \right) d\mu(x)$$

$$= e^{\langle \xi, f \rangle} = \left(S : e^{\langle \cdot, f \rangle} : \right)(\xi).$$

Hence, by the injective property of the S-transform, for all $f \in \mathcal{H}_{\mathbb{C}}$ we find

$$: e^{\langle \cdot, f \rangle} := \exp\left(\langle \cdot, f \rangle - \frac{1}{2}\langle f, f \rangle \right).$$

Besides the aforementioned properties, it turns out that the S-transform is indeed a unitary isomorphism onto the so-called Bargmann-Segal space,[26] of holomorphic functions on $\mathcal{H}_{\mathbb{C}}$.

Another transformation, important as well in applications is the so-called T-transform.

Definition 6. The T-transform of $F \in (L^2)$ is the mapping defined on $\mathcal{N}_{\mathbb{C}}$ by

$$(TF)(\xi) = \int_{\mathcal{N}'} \exp\left(i\langle x, \xi \rangle \right) F(x)\, d\mu(x), \quad \xi \in \mathcal{N}_{\mathbb{C}}.$$

In other words,

$$(TF)(\xi) = (SF)(i\xi)\exp\left(-\frac{1}{2}\langle\xi,\xi\rangle\right), \quad F \in (L^2), \xi \in \mathcal{N}_{\mathbb{C}}.$$

Therefore, the T-transform has properties similar to the S-transform and all above expressions derived for the S-transform lead easily to corresponding expressions in terms of the T-transform.

5. Test and Generalized Functions

In order to define test and generalized functions of white noise, we shall again consider the space of smooth Wick-ordered polynomials (Section 3),

$$\mathcal{P}(\mathcal{N}') = \left\{ \Phi : \Phi(x) = \sum_{n=0}^{N} \langle : x^{\otimes n} :, \varphi^{(n)} \rangle, \varphi^{(n)} \in \mathcal{N}_{\mathbb{C}}^{\widehat{\otimes} n}, x \in \mathcal{N}', N \in \mathbb{N}_0 \right\}.$$

This space can be endowed with several different topologies, but there is a natural one such that $\mathcal{P}(\mathcal{N}')$ becomes a nuclear space.[2] With respect to this topology, a sequence $(\Phi_m)_{m\in\mathbb{N}}$ of Wick-ordered polynomials $\Phi_m = \sum_{n=0}^{N(\Phi_m)} \langle : \cdot^{\otimes n} :, \varphi_m^{(n)} \rangle$ converges to $\Phi = \sum_{n=0}^{N(\Phi)} \langle : \cdot^{\otimes n} :, \varphi^{(n)} \rangle \in \mathcal{P}(\mathcal{N}')$ if and only if the sequence $(N(\Phi_m))_{m\in\mathbb{N}}$ is bounded and the sequence $(\varphi_m^{(n)})_{m\in\mathbb{N}}$ converges to $\varphi^{(n)}$ in $\mathcal{N}_{\mathbb{C}}^{\widehat{\otimes} n}$ for all $n \in \mathbb{N}_0$. Here we have set $\varphi_m^{(n)} = 0$ for all $n > N(\Phi_m)$, $m \in \mathbb{N}$, and $\varphi^{(n)} = 0$ for all $n > N(\Phi)$. It turns out that the space $\mathcal{P}(\mathcal{N}')$ endowed with this topology is densely embedded in (L^2).[2,48]

Then we can consider the dual space $\mathcal{P}'(\mathcal{N}')$ of $\mathcal{P}(\mathcal{N}')$ with respect to (L^2) and in this way we have defined the triple

$$\mathcal{P}(\mathcal{N}') \subset (L^2) \subset \mathcal{P}'(\mathcal{N}').$$

The dual pairing $\langle\!\langle \cdot, \cdot \rangle\!\rangle$ between $\mathcal{P}'(\mathcal{N}')$ and $\mathcal{P}(\mathcal{N}')$ is defined as the bilinear extension of the (sesquilinear) inner product in (L^2), that is,

$$\langle\!\langle F, \Phi \rangle\!\rangle = ((F, \bar{\Phi}))$$
$$= \int_{\mathcal{N}'} F(x)\Phi(x)\,d\mu(x), \quad F \in (L^2), \Phi \in \mathcal{P}(\mathcal{N}').$$

Remark 4. Since the function identically equal to 1 is a particular element of $\mathcal{P}(\mathcal{N}')$, we can use this equality to extend the concept of expectation to generalized functions:

$$\mathbb{E}(\Psi) := \langle\!\langle \Psi, 1 \rangle\!\rangle, \quad \Psi \in \mathcal{P}'(\mathcal{N}').$$

In order to define a space of test functions, observe that each kernel function $\varphi^{(n)}$, $n \in \mathbb{N}$ appearing in the chaos decomposition of a smooth Wick-ordered polynomial

$$\sum_{n=0}^{N} \langle : x^{\otimes n} :, \varphi^{(n)} \rangle$$

is in the space

$$\mathcal{N}_{\mathbb{C}}^{\hat{\otimes}n} = \operatorname*{prlim}_{p \in \mathbb{N}} \mathcal{H}_{p,\mathbb{C}}^{\hat{\otimes}n},$$

where $\mathcal{H}_{p,\mathbb{C}}^{\hat{\otimes}n}$, $p \in \mathbb{N}$, is the n-th symmetric tensor power of the complexified space $\mathcal{H}_{p,\mathbb{C}}$ of the Hilbert space $\mathcal{H}_p$ introduced in Section 1 (see Appendix A.1.3). Thus, $\varphi^{(n)} \in \mathcal{H}_{p,\mathbb{C}}^{\hat{\otimes}n}$ for all $p \in \mathbb{N}$, which allows to define the family of Hilbertian norms $\| \cdot \|_{p,q,\beta}$, $p, q \in \mathbb{N}$, $\beta \in [0,1]$, on $\mathcal{P}(\mathcal{N}')$ by

$$\|\Phi\|_{p,q,\beta}^2 := \sum_{n=0}^{\infty} (n!)^{1+\beta} 2^{nq} |\varphi^{(n)}|_p^2.$$

For each $p, q \in \mathbb{N}$ and each $\beta \in [0,1]$, let $(\mathcal{H}_p)_q^\beta$ be the Hilbert space obtained by completion of the space $\mathcal{P}(\mathcal{N}')$ with respect to the norm $\| \cdot \|_{p,q,\beta}$. That is,

$$(\mathcal{H}_p)_q^\beta = \left\{ \Phi = \sum_{n=0}^{\infty} \langle : \cdot^{\otimes n} :, \varphi^{(n)} \rangle \in (L^2) : \|\Phi\|_{p,q,\beta} < \infty \right\}.$$

Then we can define a family of nuclear spaces continuously and densely embedded in (L^2) (cf. e.g.[17,30]) by

$$(\mathcal{N})^\beta := \operatorname{pr} \lim_{p,q \in \mathbb{N}} (\mathcal{H}_p)_q^\beta.$$

Therefore, by the general duality theory (Appendix A.1.3), the dual space $(\mathcal{N})^{-\beta}$ of $(\mathcal{N})^\beta$ with respect to (L^2) is given by

$$(\mathcal{N})^{-\beta} = \operatorname{ind} \lim_{p,q \in \mathbb{N}} (\mathcal{H}_{-p})_{-q}^{-\beta},$$

where $(\mathcal{H}_{-p})_{-q}^{-\beta}$, $p, q \in \mathbb{N}$, $\beta \in [0,1]$, is the dual space of $(\mathcal{H}_p)_q^\beta$ with respect to (L^2). Moreover, since for all $\beta \in [0,1]$, $\mathcal{P}(\mathcal{N}') \subset (\mathcal{N})^\beta$, the spaces $(\mathcal{N})^{-\beta}$ may be regarded as subspaces of $\mathcal{P}'(\mathcal{N}')$ and hence we obtain the following extended chain of spaces:

$$\mathcal{P}(\mathcal{N}') \subset (\mathcal{N})^1 \subset (\mathcal{N})^\beta \subset (L^2) \subset (\mathcal{N})^{-\beta} \subset (\mathcal{N})^{-1} \subset \mathcal{P}'(\mathcal{N}'), \quad \beta \in [0,1).$$

The space $(\mathcal{N})^{-1}$ is the so-called *Kondratiev space*.[19-21,30] For $\mathcal{N}$ being the Schwartz space $S(\mathbb{R}^d)$, $S_d(\mathbb{R})$, $d > 1$, or $S(\mathbb{R})$ with the Hilbertian norms $|\cdot|_p$ described in Example 1, the corresponding spaces $(\mathcal{N})^0$ and $(\mathcal{N})^{-0}$ are the so-called spaces of *Hida test functions* and *Hida distributions*, respectively.[2,11,14,18,22,23,25,31,32,41] Independently of the particular choice of the Schwartz space, we shall denote the spaces $(\mathcal{N})^0$ and $(\mathcal{N})^{-0}$ by $(\mathcal{S})$ and $(\mathcal{S})'$, respectively, and the Kondratiev space by $(\mathcal{S})^{-1}$.

The chaos decomposition provides a natural decomposition of the elements in $\mathcal{P}'(\mathcal{N}')$. In fact, it turns out[28,30] that for each $\psi^{(n)} \in \mathcal{N}_{\mathbb{C}}'^{\hat{\otimes} n}$ there is a unique element in $\mathcal{P}'(\mathcal{N}')$, denoted informally by $\langle : x^{\otimes n} :, \psi^{(n)} \rangle$, acting on Wick polynomials $\Phi = \sum_{n=0}^{N} \langle : \cdot^{\otimes n} :, \varphi^{(n)} \rangle$ by

$$\langle\!\langle \langle : x^{\otimes n} :, \psi^{(n)} \rangle, \Phi \rangle\!\rangle = n! \langle \psi^{(n)}, \varphi^{(n)} \rangle.$$

Therefore, any element $\Psi \in \mathcal{P}'(\mathcal{N}')$ has a unique decomposition of the form

$$\Psi = \sum_{n=0}^{\infty} \langle : x^{\otimes n} :, \psi^{(n)} \rangle,$$

where the sum converges weakly in $\mathcal{P}'(\mathcal{N}')$, and we have

$$\langle\!\langle \Psi, \Phi \rangle\!\rangle = \sum_{n=0}^{\infty} n! \langle \psi^{(n)}, \varphi^{(n)} \rangle \tag{15}$$

for all $\Phi = \sum_{n=0}^{N} \langle : \cdot^{\otimes n} :, \varphi^{(n)} \rangle \in \mathcal{P}(\mathcal{N}')$. For more details and the proofs see,[28,30].

This internal description of the space $\mathcal{P}'(\mathcal{N}')$ allows, in particular, to describe the distributions in each space $(\mathcal{N})^{-\beta} = \bigcup_{p,q \in \mathbb{N}} (\mathcal{H}_{-p})_{-q}^{-\beta}$, $\beta \in [0,1]$. In fact, it turns out from this construction that each Hilbert space $(\mathcal{H}_{-p})_{-q}^{-\beta}$, $p,q \in \mathbb{N}$, $\beta \in [0,1]$ consists in all $\Psi = \sum_{n=0}^{\infty} \langle : x^{\otimes n} :, \psi^{(n)} \rangle \in \mathcal{P}'(\mathcal{N}')$ such that

$$\|\Psi\|_{-p,-q,-\beta}^2 := \sum_{n=0}^{\infty} (n!)^{1-\beta} 2^{-nq} \left| \psi^{(n)} \right|_{-p}^2 < \infty.$$

In[11] Hilbert spaces of smooth and generalized white noise functionals were introduced. For more details on this see Section 3.A of[14].

6. Characterization Results

Both transformations introduced in Section 4, S- and T-transform, can be extended to $(\mathcal{N})^{-\beta}$, $\beta \in [0,1]$. This yields, in particular, characterization results for the Hida and Kondratiev distributions (Subsections 6.1 and 6.2 below) as well as for any distribution in $(\mathcal{N})^{-\beta}$, $\beta \in (0,1)$ cf. e.g.[34] For $\mathcal{N} = S(\mathbb{R})$ and $\beta \in (0,1)$ such results have been obtained in[26,27].

In order to extend Definitions 5 and 6 to distributions, we first observe that for a $\xi \in \mathcal{N}_{\mathbb{C}}$ we have

$$\| : e^{\langle \cdot, \xi \rangle} : \|_{p,q,\beta}^2 = \sum_{n=0}^{\infty} (n!)^{1+\beta} 2^{nq} \left| \frac{\xi^{\otimes n}}{n!} \right|_p^2 = \sum_{n=0}^{\infty} (n!)^{\beta-1} 2^{nq} |\xi|_p^{2n}, \qquad (16)$$

which is finite if and only if $\beta < 1$ or $2^q |\xi|_p^2 < 1$ if $\beta = 1$. That is, for all $\xi \in \mathcal{N}_{\mathbb{C}}$ we have

$$: e^{\langle \cdot, \xi \rangle} : \in (\mathcal{N})^\beta, \quad \forall \beta \in [0,1).$$

Thus, Definitions 5 and 6 can be directly extended to any $\Psi \in (\mathcal{N})^{-\beta}$, $\beta \in [0,1)$, by

$$(S\Psi)(\xi) := \langle\!\langle \Psi, : e^{\langle \cdot, \xi \rangle} : \rangle\!\rangle, \quad \xi \in \mathcal{N}_{\mathbb{C}}$$

and

$$(T\Psi)(\xi) := (S\Psi)(i\xi) \exp\left(-\frac{1}{2}\langle \xi, \xi \rangle\right), \quad \xi \in \mathcal{N}_{\mathbb{C}}, \qquad (17)$$

respectively. Moreover, if $\Psi = \sum_{n=0}^{\infty} \langle: x^{\otimes n} :, \psi^{(n)} \rangle$, then

$$(S\Psi)(\xi) = \sum_{n=0}^{\infty} \langle \psi^{(n)}, \xi^{\otimes n} \rangle, \qquad (18)$$

which extends equality (13) to distributions.

Although $: e^{\langle \cdot, \xi \rangle} : \notin (\mathcal{N})^1$ for $\xi \in \mathcal{N}_{\mathbb{C}} \setminus \{0\}$, computation (16) shows that $: e^{\langle \cdot, \xi \rangle} : \in (\mathcal{H}_p)_q^1$ whenever $2^q |\xi|_p^2 < 1$. This allows to define the S-transform of Kondratiev distributions as well. Let $\Psi \in (\mathcal{N})^{-1} = \bigcup_{p,q \in \mathbb{N}} (\mathcal{H}_{-p})_{-q}^{-1}$. Then, $\Psi \in (\mathcal{H}_{-p})_{-q}^{-1}$ for some $p, q \in \mathbb{N}$. So we define the S-transform of Ψ by

$$(S\Psi)(\xi) := \langle\!\langle \Psi, : e^{\langle \cdot, \xi \rangle} : \rangle\!\rangle,$$

for all $\xi \in \mathcal{N}_{\mathbb{C}}$ such that $2^q |\xi|_p^2 < 1$. Of course, for each such a function ξ the alternative description (18) still holds. In an analogous way, we define the T-transform of $\Psi \in (\mathcal{H}_{-p})_{-q}^{-1}$ by

$$(T\Psi)(\xi) := (S\Psi)(i\xi) \exp\left(-\frac{1}{2}\langle \xi, \xi \rangle\right),$$

for all $\xi \in \mathcal{N}_{\mathbb{C}}$ such that $2^q |\xi|_p^2 < 1$.

As we have mentioned above, the Hida and Kondratiev distributions can be characterized through their S- and T-transform. Since the definition of the T-transform of those distributions is based on the S-transform, we present these characterization results, as well as their corollaries, just in terms of the S-transform.

Remark 5. For the T-transform, analogous results hold by simply replacing the S by the T-transform.

6.1. *Hida Distributions*

We recall that in this case $\mathcal{N}$ can be any Schwartz space $S(\mathbb{R}^d)$, $S_d(\mathbb{R})$, $d \geq 1$. In order to cover all these possibilities, in this subsection we shall denote all possible Schwartz test function spaces simply by $\mathcal{S}$ and the corresponding dual space by $\mathcal{S}'$.

As a first step towards the characterization of Hida distributions through its S-transform, we need the following definition (Appendix A.1.4).

Definition 7. A function $F : \mathcal{S} \to \mathbb{C}$ is called a U-functional whenever
1. for every $\xi_1, \xi_2 \in \mathcal{S}$ the mapping $\mathbb{R} \ni \lambda \mapsto F(\lambda \xi_1 + \xi_2)$ has an entire extension to $\lambda \in \mathbb{C}$,
2. there are two constants $K_1, K_2 > 0$ such that

$$|F(z\xi)| \leq K_1 \exp\left(K_2 |z|^2 |\xi|^2\right), \quad \forall z \in \mathbb{C}, \xi \in \mathcal{S}$$

for some continuous norm $|\cdot|$ on $\mathcal{S}$.

We are now ready to state the aforementioned characterization result.

Theorem 2. ([18,41]) *The S-transform defines a bijection between the space $(\mathcal{S})'$ and the space of U-functionals.*

As a consequence of Theorem 2 one may derive the next two statements. The first one concerns the convergence of sequences of Hida distributions and the second one the Bochner integration of families of Hida distributions. For more details and the proofs see[18,41] .

Corollary 1. *Let $(\Psi_n)_{n \in \mathbb{N}}$ be a sequence in $(\mathcal{S})'$ such that*
1. *for all $\xi \in \mathcal{S}$, $((S\Psi_n)(\xi))_{n \in \mathbb{N}}$ is a Cauchy sequence in $\mathbb{C}$,*
2. *there are two constants $K_1, K_2 > 0$ such that for some continuous norm $|\cdot|$ on $\mathcal{S}$ we have*

$$|(S\Psi_n)(z\xi)| \leq K_1 \exp\left(K_2|z|^2|\xi|^2\right), \quad \forall z \in \mathbb{C}, \xi \in \mathcal{S}, n \in \mathbb{N}.$$

Then $(\Psi_n)_{n \in \mathbb{N}}$ converges strongly in $(\mathcal{S})'$ to a unique Hida distribution.

Corollary 2. *Let $(\Lambda, \mathcal{B}, \nu)$ be a measure space and $\lambda \mapsto \Psi_\lambda$ be a mapping from Λ to $(\mathcal{S})'$. We assume that the S-transform of Ψ_λ fulfills the following two conditions:*
1. *the mapping $\lambda \mapsto (S\Psi_\lambda)(\xi)$ is measurable for every $\xi \in \mathcal{S}$,*
2. *all $S\Psi_\lambda$ obey the bound*

$$|(S\Psi_\lambda)(z\xi)| \leq C_1(\lambda) \exp\left(C_2(\lambda)|z|^2|\xi|^2\right), \quad z \in \mathbb{C}, \xi \in \mathcal{S},$$

for some continuous norm $|\cdot|$ on $\mathcal{S}$ and for some $C_1 \in L^1(\Lambda, \mathcal{B}, \nu)$, $C_2 \in L^\infty(\Lambda, \mathcal{B}, \nu)$.

Then the Bochner integral

$$\int_\Lambda \Psi_\lambda \, d\nu(\lambda)$$

exists in $(\mathcal{S})'$ and

$$S\left(\int_\Lambda \Psi_\lambda \, d\nu(\lambda)\right) = \int_\Lambda (S\Psi_\lambda) \, d\nu(\lambda).$$

Example 3. Given the one-dimensional Brownian motion $B_t = \langle \cdot, \mathbb{1}_{[0,t)} \rangle$, $t \geq 0$ defined in Example 2 and the Dirac delta function $\delta_a \in S'(\mathbb{R})$ with mass at $a \in \mathbb{R}$, consider the informal composition

$$\delta_a(B_t) = \delta_0(B_t - a).$$

Based on an approximation procedure by Hida distributions (Corollary 1) and Corollary 2, a rigorous meaning of the Donsker's delta function $\delta_0(B_t - a)$ as the Bochner integral in $(\mathcal{S})'$

$$\delta_0(B_t - a) := \frac{1}{2\pi} \int_\mathbb{R} e^{iu(B_t - a)} \, du$$

has been given in.[35] Its S-transform is given (see e.g.[13,14,33,35]) by

$$(S\delta_0(B_t - a))(\xi) = \frac{1}{\sqrt{2\pi t}} \exp\left(-\frac{1}{2t}\left(a - \int_0^t \xi(u)\,du\right)^2\right), \quad \xi \in S(\mathbb{R}),$$

which is obviously a U-functional.

Among Hida distributions the positive ones have particular characteristics. We recall that a $\Psi \in (\mathcal{S})'$ is said to be *positive* whenever $\langle\langle \Psi, \Phi \rangle\rangle \geq 0$ for all $\Phi \in (\mathcal{S})$ μ-a.e. positive (being μ the Gaussian measure on $(\mathcal{S}', \mathcal{C}_\sigma(\mathcal{S}'))$). As shown independently in[24] and in[51], we have the following result.

Theorem 3. *If $\Psi \in (\mathcal{S})'$ belongs to the cone $(\mathcal{S})'_+$ of positive Hida distributions, then there is a unique (positive) finite measure ν_Ψ on $(\mathcal{S}', \mathcal{C}_\sigma(\mathcal{S}'))$ such that*

$$\langle\langle \Psi, \Phi \rangle\rangle = \int_{\mathcal{S}'} \Phi(\omega) \, d\nu_\Psi(\omega)$$

for all $\Phi \in (\mathcal{S})$.

Example 4. Coming back to Example 3, in[49] the authors have proved that $\delta_0(B_t - a) \in (\mathcal{S})'_+$. On the other hand, Y. Yokoi has shown in[51] that $\Psi \in (\mathcal{S})'$ is positive if and only if $T\Psi$ is positive definite. By Example 3 and (17), it is clear that the latter condition holds for $\Psi = \delta_0(B_t - a)$. Thus, according to Theorem 3, the Donsker's delta function $\delta_0(B_t - a)$ defines a finite measure on $(S'(\mathbb{R}), \mathcal{C}_\sigma(S'(\mathbb{R})))$.

6.2. Kondratiev Distributions

Theorem 4. (^{19}) *Let $0 \in U \subset \mathcal{N}_{\mathbb{C}}$ be an open set and $F : U \to \mathbb{C}$ be a holomorphic function on U. Then there is a unique $\Psi \in (\mathcal{N})^{-1}$ such that $S\Psi = F$. Conversely, given a $\Psi \in (\mathcal{N})^{-1}$ the function $S\Psi$ is holomorphic on some open set in $\mathcal{N}_{\mathbb{C}}$ containing 0.*

The correspondence between F and Ψ is a bijection if one identifies holomorphic functions which coincide on some open neighborhood of 0 in $\mathcal{N}_{\mathbb{C}}$.

We shall do so. As a consequence, we can derive the next two statements. The first one concerns the convergence of sequences of Kondratiev distributions and the second one the Bochner integration of families of the same type of generalized functions.

Corollary 3 (^{19}). *Let $(\Psi_n)_{n\in\mathbb{N}}$ be a sequence in $(\mathcal{N})^{-1}$ such that there are $p, q \in \mathbb{N}$ so that*
1. *all $S\Psi_n$ are holomorphic on the open neighborhood $U_{p,q} := \{\xi \in \mathcal{N}_{\mathbb{C}} : 2^q|\xi|_p^2 < 1\}$ of $0 \in \mathcal{N}_{\mathbb{C}}$,*
2. *there is a $C > 0$ such that $|S\Psi_n(\xi)| \leq C$ for all $\xi \in U_{p,q}$ and all $n \in \mathbb{N}$,*
3. *$(S\Psi_n(\xi))_{n\in\mathbb{N}}$ is a Cauchy sequence in $\mathbb{C}$ for all $\xi \in U_{p,q}$.*
   Then $(\Psi_n)_{n\in\mathbb{N}}$ converges strongly in $(\mathcal{N})^{-1}$.

Corollary 4 (^{19}). *Let $(\Lambda, \mathcal{B}, \nu)$ be a measure space and $\lambda \mapsto \Psi_\lambda$ be a mapping from Λ to $(\mathcal{N})^{-1}$. We assume that there is a $U_{p,q} = \{\xi \in \mathcal{N}_{\mathbb{C}} : 2^q|\xi|_p^2 < 1\}$, $p, q \in \mathbb{N}$, such that*
1. *$S\Psi_\lambda$ is holomorphic on $U_{p,q}$ for every $\lambda \in \Lambda$,*
2. *the mapping $\lambda \mapsto (S\Psi_\lambda)(\xi)$ is measurable for every $\xi \in U_{p,q}$,*
3. *there is a $C \in L^1(\Lambda, \mathcal{B}, \nu)$ such that*

$$|(S\Psi_\lambda)(\xi)| \leq C(\lambda), \quad \forall \xi \in U_{p,q}, \, \nu - a.a. \, \lambda \in \Lambda.$$

Then there are $p', q' \in \mathbb{N}$, which only depend on p, q, such that

$$\int_\Lambda \Psi_\lambda \, d\nu(\lambda)$$

exists as a Bochner integral in $(\mathcal{H}_{-p'})_{-q'}^{-1}$. In particular, $S\left(\int_\Lambda \Phi_\lambda \, d\nu(\lambda)\right)$ is holomorphic on $U_{p',q'} = \{\xi \in \mathcal{N}_{\mathbb{C}} : 2^{q'}|\xi|_{p'}^2 < 1\}$ and

$$\left\langle\!\left\langle \int_\Lambda \Psi_\lambda \, d\nu(\lambda), \Phi \right\rangle\!\right\rangle = \int_\Lambda \langle\!\langle \Psi_\lambda, \Phi \rangle\!\rangle \, d\nu(\lambda), \quad \forall \Phi \in (\mathcal{N})^1.$$

Positive Kondratiev distributions are defined similarly to the Hida case. For such a particular class of generalized functions in $(\mathcal{N})^{-1}$ the following characterization result holds.

Theorem 5.[29] *If* $\Psi \in (\mathcal{N})^{-1}$ *belongs to the cone* $(\mathcal{N})^{-1}_+$ *of positive Kondratiev distributions, then there is a unique (positive) finite measure* $\nu = \nu_\Psi$ *on* $(\mathcal{N}', \mathcal{C}_\sigma(\mathcal{N}'))$ *such that for all* $\Phi \in (\mathcal{N})^1$

$$\langle\!\langle \Psi, \Phi \rangle\!\rangle = \int_{\mathcal{N}'} \Phi(x)\, d\nu(x) \tag{19}$$

and, moreover, there are $p \in \mathbb{N}$, $K, C > 0$ *so that*

$$\left| \int_{\mathcal{N}'} \langle x, \xi \rangle^n\, d\nu(x) \right| = KC^n n! |\xi|_p^n \tag{20}$$

for all $\xi \in \mathcal{N}$ *and all* $n \in \mathbb{N}_0$. *Vice versa, any (positive) measure* ν *that obeys* (20) *defines a positive distribution* $\Psi \in (\mathcal{N})^{-1}_+$ *by* (19).

7. Wick Product and $*$-Convolution

According to the Characterization Theorem 4, the S-transform on $(\mathcal{N})^{-1}$ is a bijection if we consider *germs* of holomorphic functions at zero, that is, if we identify holomorphic functions which coincide on some open neighborhood of 0 in $\mathcal{N}_\mathbb{C}$. Thus, we define $\mathrm{Hol}_0(\mathcal{N}_\mathbb{C})$ as the algebra of germs of holomorphic functions at zero. Algebraically, it is clear that $\mathrm{Hol}_0(\mathcal{N}_\mathbb{C})$ endowed with the pointwise multiplication of functions is an algebra. Therefore, by Theorem 4, for each pair $\Psi_1, \Psi_2 \in (\mathcal{N})^{-1}$ we can define the so-called *Wick product* $\Psi_1 \diamond \Psi_2 \in (\mathcal{N})^{-1}$ *of* Ψ_1 *and* Ψ_2 by

$$\Psi_1 \diamond \Psi_2 := S^{-1}\left((S\Psi_1)(S\Psi_2) \right).$$

It turns out that the space $(\mathcal{N})^{-1}$ endowed with the Wick product is a commutative algebra with unit element $:e^{\langle x, 0 \rangle}: \equiv 1$[14,15,19].

The Wick product can be described in terms of chaos decomposition as well. If $\Psi_i = \sum_{n=0}^{\infty} \langle : x^{\otimes n} :, \psi_i^{(n)} \rangle \in (\mathcal{N})^{-1}$, $i = 1, 2$, then

$$\Psi_1 \diamond \Psi_2 = \sum_{n=0}^{\infty} \left\langle : x^{\otimes n} :, \sum_{k=0}^{n} \psi_1^{(k)} \widehat{\otimes} \psi_2^{(n-k)} \right\rangle.$$

Clearly, we can also define

$$\Psi^{\diamond n} := S^{-1}((S\Psi)^n) = \Psi \diamond \dots \diamond \Psi \ (n \text{ times}), \quad \Psi \in (\mathcal{N})^{-1}$$

and thus finite linear combinations of the form $\sum_{n=0}^{N} a_n \Psi^{\diamond n}$ ($\Psi^{\diamond 0} := 1$). Moreover, given a function $g : \mathbb{C} \to \mathbb{C}$ analytic on some neighborhood of $(S\Psi)(0) = \mathbb{E}(\Psi) \in \mathbb{C}$ (Remark 4), we can define $g^\diamond(\Psi) \in (\mathcal{N})^{-1}$ by $g^\diamond(\Psi) := S^{-1}(g(S\Psi))$[19,30]. In particular, we can define $\exp^\diamond \Psi \in (\mathcal{N})^{-1}$:

$$\exp^\diamond \Psi := S^{-1}(\exp(S\Psi)) = \sum_{n=0}^{\infty} \frac{1}{n!} \Psi^{\diamond n}.$$

In fact, if the power series representation of an analytic function g has the form

$$g(z) = \sum_{n=0}^{\infty} a_n(z - \mathbb{E}(\Psi))^n, \quad z \in \mathbb{C},$$

then the Wick series

$$\sum_{n=0}^{\infty} a_n(\Psi - \mathbb{E}(\Psi))^{\diamond n}$$

converges in $(\mathcal{N})^{-1}$ and, moreover,

$$g^\diamond(\Psi) = \sum_{n=0}^{\infty} a_n(\Psi - \mathbb{E}(\Psi))^{\diamond n}.$$

For more details concerning the Wick product see e.g.[14,15,19,30].

By Theorem 4 and Remark 5, the T-transform also yields the definition of an algebraic structure on $(\mathcal{N})^{-1}$. For each pair $\Psi_1, \Psi_2 \in (\mathcal{N})^{-1}$ we define the so-called *-*convolution* $\Psi_1 * \Psi_2 \in (\mathcal{N})^{-1}$ *of* Ψ_1 *and* Ψ_2 by

$$\Psi_1 * \Psi_2 := T^{-1}((T\Psi_1)(T\Psi_2)).$$

Due to the close relation between the T- and the S-transform on $(\mathcal{N})^{-1}$, all the above results quoted for the Wick product straightforwardly hold for the *-convolution. Moreover, through the so-called *Fourier transform on* $(\mathcal{N})^{-1}$,

$$\mathcal{F}\Psi := T^{-1}(S\Psi), \quad \Psi \in (\mathcal{N})^{-1},$$

both algebraic convolutions are related[19] by

$$\mathcal{F}(\Psi_1 \diamond \Psi_2) = (\mathcal{F}\Psi_1) * (\mathcal{F}\Psi_2), \quad \Psi_1, \Psi_2 \in (\mathcal{N})^{-1}.$$

For more details concerning this Fourier transform $\mathcal{F}$ see e.g.[14,33] and the references therein.

8. Annihilation, Creation and Second Quantization Operators

The Itô-Segal-Wiener isomorphism between the space (L^2) and the symmetric Fock space provides natural operators on (L^2) by carrying over standard Fock spaces operators, namely, the annihilation, creation and second quantizations operators cf. e.g.[2,44].

8.1. *Annihilation Operators*

Let $h \in \mathcal{H}$ be given. We recall that the so-called *annihilation operator* $a(h)$ is the operator acting on $f^{(n)} \in \mathrm{Exp}_n(\mathcal{H}_{\mathbb{C}})$, $n \in \mathbb{N}$, of the form (Appendix A.1.2)

$$f^{(n)} = f_1 \widehat{\otimes} \ldots \widehat{\otimes} f_n \in \mathcal{H}_{\mathbb{C}}^{\widehat{\otimes}n}, \quad f_i \in \mathcal{H}_{\mathbb{C}}, i = 1, \ldots, n \tag{21}$$

by

$$(a(h))f^{(n)} := \sum_{j=1}^{n} (h, f_j) f_1 \widehat{\otimes} \ldots \widehat{\otimes} f_{j-1} \widehat{\otimes} f_{j+1} \widehat{\otimes} \ldots \widehat{\otimes} f_n \in \mathcal{H}_{\mathbb{C}}^{\widehat{\otimes}(n-1)}.$$

This definition can be linearly extended to the dense space in $\mathrm{Exp}_n(\mathcal{H}_{\mathbb{C}})$ spanned by elements of the form (21). Moreover, for such elements the following equality of norms holds (cf. e.g.[44]),

$$|(a(h))f^{(n)}|_{\mathrm{Exp}_{n-1}(\mathcal{H}_{\mathbb{C}})} \leq \sqrt{n} |h| |f^{(n)}|_{\mathrm{Exp}_n(\mathcal{H}_{\mathbb{C}})}, \tag{22}$$

which allows to extend $a(h)$ to a bounded operator $a(h) : \mathrm{Exp}_n(\mathcal{H}_{\mathbb{C}}) \to \mathrm{Exp}_{n-1}(\mathcal{H}_{\mathbb{C}})$.

Therefore, in terms of the space (L^2), the Itô-Segal-Wiener isomorphism yields an operator, also denoted by $a(h)$, such that for all $x \in \mathcal{N}'$ and all $f_1, \ldots, f_n \in \mathcal{H}_{\mathbb{C}}$, $n \in \mathbb{N}$,

$$(a(h)) \langle : x^{\otimes n} :, f_1 \widehat{\otimes} \ldots \widehat{\otimes} f_n \rangle$$
$$= \sum_{j=1}^{n} (h, f_j) \langle : x^{\otimes(n-1)} :, f_1 \widehat{\otimes} \ldots \widehat{\otimes} f_{j-1} \widehat{\otimes} f_{j+1} \widehat{\otimes} \ldots \widehat{\otimes} f_n \rangle.$$

Due to (22), observe that for $x \in \mathcal{N}'$, $f_1, \ldots, f_n \in \mathcal{H}$, $n \in \mathbb{N}$, fixed, the linear functional on $\mathcal{H}$

$$\mathcal{H} \ni h \mapsto (a(h))P(x), \quad P(x) := \langle : x^{\otimes n} :, f_1 \widehat{\otimes} \ldots \widehat{\otimes} f_n \rangle$$

is bounded. Therefore, by the Riesz representation theorem, it is given by an inner product

$$(h, \nabla P(x)), \quad \forall h \in \mathcal{H}$$

for some $\nabla P(x) \in \mathcal{H}$. In particular, for $\mathcal{N} = S(\mathbb{R})$ and $\mathcal{H} = L^2(\mathbb{R})$, we have

$$(a(h))P(\omega) = (h, \nabla P(\omega)) = \int_{-\infty}^{+\infty} h(t) \partial_t P(\omega) \, dt, \quad h \in L^2(\mathbb{R}),$$

where ∂_t, $t \in \mathbb{R}$ is the Hida derivative introduced in[11].

The reason for this name lies on the fact that the Hida derivative is indeed a (Gâteaux) derivative. For $\omega, \omega_0 \in S'(\mathbb{R})$ fixed, let F be a real or complex-valued function pointwisely defined on $S'(\mathbb{R})$. We say that F is *Gâteaux differentiable at ω in direction ω_0* if the function $\mathbb{R} \ni \lambda \mapsto F(\omega + \lambda \omega_0)$ is differentiable at $\lambda = 0$. In this case, we shall use the notation

$$D_{\omega_0} F(\omega) := \frac{d}{d\lambda} F(\omega + \lambda \omega_0)\Big|_{\lambda=0}. \tag{23}$$

In particular, for P defined as before we easily find

$$D_{\omega_0} P(\omega) = \sum_{j=1}^{n} \langle \omega_0, f_j \rangle \langle : \omega^{\otimes(n-1)} :, f_1 \widehat{\otimes} \ldots \widehat{\otimes} f_{j-1} \widehat{\otimes} f_{j+1} \widehat{\otimes} \ldots \widehat{\otimes} f_n \rangle.$$

This shows that $a(h)P(\omega)$, $h \in L^2(\mathbb{R})$ is, in particular, a Gâteaux derivative at the point $\omega \in S'(\mathbb{R})$ in direction $h \in L^2(\mathbb{R})$. Therefore, we can regard definition (23) as an extension of $a(h)$, $h \in L^2(\mathbb{R})$, to tempered distributions directions.

In particular, for the Dirac delta function $\delta_t \in S'(\mathbb{R})$, $t \in \mathbb{R}$, it turns out cf. e.g.[14] that

$$D_{\delta_t} P = \partial_t P,$$

where $\partial_t P$ is the Hida derivative of P. In fact,

$$\partial.\Phi(\omega) = \nabla\Phi(\omega)$$

defines a Fréchet derivative on $(\mathcal{S})$, see e.g.,[14] and for suitable positive $\Psi \in (\mathcal{S})'$

$$\varepsilon(\Phi) := \langle\langle \Psi, |\nabla\Phi|^2 \rangle\rangle$$

will give rise to (pre-)Dirichlet forms, see[14].

8.2. Creation Operators

In order to recall the definition of creation operators on the Fock space, we come back to the bounded operator $a(h) : \mathrm{Exp}_n(\mathcal{H}_{\mathbb{C}}) \to \mathrm{Exp}_{n-1}(\mathcal{H}_{\mathbb{C}})$, $h \in \mathcal{H}$, defined at the beginning of Subsection 8.1. Clearly, we can extend componentwise $a(h)$ to the dense subspace of $\mathrm{Exp}(\mathcal{H}_{\mathbb{C}})$ consisting of all sequences $(f^{(n)})_{n \in \mathbb{N}_0}$ such that all terms vanish except finitely many ones. Therefore, the adjoint $a^*(h)$ of $a(h)$ is a well-defined operator on $\mathrm{Exp}(\mathcal{H}_{\mathbb{C}})$. A straightforward computation shows that its action on $f^{(n)} \in \mathrm{Exp}_n(\mathcal{H}_{\mathbb{C}})$, $n \in \mathbb{N}$, is given by

$$(a^*(h))f^{(n)} = h\widehat{\otimes}f^{(n)} \in \mathrm{Exp}_{n+1}(\mathcal{H}_{\mathbb{C}})$$

and, moreover,

$$|(a^*(h))f^{(n)}|_{\mathrm{Exp}_{n+1}(\mathcal{H}_{\mathbb{C}})} \leq \sqrt{n+1}|h||f^{(n)}|_{\mathrm{Exp}_n(\mathcal{H}_{\mathbb{C}})},$$

see e.g.[2,44]. Since $a(h)$ and $a^*(h)$ are densely defined, they are closable, and thus both operators can be extended to their closures. We shall denote both extended operators also by $a(h)$ and $a^*(h)$, respectively. The following equalities hold

$$[a(f), a(h)] = [a^*(f), a^*(h)] = 0, \quad [a(f), a^*(h)] \subseteq (f, h), \qquad (24)$$

which are well-known as the canonical commutation relations. Here $[\cdot, \cdot]$ is the usual *commutator* between two operators, $[A, B] := AB - BA$.

Concerning $a^*(h)$, that is, the so-called *creation operator*, its image under the Itô-Segal-Wiener isomorphism leads as before to an operator on (L^2), also denoted by $a^*(h)$,

$$(a^*(h))\langle: x^{\otimes n} :, f^{(n)}\rangle = \langle: x^{\otimes(n+1)} :, h\widehat{\otimes}f^{(n)}\rangle, \quad f^{(n)} \in \mathcal{H}_{\mathbb{C}}^{\widehat{\otimes}n}, n \in \mathbb{N}.$$

Of course, by construction, $a^*(h)$ is the adjoint operator of $a(h)$ on (L^2) and relations corresponding to (24) hold.

In order to proceed towards distributions, let us consider $\mathcal{H} = L^2(\mathbb{R})$ and $\mathcal{N} = S(\mathbb{R})$. Concerning the corresponding space $(\mathcal{S})$ of Hida test functions, it turns out cf. e.g.[14] that each $\sum_{n=0}^{\infty}\langle: \cdot^{\otimes n} :, \varphi^{(n)}\rangle \in (\mathcal{S})$ has a continuous version. That is, each kernel $\varphi^{(n)}$, $n \in \mathbb{N}$, has a continuous version in $(S_{\mathbb{C}}(\mathbb{R}))^{\widehat{\otimes}n}$. For technical reasons, in what follows we shall always consider the continuous version of a Hida test function. Of course, it follows from the previous subsection that $D_{\omega_0}\langle: \omega^{\otimes n} :, \varphi^{(n)}\rangle$ exists for all $\omega_0 \in S'(\mathbb{R})$ and

$$D_{\omega_0}\langle: \omega^{\otimes n} :, \varphi^{(n)}\rangle = n\langle\omega_0\widehat{\otimes} : \omega^{\otimes(n-1)} :, \varphi^{(n)}\rangle = n\left\langle: \omega^{\otimes(n-1)} :, \langle\omega_0, \varphi^{(n)}\rangle\right\rangle, \qquad (25)$$

where $\langle\omega_0, \varphi^{(n)}\rangle$ means that ω_0 is evaluated on $\varphi^{(n)}$ in the first argument. Moreover, for a fixed $\omega_0 \in S'(\mathbb{R})$, D_{ω_0} is a continuous linear operator from $(\mathcal{S})$ into itself. This is a consequence of the fact that given a $\omega_0 \in \mathcal{H}_{-q}$, being $\mathcal{H}_{-q}$, $p \in \mathbb{N}$ the Hilbert spaces introduced in Example 1 (i) with the corresponding norm $|\cdot|_{-q}$, for every $\Phi \in (\mathcal{S})$ and every $p, r \in \mathbb{N}$ we find cf.,[14,39]

$$\|D_{\omega_0}\Phi\|_{p,r,0} \leq 2^{q-p}|\omega_0|_{-q}\|\Phi\|_{\max\{p,q\},r,0},$$

where the first and the last norms are the ones on $(\mathcal{H}_p)_r^0$, $p, r \in \mathbb{N}$, $\mathrm{prlim}_{p,r\in\mathbb{N}}(\mathcal{H}_p)_r^0 = (\mathcal{S})$ (Section 5). Therefore, we can consider the adjoint operator $D_{\omega_0}^*$ of the Gâteaux derivative D_{ω_0}, $\omega_0 \in S'(\mathbb{R})$, which (strongly) continuously maps the space of Hida distributions $(\mathcal{S})'$ into itself. Due to (15) and (25), the action of $D^*_{\omega_0}$ on a $\Psi = \sum_{n=0}^{\infty} \langle : \omega^{\otimes n} :, \psi^{(n)} \rangle \in (\mathcal{S})'$, thus with symmetric tempered distribution kernels $\psi^{(n)}$, is given for all $\Phi = \sum_{n=0}^{\infty} \langle : \cdot^{\otimes n} :, \varphi^{(n)} \rangle \in (\mathcal{S})$ by

$$\langle\langle D^*_{\omega_0} \Psi, \Phi \rangle\rangle = \langle\langle \Psi, D_{\omega_0} \Phi \rangle\rangle = \sum_{n=0}^{\infty} n!(n+1) \left\langle \psi^{(n)}, \langle \omega_0, \varphi^{(n+1)} \rangle \right\rangle$$

$$= \sum_{n=0}^{\infty} (n+1)! \langle \omega_0 \widehat{\otimes} \psi^{(n)}, \varphi^{(n+1)} \rangle.$$

That is,

$$D^*_{\omega_0} \Psi = \sum_{n=1}^{\infty} \langle : \omega^{\otimes n} :, \omega_0 \widehat{\otimes} \psi^{(n-1)} \rangle. \tag{26}$$

It extends to tempered distributions the operator $a^*(h)$, $h \in L^2(\mathbb{R})$, above defined on (L^2). As we can expect, relations similar to (24) can be stated for D_{ω_0} and $D^*_{\omega_0}$, $\omega_0 \in S'(\mathbb{R})$, as well. From those, we give particular attention to the case $\omega_0 = \delta_t$, which leads to the *canonical commutation relations* used in quantum field theory. Informally, for $s, t \in \mathbb{R}$,

$$[\partial_s, \partial_t] = [\partial^*_s, \partial^*_t] = 0, \quad [\partial_s, \partial^*_t] = \delta(s - t),$$

with $\partial^*_t = D^*_{\delta_t}$, $t \in \mathbb{R}$. Furthermore, given the white noise $\omega(t)$ (Example 2), for all $\Phi \in (\mathcal{S})$ we find

$$\omega(t)\Phi = (\partial_t + \partial^*_t)\Phi,$$

where the multiplication appearing in the left-hand side is a Hida distribution. Indeed, since $(\mathcal{S})$ is closed under the pointwise multiplication and the multiplication of Hida test functions is a continuous bilinear mapping on $(\mathcal{S})$ cf. e.g.,[39] the multiplication $\omega(t)\Phi \in (\mathcal{S})'$ is well-defined by

$$\langle\langle \omega(t)\Phi, \Phi_0 \rangle\rangle := \langle\langle \omega(t), \Phi\Phi_0 \rangle\rangle, \quad \forall \Phi_0 \in (\mathcal{S}).$$

For more details and the proofs see e.g.[14,39].

Another application of this particular case concerns the definition of the so-called Hitsuda-Skorohod integral, related to the Skorohod and the Itô integrals, both well-known in stochastic analysis.

From now on, let $T \subset \mathbb{R}$ be a bounded interval with the Borel σ-algebra $\mathcal{B}$ and the Lebesgue measure. Since $\partial^*_t : (\mathcal{S})' \to (\mathcal{S})'$, given a mapping $\Psi : T \to (\mathcal{S})'$ defined a.e., we can then consider the mapping defined for a.a. $t \in T$ by $t \mapsto \partial_t^* \Psi(t)$. If, in addition, for every $\Phi \in (\mathcal{S})$, $\langle\!\langle \Psi(\cdot), \Phi \rangle\!\rangle \in L^1(T, \mathcal{B}, dt)$, then we have a well-defined *Pettis integral of* $\partial_t^* \Psi$,

$$\int_T \partial_t^* \Psi(t) \, dt \in (\mathcal{S})'. \tag{27}$$

That is, (27) is the unique element in $(\mathcal{S})'$ such that for all $\Phi \in (\mathcal{S})$

$$\left\langle\!\!\left\langle \int_T \partial_t^* \Psi(t) \, dt, \Phi \right\rangle\!\!\right\rangle = \int_T \langle\!\langle \partial_t^* \Psi(t), \Phi \rangle\!\rangle \, dt.$$

In particular, for $\Phi =: e^{\langle \cdot, \xi \rangle} :\in (\mathcal{S})$, $\xi \in S(\mathbb{R})$ (Section 6), it follows immediately from this definition that the S-transform of (27) is given by

$$S\left(\int_T \partial_t^* \Psi(t) \, dt \right)(\xi) = \int_T S\left(\partial_t^* \Psi(t) \right)(\xi) \, dt.$$

We call (27) the *Hitsuda-Skorohod integral of* Ψ.

In terms of chaos decomposition, observe that if $\Psi(t) = \sum_{n=0}^\infty \langle : \omega^{\otimes n} :, \psi^{(n)}(t)\rangle$ for a.a. $t \in T$, then, by (26),

$$\partial_t^* \Psi(t) = \sum_{n=1}^\infty \langle : \omega^{\otimes n} :, \delta_t \widehat{\otimes} \psi^{(n-1)}(t)\rangle = \omega(t) \diamond \sum_{n=0}^\infty \langle : \omega^{\otimes n} :, \psi^{(n)}(t)\rangle = \omega(t) \diamond \Psi(t),$$

where $\diamond$ is the Wick product introduced in Section 7. This yields another approach to introduce the Skorohod and the Itô integrals[15].

Now let $T = [0, 1]$ and let $X : [0, 1] \to (L^2)$ be a square integrable function with chaos decomposition

$$X(t) = \sum_{n=0}^\infty \langle : \omega^{\otimes n} :, f^{(n)}(t)\rangle \quad a.a. \ t \in [0, 1]$$

such that for a.a. $t \in [0, 1]$

$$\sum_{n=0}^\infty n n! |f^{(n)}(t)|^2 < \infty.$$

Then, it can be shown cf. e.g.[14] that the Hitsuda-Skorohod integral of X belongs to (L^2) and coincides with its Skorohod integral.

In order to recover the definition of the Itô integral, let $X : [0, 1] \to (L^2)$ be a square integrable function which we assume to be adapted to the filtration generated by Brownian motion. In terms of chaos decomposition this means that if X is given as before, then for all $n \in \mathbb{N}$ and for a.a. $(u_1, \ldots, u_n) \in \mathbb{R}^n \setminus [0, t]^n$, $f^{(n)}(t; u_1, \ldots, u_n) = 0$, $t \in [0, 1]$. Then, it turns out cf. e.g.[14] that the Hitsuda-Skorohod integral of X coincides with its Itô integral:

$$\int_0^1 \partial_t^* X(t) \, dt = \int_0^1 X(t) \, dB(t).$$

In accordance with the definition of Itô integral in stochastic analysis, here we should choose a continuous version of the Brownian motion defined in Example 2.

For more details and the proofs see e.g.[14,34].

8.3. *Second Quantization Operators*

Given a contraction operator B on $\mathcal{H}_{\mathbb{C}}$, we can define a contraction operator $\mathrm{Exp}B$ on the Fock space $\mathrm{Exp}(\mathcal{H}_{\mathbb{C}})$ defined on each space $\mathrm{Exp}_n(\mathcal{H}_{\mathbb{C}})$, $n \in \mathbb{N}$, by $B^{\otimes n}$, $\mathrm{Exp}B \restriction \mathrm{Exp}_0(\mathcal{H}_{\mathbb{C}}) := 1$. Therefore, for any coherent state $e(f)$, $f \in \mathcal{H}_{\mathbb{C}}$,

$$\mathrm{Exp}B\,(e(f)) = e(Bf).$$

In particular, given a positive self-adjoint operator A on $\mathcal{H}_{\mathbb{C}}$ and the contraction semigroup e^{-tA}, $t \geq 0$, we have defined a contraction semigroup $\mathrm{Exp}\left(e^{-tA}\right)$, $t \geq 0$, on $\mathrm{Exp}(\mathcal{H}_{\mathbb{C}})$. The generator of this semigroup is the so-called *second quantization operator corresponding to* A and we shall denote it by $d\mathrm{Exp}A$, i.e.,

$$\mathrm{Exp}\left(e^{-tA}\right) = \exp\left(-td\mathrm{Exp}A\right), \quad t \geq 0.$$

For more details see e.g.[2,43]. The action of $d\mathrm{Exp}A$ on coherent states $e(f)$ with f in the domain of A is given by

$$d\mathrm{Exp}A\,(e(f)) = \left((Af)\,\widehat{\otimes}\,\frac{f^{\otimes(n-1)}}{(n-1)!}\right)_{n \in \mathbb{N}_0}$$

with $(Af)\,\widehat{\otimes}\,\frac{f^{\otimes(n-1)}}{(n-1)!} = 0$ if $n = 0$, cf. e.g.[2].

The definition of second quantization operators leads, for instance, to the construction of countably Hilbert spaces and triples on Fock spaces and thus to the definition of new nuclear triples cf. e.g.[2,14,22,39]. In particular, for A being the Hamiltonian H of the quantum harmonic oscillator on $L^2(\mathbb{R})$ (Example 1),

$$(H\xi)(u) = -\frac{d^2\xi}{du^2}(u) + (u^2 + 1)\xi(u), \quad u \in \mathbb{R},$$

we rediscover the Gelfand triple

$$(\mathcal{S}) \subset (L^2) \subset (\mathcal{S})'.$$

For more details concerning this constructive scheme and examples see e.g.[2,14,22].

A.1. Appendices

A.1.1. *Tensor Powers of Hilbert Spaces*

Instead of reproducing the abstract construction of tensor powers of general topological vector spaces like e.g. in[45,50], we follow closely the direct approach of[43] for Hilbert spaces.

Let $\mathcal{H}$ be a (real or complex) Hilbert space with inner product $(\cdot,\cdot)$. For each $n \in \mathbb{N}$, $n \geq 2$, and every $g_1, \ldots, g_n \in \mathcal{H}$, we consider the following n-linear form $g_1 \otimes \ldots \otimes g_n := \otimes_{i=1}^n g_i$ defined on $\mathcal{H}^n$ by

$$(g_1 \otimes \ldots \otimes g_n)(h_1, \ldots, h_n) := \prod_{i=1}^n (h_i, g_i), \quad h_1, \ldots, h_n \in \mathcal{H}.$$

We shall call the linear space spanned by such n-linear forms the *algebraic n-th tensor power of* $\mathcal{H}$ and denote it by $\mathcal{H}^{\otimes_a n}$. In order to introduce a topological structure on $\mathcal{H}^{\otimes_a n}$, we define an inner product on $\mathcal{H}^{\otimes_a n}$, also denoted by $(\cdot,\cdot)$, acting on elements $\otimes_{i=1}^n g_{1i}$, $\otimes_{j=1}^n g_{2j} \in \mathcal{H}^{\otimes_a n}$ by

$$\left(\otimes_{i=1}^n g_{1i}, \otimes_{j=1}^n g_{2j} \right) := \prod_{k=1}^n (g_{1k}, g_{2k}). \tag{A.1}$$

Remark 6. It turns out (cf. e.g.[43]) that the value of (G_1, G_2), G_1, $G_2 \in \mathcal{H}^{\otimes_a n}$ is independent of the linear combinations used to express G_1 and G_2, and thus $(\cdot,\cdot)$ is well-defined.

Definition 8. The completion of $\mathcal{H}^{\otimes_a n}$ with respect to the norm induced by the inner product (A.1) is called the (*topological*) *n-th tensor power of* $\mathcal{H}$. We shall denoted it by $\mathcal{H}^{\otimes n}$.

We observe that if the Hilbert space $\mathcal{H}$ is separable and $\{e_k\}_{k\in\mathbb{N}}$ is an orthonormal basis of $\mathcal{H}$, then $\mathcal{H}^{\otimes n}$ is also a separable Hilbert space and the set of elements of the form $e_{\mathbf{k}} := \otimes_{i=1}^n e_{k_i}$ indexed by $\mathbf{K} := (k_1, \ldots, k_n) \in \mathbb{N}^n$ is a Hilbertian basis of $\mathcal{H}^{\otimes n}$.

In particular, we can consider the Hilbert space $L^2(\mathbb{R})$ or its complexified space $L_{\mathbb{C}}^2 := \{f_1 + if_2 : f_1, f_2 \in L^2(\mathbb{R})\}$ (see Appendix A.1.2). In these cases the tensor powers $(L^2(\mathbb{R}))^{\otimes n}$, $(L_{\mathbb{C}}^2(\mathbb{R}))^{\otimes n}$ can be identified with $L^2(\mathbb{R}^n)$, $L_{\mathbb{C}}^2(\mathbb{R}^n)$, respectively.

Proposition 4. *The spaces* $(L^2(\mathbb{R}))^{\otimes n}$ *and* $(L_{\mathbb{C}}^2(\mathbb{R}))^{\otimes n}$ *are unitarily isomorphic to* $L^2(\mathbb{R}^n)$ *and* $L_{\mathbb{C}}^2(\mathbb{R}^n)$, *respectively.*

Proof. (Sketch) Given an orthonormal basis $\{e_k\}_{k\in\mathbb{N}}$ of $L^2(\mathbb{R})$, consider the orthonormal basis $R_{\mathbf{K}}$, $\mathbf{K} := (k_1, \ldots, k_n) \in \mathbb{N}^n$, of $L^2(\mathbb{R}^n)$,

$$R_{\mathbf{K}}(x_1, \ldots, x_n) := e_{k_1}(x_1) \ldots e_{k_n}(x_n),$$

and the linear mapping R which maps the orthonormal basis $\{e_{\mathbf{k}}\}_{\mathbf{k}\in\mathbb{N}^n}$ of $(L^2(\mathbb{R}))^{\otimes n}$ onto $\{R_{\mathbf{k}}\}_{\mathbf{k}\in\mathbb{N}^n}$,

$$R : e_{k_1} \otimes \ldots \otimes e_{k_n} \mapsto R_{\mathbf{K}}, \quad \mathbf{K} = (k_1, \ldots, k_n).$$

Clearly the following equality of norms holds:

$$|R_{\mathbf{K}}|_{L^2(\mathbb{R}^n)} = |e_{k_1} \otimes \ldots \otimes e_{k_n}|_{(L^2(\mathbb{R}))^{\otimes n}}, \quad \forall \mathbf{K} = (k_1, \ldots, k_n) \in \mathbb{N}^n,$$

which leads to the existence of a unique extension of R to a unitary isomorphism of $(L^2(\mathbb{R}))^{\otimes n}$ onto $L^2(\mathbb{R}^n)$, also denoted by R. The assertion for the complex case follows by simply replacing each L^2-space by the corresponding $L^2_{\mathbb{C}}$ complexified space. $\qquad\square$

Remark 7. In view of this proof, we shall identify each n-linear form $g_1 \otimes \ldots \otimes g_n$ with $R(g_1 \otimes \ldots \otimes g_n) \in L^2(\mathbb{R}^n)$, that is,

$$(g_1 \otimes \ldots \otimes g_n)(x_1, \ldots, x_n) := g_1(x_1) \ldots g_n(x_n),$$

and $g_1 \otimes \ldots \otimes g_n \in L^2(\mathbb{R}^n)$ will be called the *tensor product of* $g_1, \ldots, g_n$. The latter equality is adopted to define $g_1 \otimes \ldots \otimes g_n$ with $g_1, \ldots, g_n$ being elements of a generic space of functions.

A.1.2. *Fock Space*

The definition and main properties of the symmetric Fock spaces are described below. For more details see e.g.[8,43,46].

Given a real separable Hilbert space $\mathcal{H}$, let $\mathcal{H}_{\mathbb{C}}$ be the complexified space of $\mathcal{H}$,

$$\mathcal{H}_{\mathbb{C}} := \{f + ig : f, g \in \mathcal{H}\}$$

with the inner product

$$(f_1 + ig_1, f_2 + ig_2) := (f_1, f_2) + (g_1, g_2) + i(g_1, f_2) - i(f_1, g_2)$$

(antilinear in the second factor) and the corresponding norm $|\cdot|$.

For each $n \in \mathbb{N}$ fixed and any $\iota \in S_n$ ($S_n :=$ the permutation group over $\{1, \ldots, n\}$), we consider the unitary isomorphism $U_{\iota,n}$ defined on the total set[a] of elements of the form $g_1 \otimes \ldots \otimes g_n \in \mathcal{H}_{\mathbb{C}}^{\otimes n}$, $g_i \in \mathcal{H}_{\mathbb{C}}$, $i = 1, \ldots, n$, by

$$U_{\iota,n}(g_1 \otimes \ldots \otimes g_n) := g_{\iota(1)} \otimes \ldots \otimes g_{\iota(n)}.$$

---

[a]A subset A of a Hilbert space is said to be *total* whenever the closure of the space spanned by A coincides with the whole space.

Then, given the family of unitary isomorphisms $U_{\iota,n}$ with $\iota \in S_n$, we define the operator P_n on $\mathcal{H}_{\mathbb{C}}^{\otimes n}$ by

$$P_n := \frac{1}{n!} \sum_{\iota \in S_n} U_{\iota,n}.$$

It is easy to check that $P_n \circ P_n = P_n$ and the adjoint operator of P_n coincides with P_n itself. That is, P_n is an orthogonal projection. We shall call the image of $\mathcal{H}_{\mathbb{C}}^{\otimes n}$ under P_n the *n-th symmetric tensor power* of $\mathcal{H}_{\mathbb{C}}$ and denote it by $\mathcal{H}_{\mathbb{C}}^{\widehat{\otimes} n}$. We shall denote each $P_n(g_1 \otimes \ldots \otimes g_n)$ by $g_1 \widehat{\otimes} \ldots \widehat{\otimes} g_n$.

Due to Proposition 4 and its proof, it is clear that in particular for $\mathcal{H}_{\mathbb{C}} = L_{\mathbb{C}}^2(\mathbb{R})$, the n-th symmetric tensor power $(L_{\mathbb{C}}^2(\mathbb{R}))^{\widehat{\otimes} n}$ is unitarily isomorphic to the subspace $\widehat{L}_{\mathbb{C}}^2(\mathbb{R}^n) \subset L_{\mathbb{C}}^2(\mathbb{R}^n)$ of all symmetric square integrable functions. For this reason, in accordance with Remark 7 we shall identify the space $(L_{\mathbb{C}}^2(\mathbb{R}))^{\widehat{\otimes} n}$ with the space $\widehat{L}_{\mathbb{C}}^2(\mathbb{R}^n)$.

Definition 9. The Bose or symmetric Fock space $\mathrm{Exp}\mathcal{H}_{\mathbb{C}}$ over $\mathcal{H}_{\mathbb{C}}$ is the Hilbert space defined by the Hilbertian direct sum

$$\mathrm{Exp}(\mathcal{H}_{\mathbb{C}}) := \bigoplus_{n=0}^{\infty} \mathrm{Exp}_n(\mathcal{H}_{\mathbb{C}}),$$

where $\mathrm{Exp}_0(\mathcal{H}_{\mathbb{C}}) := \mathbb{C}$ is the so-called *vacuum subspace*, and each $\mathrm{Exp}_n(\mathcal{H}_{\mathbb{C}})$, $n \in \mathbb{N}$, defined by the space $\mathcal{H}_{\mathbb{C}}^{\widehat{\otimes} n}$ endowed with the inner product $n!(\cdot,\cdot)_{\mathcal{H}_{\mathbb{C}}^{\otimes n}}$, is the so-called *n-particle subspace*.

In other words, a generic element $F \in \mathrm{Exp}(\mathcal{H}_{\mathbb{C}})$ is a sequence $F = (f^{(n)})_{n \in \mathbb{N}_0}$ with $f^{(n)} \in \mathcal{H}_{\mathbb{C}}^{\widehat{\otimes} n}$, $n \in \mathbb{N}$, and

$$\|F\|_{\mathrm{Exp}(\mathcal{H}_{\mathbb{C}})}^2 := \sum_{n=0}^{\infty} n! |f^{(n)}|_{\mathcal{H}_{\mathbb{C}}^{\otimes n}}^2 < \infty.$$

Among the elements in $\mathrm{Exp}(\mathcal{H}_{\mathbb{C}})$ we distinguish the so-called *exponential vectors* or *coherent states* $e(f) \in \mathrm{Exp}(\mathcal{H}_{\mathbb{C}})$ corresponding to the one-particle vector $f \in \mathcal{H}_{\mathbb{C}}$:

$$e(f) := \left(1, f, \frac{1}{2!}f^{\otimes 2}, \ldots, \frac{1}{n!}f^{\otimes n}, \ldots\right), \quad f^{\otimes n} := f \otimes \ldots \otimes f \ (n \text{ times}).$$

According to the definition of $\mathrm{Exp}(\mathcal{H}_{\mathbb{C}})$, we have

$$(e(f_1), e(f_2))_{\mathrm{Exp}(\mathcal{H}_{\mathbb{C}})} = \exp\left((f_1, f_2)\right), \quad f_1, f_2 \in \mathcal{H}_{\mathbb{C}},$$

and thus

$$\|e(f)\|_{\mathrm{Exp}(\mathcal{H}_{\mathbb{C}})}^2 = \exp\left(|f|^2\right), \quad f \in \mathcal{H}_{\mathbb{C}}.$$

The next statement emphasizes the role of coherent states (see e.g.[2]).

Proposition 5. *Given a linear subspace $\mathcal{L} \subset \mathcal{H}_{\mathbb{C}}$, the family of coherent states $\{e(f) : f \in \mathcal{L}\}$ is total in $\mathrm{Exp}(\mathcal{H}_{\mathbb{C}})$ whenever $\mathcal{L}$ is dense in $\mathcal{H}_{\mathbb{C}}$.*

Remark 8. In view of Appendix A.1.3 below, let us mention that in an analogous way we can define symmetric tensor powers of real Hilbert spaces as well.

A.1.3. *Tensor Powers of Nuclear Spaces*

As in Section 1, let $\mathcal{N} \subset \mathcal{H} \subset \mathcal{N}'$ be a nuclear triple,

$$\mathcal{N} = \mathop{\mathrm{pr\,lim}}_{p \in \mathbb{N}} \mathcal{H}_p, \quad \mathcal{N}' = \mathop{\mathrm{ind\,lim}}_{p \in \mathbb{N}} \mathcal{H}_{-p}.$$

In order to define tensor powers $\mathcal{N}^{\otimes n}$ and symmetric tensor powers $\mathcal{N}^{\widehat{\otimes} n}$, $n \in \mathbb{N}$, $n \geq 2$, of the nuclear space $\mathcal{N}$, we consider the families of tensor powers of the Hilbert spaces $\mathcal{H}_p^{\otimes n}$ and $\mathcal{H}_p^{\widehat{\otimes} n}$, both indexed by $p \in \mathbb{N}$. Since there is no risk of confusion, we shall use the notation $|\cdot|_p$ also for the Hilbertian norm on $\mathcal{H}_p^{\otimes n}$. The n-th tensor power $\mathcal{N}^{\otimes n}$ of $\mathcal{N}$ and the n-th symmetric tensor power $\mathcal{N}^{\widehat{\otimes} n}$ of $\mathcal{N}$ are the nuclear Fréchet spaces defined by

$$\mathcal{N}^{\otimes n} := \mathop{\mathrm{pr\,lim}}_{p \in \mathbb{N}} \mathcal{H}_p^{\otimes n}, \quad \mathcal{N}^{\widehat{\otimes} n} := \mathop{\mathrm{pr\,lim}}_{p \in \mathbb{N}} \mathcal{H}_p^{\widehat{\otimes} n},$$

respectively.[2]

Moreover, if each $\mathcal{H}_{-p}^{\otimes n}$ (resp., $\mathcal{H}_{-p}^{\widehat{\otimes} n}$) is the dual space of $\mathcal{H}_p^{\otimes n}$ (resp., $\mathcal{H}_p^{\widehat{\otimes} n}$) with respect to $\mathcal{H}^{\otimes n}$ (resp., $\mathcal{H}^{\widehat{\otimes} n}$), then the dual space $\mathcal{N}'^{\otimes n}$ of $\mathcal{N}^{\otimes n}$ with respect to $\mathcal{H}^{\otimes n}$ and the dual space $\mathcal{N}'^{\widehat{\otimes} n}$ of $\mathcal{N}^{\widehat{\otimes} n}$ with respect to $\mathcal{H}^{\widehat{\otimes} n}$ can be written as

$$\mathcal{N}'^{\otimes n} = \mathop{\mathrm{ind\,lim}}_{p \in \mathbb{N}} \mathcal{H}_{-p}^{\otimes n} \quad \text{and} \quad \mathcal{N}'^{\widehat{\otimes} n} = \mathop{\mathrm{ind\,lim}}_{p \in \mathbb{N}} \mathcal{H}_{-p}^{\widehat{\otimes} n},$$

respectively. As before, in this work we shall also use the notation $|\cdot|_{-p}$ for the norm on $\mathcal{H}_{-p}^{\otimes n}$, $p \in \mathbb{N}$, and $\langle \cdot, \cdot \rangle$ for the dual pairing between $\mathcal{N}'^{\otimes n}$ and $\mathcal{N}^{\otimes n}$.

Thus we have defined the nuclear triples

$$\mathcal{N}^{\otimes n} \subset \mathcal{H}^{\otimes n} \subset \mathcal{N}'^{\otimes n} \quad \text{and} \quad \mathcal{N}^{\widehat{\otimes} n} \subset \mathcal{H}^{\widehat{\otimes} n} \subset \mathcal{N}'^{\widehat{\otimes} n}.$$

Remark 9. All the above results quoted still hold for complex spaces. In that case, we shall use the same notation as above.

A.1.4. *Holomorphy on Locally Convex Spaces*

In this part we generalize the notion of holomorphic or analytic functions in complex analysis to complex-valued functions defined on a locally convex topological vector space $\mathcal{E}$ over the complex field $\mathbb{C}$. For more details and the proofs see e.g.[1,4,5].

A function $F : U \to \mathbb{C}$ defined on an open set $U \subset \mathcal{E}$ is called *G-holomorphic* or *Gâteaux-holomorphic* if for each $\xi_0 \in \mathcal{U}$ and each $\xi \in \mathcal{E}$ the complex-valued function

$$\mathbb{C} \ni z \mapsto F(\xi_0 + z\xi) \in \mathbb{C}$$

is analytic on some neighborhood of $0 \in \mathbb{C}$. Hence, given a G-holomorphic function $F : U \to \mathbb{C}$, it turns out by the general theory of complex analysis that for each $\xi_0 \in U$ we can write

$$F(\xi_0 + \xi) = \sum_{n=0}^{\infty} \frac{1}{n!} d^n F(\xi_0; \xi) \tag{A.2}$$

for all ξ in some open neighborhood of zero, where $d^n F(\xi_0; \xi)$ is the differential $d^n F(\xi_0)(\xi, \ldots, \xi)$ of n-th order of F at the point ξ_0 along the direction $(\xi, \ldots, \xi)$.

A G-holomorphic function $F : U \to \mathbb{C}$ is called *holomorphic* whenever for all $\xi_0 \in U$ there is an open neighborhood V of zero such that the series in (A.2) converges uniformly on V. It turns out (cf. e.g. [5, Lemma 2.8]) that a G-holomorphic function is holomorphic if and only if it is locally bounded.

A function F is said to be *holomorphic at a point* $\xi_0 \in \mathcal{E}$ if there is an open neighborhood $U \subset \mathcal{E}$ of ξ_0 such that $F : U \to \mathbb{C}$ is holomorphic.

References

1. J. A. Barroso. *Introduction to Holomorphy*, volume 106 of *Mathematics Studies*. North-Holland Publ. Co., Amsterdam, 1985.
2. Yu. M. Berezansky and Yu. G. Kondratiev. *Spectral Methods in Infinite-Dimensional Analysis*. Naukova Dumka, Kiev, 1988. (in Russian). English translation, Kluwer Academic Publishers, Dordrecht, 1995.
3. Yu. M. Berezansky, Z. G. Sheftel, and G. F. Us. *Functional Analysis*, volume II. Birkhäuser, Boston, Basel and Berlin, 1996.
4. J. F. Colombeau. *Differential Calculus and Holomorphy: Real and Complex Analysis in Locally Convex Spaces*, volume 64 of *Mathematics Studies*. North-Holland Publ. Co., Amsterdam, 1982.

5. S. Dineen. *Complex Analysis in Locally Convex Spaces*, volume 57 of *Mathematics Studies*. North-Holland Publ. Co., Amsterdam, 1981.

6. M. Grothaus, Yu. G. Kondratiev, and L. Streit. Complex Gaussian analysis and the Bargmann-Segal space. *Methods Funct. Anal. Topology*, 3(4):46–64, 1997.

7. A. Grothendieck. *Produits Tensoriels Topologiques et Espaces Nucléaires*, volume 16 of *Mem. Amer. Math. Soc.* American Mathematical Society, 1955.

8. A. Guichardet. *Symmetric Hilbert Spaces and Related Topics*, volume 261 of *Lecture Notes in Math.* Springer Verlag, Berlin, Heidelberg and New York, 1972.

9. I. M. Gelfand and N. Ya. Vilenkin. *Generalized Functions.* Academic Press, New York and London, 1968.

10. T. Hida. *Stationary Stochastic Processes.* Princeton University Press, 1970.

11. T. Hida. *Analysis of Brownian Functionals*, volume 13 of *Carleton Mathematical Lecture Notes.* Carleton University, Ottawa, 1975.

12. T. Hida. *Brownian Motion*, volume 11 of *Appl. Math.* Springer Verlag, New York, 1980.

13. T. Hida. Generalized Brownian functionals. In G. Kallianpur, editor, *Theory and Applications of Random Fields*, pages 89–95, Berlin, Heidelberg and New York, 1983. Springer.

14. T. Hida, H. H. Kuo, J. Potthoff, and L. Streit. *White Noise. An Infinite Dimensional Calculus.* Kluwer Academic Publishers, Dordrecht, 1993.

15. H. Holden, B. Øksendal, J. Ubøe, and T. Zhang. *Stochastic Partial Differential Equations: A Modeling, White Noise Functional Approach.* Birkhäuser, Boston, Basel, and Berlin, 1996.

16. Z. Y. Huang and J. A. Yan. *Introduction to Infinite Dimensional Stochastic Analysis.* Kluwer Academic Publishers, Beijing and New York, 1997.

17. N. A. Kachanovsky and S. V. Koshkin. Minimality of Appell-like systems and embeddings of test function spaces in a generalization of white noise analysis. *Methods Funct. Anal. Topology*, 5(3):13–25, 1999.

18. Yu. G. Kondratiev, P. Leukert, J. Potthoff, L. Streit, and W. Westerkamp. Generalized functionals in Gaussian spaces: The characterization theorem revisited. *J. Funct. Anal.*, 141:301–318, 1996.

19. Yu. G. Kondratiev, P. Leukert, and L. Streit. Wick calculus in Gaussian analysis. *Acta Appl. Math.*, 44:269–294, 1996.

20. Yu. G. Kondratiev. Spaces of Test and Generalized Functions of an Infinite Number of Variables. Master's thesis, University of Kiev, 1975.

21. Yu. G. Kondratiev. *Generalized Functions in Problems of Infinite Dimensional Analysis.* PhD thesis, University of Kiev, 1978.

22. Yu. G. Kondratiev. Nuclear spaces of entire functions in problems of infinite dimensional analysis. *Soviet Math. Dokl.*, 22:588–592, 1980.

23. Yu. G. Kondratiev. Spaces of entire functions of an infinite number of variables, connected with the rigging of a Fock space. In *Spectral Analysis of Differential Operators*, pages 18–37. Institute of Mathematics of the National Academy of Sciences of Ukraine, SSR, Kiev, 1980. (in Russian). English translation in *Selecta Mathematica Sovietica*, 10:165–180, 1991.

24. Yu. G. Kondratiev and Yu. S. Samoilenko. Integral representation of generalized positive definite kernels of an infinite number of variables. *Soviet Math. Dokl.*, 17:517–521, 1976.

25. Yu. G. Kondratiev and Yu. S. Samoilenko. Spaces of trial and generalized functions of an infinite number of variables. *Rep. Math. Phys.*, 14(3):325–350, 1978.

26. Yu. G. Kondratiev and L. Streit. A remark about a norm estimate for white noise distributions. *Ukrainian Math. J.*, 44:832–835, 1992.

27. Yu. G. Kondratiev and L. Streit. Spaces of white noise distributions: constructions, descriptions, applications I. *Rep. Math. Phys.*, 33:341–366, 1993.

28. Yu. G. Kondratiev, J. L. Silva, and L. Streit. Generalized Appell systems. *Methods Funct. Anal. Topology*, 3(2):28–61, 1997.

29. Yu. G. Kondratiev, L. Streit, and W. Westerkamp. A note on positive distributions in Gaussian analysis. *Ukrainian Math. J.*, 47(5):749–759, 1995.

30. Yu. G. Kondratiev, L. Streit, W. Westerkamp, and J. Yan. Generalized functions in infinite dimensional analysis. *Hiroshima Math. J.*, 28(2):213–260, 1998.

31. I. Kubo and S. Takenaka. Calculus on Gaussian white noise I. *Proc. Japan Acad. Ser. A Math. Sci.*, 56:376–380, 1980.

32. I. Kubo and S. Takenaka. Calculus on Gaussian white noise II. *Proc. Japan Acad. Ser. A Math. Sci.*, 56:411–416, 1980.

33. H. H. Kuo. Lectures on white noise analysis. *Soochow J. Math.*, 18:229–300, 1992.

34. H. H. Kuo. *White Noise Distribution Theory*. CRC Press, Boca Raton, New York, London, and Tokyo, 1996.

35. A. Lascheck, P. Leukert, L. Streit, and W. Westerkamp. More about Donsker's delta function. *Soochow J. Math.*, 20:401–418, 1994.

36. P. Malliavin. *Stochastic Analysis*, volume 313 of *Grundlehren der Mathematischen Wissenschaften*. Springer Verlag, Berlin, 1997.

37. R. Minlos. Generalized random processes and their extension to a measure. *Trudy Moskov. Mat. Obšč.*, 8:497–518, 1959. (in Russian). English translation, Selected Transl. Math. Statist. and Prob. 3 (1963), 291–313.

38. D. Nualart. *The Malliavin Calculus and Related Topics*. Prob. Appl. Springer, Berlin and New York, 1995.

39. N. Obata. *White Noise Calculus and Fock Space*, volume 1577 of *Lecture Notes in Math.* Springer Verlag, Berlin, Heidelberg and New York, 1994.

40. A. Pietsch. *Nuclear Locally Convex Spaces*, volume 66 of *Ergebnisse der Mathematik und ihrer Grenzgebiete*. Springer Verlag, Berlin, Heidelberg and New York, 1972.

41. J. Potthoff and L. Streit. A characterization of Hida distributions. *J. Funct. Anal.*, 101:212–229, 1991.

42. A. Robertson and W. Robertson. *Topological Vector Spaces*. Cambridge University Press, Cambridge, 1973.

43. M. Reed and B. Simon. *Methods of Modern Mathematical Physics*, volume I. Academic Press, New York and London, 1972.

44. M. Reed and B. Simon. *Methods of Modern Mathematical Physics*, volume II.

Academic Press, New York and London, 1975.

45. H. H. Schaefer. *Topological Vector Spaces*. Springer Verlag, Berlin, Heidelberg and New York, 1971.

46. I. E. Segal. Tensor algebras over Hilbert spaces. *Trans. Amer. Math. Soc.*, 81:106–134, 1956.

47. B. Simon. Distributions and their Hermite expansions. *J. Math. Phys.*, 12:140–148, 1971.

48. A. V. Skorohod. *Integration in Hilbert Space*. Springer Verlag, Berlin, Heidelberg and New York, 1974.

49. N. R. Shieh and Y. Yokoi. Positivity of Donsker's delta function. In T. Hida et al, editor, *White Noise Analysis - Mathematics and Applications*, pages 374–382. World Scientific, Singapore, 1990.

50. F. Treves. *Topological Vector Spaces, Distributions and Kernels*. Academic Press, New York and London, 1967.

51. Y. Yokoi. Positive generalized white noise functionals. *Hiroshima Math. J.*, 20:137–157, 1990.

Chapter 2

Quantum Fields

Sergio Albeverio[1], Michael Röckner[2] and Minoru W. Yoshida[3]

[1] *Inst. Angewandte Mathematik, and HCM, Universität Bonn,*
[2] *Dept. Math., Univ. Bielefeld,*
[3] *Faculty of Engineering, Dept. Information Sci., Lab. of Math. Analysis, Kanagawa Univ.*

1. Introduction and Preliminaries

From a functional analytic point of view, the **Gaussian** white noise analysis operates on a symmetric Fock space over the *real* Hilbert space $\mathcal{H} = L^2(\mathbb{R}^d, \mathbb{R})$ or the *complex* Hilbert space $\mathcal{H}_{\mathbb{C}} = L^2(\mathbb{R}^d, \mathbb{C})$ with $d \in \mathbb{N}$. Precisely, in Chapter 1, we have seen that the **Gaussian** white noise analysis gives structures of Sobolev spaces to the symmetric Fock space resp. its dual space, that are expressed by $(\mathcal{S})$, $(\mathcal{S})'$, $(\mathcal{H}_p)_q^\beta$, $(\mathcal{H}_{-p})_{-q}^{-\beta}$ and other symbols (cf. Chapter 1). Correspondingly, we have seen that the T-transform is a natural extension of the traditional Fourier transform on these Sobolev spaces.

Since the Fock spaces are fundamental structures of quantum field theory (cf., e.g.,[18,19,29,41,46,59,70,100] and references therein), it is natural that we consider the (constructive) quantum field theory through the white noise analysis.

In this chapter, we introduce and review, mainly, three developments on constructive quantum field theory through the white noise analysis. Two of them are considerations by means of the **Gaussian** white noise analysis (cf. sections 2 and 3) that depend on the structures of **Gaussian** white noise (or **Gaussian** random fields), and the other one is a result derived through generalized (**non-Gaussian**) white noise analysis (cf. section 4). (The mathematical analysis based on **non-Gaussian** has its own special importance, we refer to[102,103], and references therein, on this concern).

Generally and roughly speaking, every real valued random field (or equivalently a set of real valued random variables) indexed by $\mathcal{S}(\mathbb{R}^d, \mathbb{R})$, the space of Schwartz rapidly decreasing test functions, possessing a continuity with respect to some Schwartz norm and expectation, can be expressed as an element of *the space of algebraic (i.e., incomplete) direct sum of spaces of the form* $\bigoplus_{n=0}^{\infty} (\mathcal{S}'(\mathbb{R}^d, \mathbb{R}))^{\hat{\otimes} n}$. This direct sum is sufficiently large, and includes the spaces of S-transforms of $(\mathcal{H}_{-p})_{-q}^{-\beta}$ and $(\mathcal{S})'$, and the symmetric Fock space (i.e., *the completed direct sum having*) $\bigoplus_{n=0}^{\infty} (L^2(\mathbb{R}^d))^{\hat{\otimes} n}$ as its subspaces, cf. (10) in the next section. Every example introduced in the present chapter, build using **Gaussian** and **generalized (non-Gaussian)** white noise (or random field), is understood relative to this general mathematical structure (cf. (10), Theorem 1, Remark 2, Remark 2-2°), (20), (33), (50) and (51)).

Before proceeding to the next sections, we recall the structure of the spaces of **Gaussian** white noise, e.g., the spaces $(\mathcal{S})$ and $(\mathcal{S})'$ and others, and the corresponding symmetric Fock spaces.

Below (throughout this chapter), in order to avoid confusions of the symbols such as S-transform, the space of Schwartz distributions and the spaces of generalized random variables $(\mathcal{S})$ and $(\mathcal{S})'$, we denote the space of test functions of Schwartz distributions (the space of rapidly decreasing functions) on $\mathbb{R}^d$ (taking values in $\mathbb{R}$ resp. $\mathbb{C}$) and their dual spaces, the space of Schwartz distributions, by using the corresponding *exact* notations such as $\mathcal{S}(\mathbb{R}^d, \mathbb{R})$, resp. $\mathcal{S}(\mathbb{R}^d, \mathbb{C})$ and $\mathcal{S}'(\mathbb{R}^d, \mathbb{R})$, resp. $\mathcal{S}'(\mathbb{R}^d, \mathbb{C})$;

Throughout this chapter, let $d \in \mathbb{N}$ denote a given space (resp. space-time) dimension on which our discussions are carried out, and let μ be the canonical **Gaussian** measure on $\mathcal{S}'(\mathbb{R}^d, \mathbb{R})$ such that

$$\int_{\mathcal{S}'(\mathbb{R}^d, \mathbb{R})} e^{i<\omega, f>} d\mu(\omega) = e^{-\frac{1}{2}|f|^2}, \quad \forall f \in \mathcal{S}(\mathbb{R}^d, \mathbb{R}),$$

$$\text{with} \quad |f|^2 := \int_{\mathbb{R}^d} |f(x)|^2 dx. \tag{1}$$

Let $\mathcal{P}(\mathcal{S}'(\mathbb{R}^d, \mathbb{R}))$ be the space of smooth Wick polynomials on $\mathcal{S}'(\mathbb{R}^d, \mathbb{R})$, that is a subspace of $L^2(\mu)$ random variables (i.e., a subspace of (L^2), cf. Chapter 1) such that

$$\mathcal{P}(\mathcal{S}'(\mathbb{R}^d, \mathbb{R})) \equiv \left\{ \varphi \; \middle| \; \varphi(\omega) = \sum_{n=0}^{N} <: \omega^{\otimes n} :, \varphi^{(n)} >, \right.$$

$$\left. \varphi^{(n)} \in (\mathcal{S}(\mathbb{R}^d, \mathbb{C}))^{\hat{\otimes} n}, \omega \in \mathcal{S}'(\mathbb{R}^d, \mathbb{R}), N \in \mathbb{N}_0 \right\} \subset (L^2), \tag{2}$$

where $\mathbb{N}_0 := \mathbb{N} \cup \{0\}$.

By Chapter 1, (L^2) is unitary equivalent with the symmetric Fock space $Exp(\mathcal{H}_\mathbb{C})$ defined by

$$Exp(\mathcal{H}_\mathbb{C}) = \text{the complete direct sum of } \bigoplus_{n=0}^{\infty} (\mathcal{H}_\mathbb{C})^{\hat{\otimes} n}, \tag{3}$$

where $\mathcal{H}_\mathbb{C} \equiv L^2$ space (with respect to the Lebesgue measure on $\mathbb{R}^d$) of complex valued functions, and the norm of which is

$$\|\Phi\|_{Exp(\mathcal{H}_\mathbb{C})}^2 \equiv \sum_{n=0}^{\infty} n! \, \|\Phi^{(n)}\|_{L^2}^2,$$

for $\Phi = \bigoplus_{n=0}^{\infty} \Phi^{(n)}$, $\Phi^{(n)} \in (\mathcal{H}_\mathbb{C})^{\hat{\otimes} n}$, $n \in \mathbb{N}_0$. We have in fact the equivalence given by the S-transform:

$$S\big((L^2)\big) = \text{the complete direct sum of } \bigoplus_{n=0}^{\infty} (\mathcal{H}_\mathbb{C})^{\hat{\otimes} n} := Exp(\mathcal{H}). \tag{4}$$

Correspondingly, by (2) and (4), we have the following expression

$$S\big(\mathcal{P}(\mathcal{S}'(\mathbb{R}^d, \mathbb{R}))\big) = \Big\{ \Phi = \bigoplus_{n=0}^{\infty} \Phi^{(n)} \,\Big|\, \Phi^{(n)} \in (\mathcal{S}(\mathbb{R}^d, \mathbb{R}))^{\hat{\otimes} n}, \; n \in \mathbb{N}_0,$$

$$\text{with } \Phi^{(n)} = 0 \text{ except for a finite number of } n\text{'s}\Big\}$$

$$\subset S\big((\mathcal{S})\big) \subset S\big((L^2)\big) = Exp(\mathcal{H}). \tag{5}$$

If we let $S(\mathcal{P}(\mathcal{S}'(\mathbb{R}^d, \mathbb{R})))$ be a topological space equipped with a *sufficiently strong* topology, denoting this *topological vector space* by $\tilde{\mathcal{P}}$, with the open base $\mathcal{O}$ such that

$$\mathcal{O} := \bigoplus_{n=0}^{\infty} \mathcal{O}^{(n)}, \quad \mathcal{O}^{(n)} \text{ an open subset of } (\mathcal{S}(\mathbb{R}^d, \mathbb{R}))^{\hat{\otimes} n}, \; n \in \mathbb{N}_0,$$

$$\text{and } \mathcal{O}^{(n)} = \{\mathbf{0}\} \text{ except for a finite number of } n\text{'s},$$

$$\text{with } \mathbf{0} = \text{the 0 vector of } (\mathcal{S}(\mathbb{R}^d, \mathbb{R}))^{\hat{\otimes} n}, \tag{6}$$

then we have the following continuous embedding (cf. Chapter 1, for the notations):

$$\tilde{\mathcal{P}} = S\big(\mathcal{P}(\mathcal{S}'(\mathbb{R}^d, \mathbb{R}))\big) \hookrightarrow S\big((\mathcal{S})\big) \hookrightarrow S\big((L^2)\big) = Exp(\mathcal{H}). \tag{7}$$

For the dual spaces (i.e., the *linear* spaces of continuous linear functionals), since $S\big((\mathcal{S})\big)' = S\big((\mathcal{S})'\big)$ and $S\big((L^2)\big)' = S\big((L^2)\big)$, by (1.7) we have

$$\tilde{\mathcal{P}}' \supset S\big((\mathcal{S})'\big) \supset S\big((L^2)\big). \tag{8}$$

Moreover, by (1.6) obviously $\bigoplus_{n=0}^{\infty}(\mathcal{S}'(\mathbb{R}^d,\mathbb{R}))^{\hat{\otimes}n} \subset \tilde{\mathcal{P}}'$. Moreover, by (1.5) any element of $\tilde{\mathcal{P}}'$ must be an element of $\bigoplus_{n=0}^{\infty}(\mathcal{S}'(\mathbb{R}^d,\mathbb{R}))^{\hat{\otimes}n}$, and from this we see that

$$\tilde{\mathcal{P}}' = \text{the algebraic direct sum of } \bigoplus_{n=0}^{\infty}(\mathcal{S}'(\mathbb{R}^d,\mathbb{R}))^{\hat{\otimes}n}. \tag{9}$$

Consequently, by (3),(4),(8) and (9) and also by the definition of the spaces $(\mathcal{H}_{-p})_{-q}^{-\beta}$ for any p, q, $\beta \in \mathbb{N}$, and $(\mathcal{S})^{-1}$ (cf. Chapter 1 and section 3 of this chapter) we have

$$Exp(\mathcal{H}_{\mathbb{C}}) = \text{the complete direct sum of } \bigoplus_{n=0}^{\infty}(\mathcal{H}_{\mathbb{C}})^{\hat{\otimes}n} = S\big((L^2)\big)$$
$$\subset S\big((\mathcal{S})'\big) \subset S\big((\mathcal{H}_{-p})_{-q}^{-1}\big) \subset S\big((\mathcal{S})^{-1}\big)$$
$$\subset \text{ the algebraic direct sum of } \bigoplus_{n=0}^{\infty}(\mathcal{S}'(\mathbb{R}^d,\mathbb{R}))^{\hat{\otimes}n}. \tag{10}$$

2. Applications of white noise analysis to canonical quantum field theory with ϕ-bound

2.1. *A general formulation*

Canonical quantum field theory (cf., e.g.[20,23,41,45,59] and references therein) in d-dimensional space-time, $d \in \mathbb{N}$, is based on the consideration of *canonical commutation relations* (CCR).

Namely, in each canonical quantum field theory, we study a *complex* Hilbert space $\mathcal{H}_{\mathbb{C}}$ with an inner product $(\cdot, \cdot)$, self-adjoint operator valued distributions ϕ, π (*the field operators*) such that for $s := d - 1$,

$$\phi : \mathcal{S}(\mathbb{R}^s,\mathbb{R}) \ni \xi \longmapsto \phi(\xi) \in \text{ a family of self-adjoint operators on } \mathcal{H}_{\mathbb{C}},$$

$$\pi : \mathcal{S}(\mathbb{R}^s,\mathbb{R}) \ni \xi \longmapsto \pi(\xi) \in \text{ a family of self-adjoint operators on } \mathcal{H}_{\mathbb{C}},$$

and a self-adjoint operator $A \geq 0$ (*the Hamiltonian*), which satisfy (on a suitable dense domain) the following *canonical commutation relations* (CCR):

$$[\phi(\xi),\,\phi(\eta)] = [\pi(\xi),\,\pi(\eta)] = 0, \qquad [\phi(\xi),\,\pi(\eta)] = i\,(\xi,\,\eta), \tag{11}$$

and

$$[A,\,\phi(\eta)] = -i\,\pi(\xi), \qquad\qquad \forall \xi,\,\eta \in \mathcal{S}(\mathbb{R}^s,\mathbb{R}), \tag{12}$$

where $[A_1, A_2] \equiv A_1 A_2 - A_2 A_1$ for linear operators A_1, A_2 acting in $\mathcal{H}_{\mathbb{C}}$ (on a domain where all operators are defined). Moreover, usually it is assumed that one has the existence of a unique normalized eigen vector (*vacuum*) $\Omega \in \mathcal{H}_{\mathbb{C}}$ of A such that

$$A\Omega = 0. \tag{13}$$

A system $\mathcal{H}_{\mathbb{C}}$, ϕ, π, A and Ω that satisfy (11), (12) and (13) defines a canonical quantum field theory. If such a system satisfies, in addition, an assumption denoted as the ϕ-bound, then the canonical quantum field theory admits an expression in the framework of *Gaussian* white noise analysis (cf. Theorem 2.1 given below).

In order to explain this, we prepare some notions.
Let

$$U(\xi) \equiv e^{i\phi(\xi)}, \qquad V(\xi) \equiv e^{i\pi(\xi)}, \qquad \xi \in \mathcal{S}(\mathbb{R}^s, \mathbb{R}),$$

the unitary operator associated to $\phi(\xi)$ resp. $\pi(\xi)$, then (2.1) can be rewritten as the following Weyl (commutation) relation:

$$U(\xi)U(\eta) = U(\xi + \eta), \quad V(\xi)V(\eta) = V(\xi + \eta), \tag{14}$$
$$U(\xi)V(\eta) = V(\eta)U(\xi)e^{i(\xi,\eta)}, \forall \xi, \eta \in \mathcal{S}(\mathbb{R}^s, \mathbb{R}),$$

and (2.2) is then replaced by (cf.[45])

$$[A, U(\xi)] = U(\xi)\pi(\xi) + \frac{1}{2}(\xi, \xi)U(\xi), \tag{15}$$

on a suitable dense domain. Under the additional assumption that the Hamiltonian A is invariant under time reversal, from (14) and (15) one deduces the following *Araki relation*:

$$\Big(U(\xi)\Omega, \ AU(\eta)\Omega\Big) = \frac{1}{2}(\xi, \eta)\Big(U(\xi)\Omega, \ U(\eta)\Omega\Big), \qquad \forall \xi, \eta \in \mathcal{S}(\mathbb{R}^s, \mathbb{R}) \tag{16}$$

(2.6) implies the following lemma (cf.[20–22,66,70]):

Lemma 1. *On a given Hilbert space $\mathcal{H}_{\mathbb{C}}$, assume that for (ϕ, A, Ω) the relation (16) holds. For $\xi_1, \ldots, \xi_n \in \mathcal{S}(\mathbb{R}^s, \mathbb{R})$ $(n \in \mathbb{N})$, denote $\phi(\xi) \equiv (\phi(\xi_1), \ldots, \phi(\xi_n))$. Suppose that $F \in C^1(\mathbb{R}^s, \mathbb{R})$ so that $F(\phi(\xi))\Omega$, $(D_j F)(\phi(\xi))\Omega \in \mathcal{H}_{\mathbb{C}}$, $j = 1, 2, \ldots, n$, where $D_j F$ denotes the j-th partial derivative of F. Then, denoting $\mathcal{D}(A^{\frac{1}{2}})$ the domain of the operator $A^{\frac{1}{2}}$ for the non-negative self-adjoint operator A, it holds that $F(\phi(\xi))\Omega \in \mathcal{D}(A^{\frac{1}{2}})$ and*

$$\Big(F(\phi(\xi))\Omega, \ AF(\phi(\xi))\Omega\Big)_{\mathcal{H}_{\mathbb{C}}} = \frac{1}{2}\sum_{i,j=1}^{n}(\xi_i, \xi_j)_{L^2(\mathbb{R}^s)}\Big((D_iF)(\phi(\xi))\Omega,$$
$$(D_jF)(\phi(\xi))\Omega\Big)_{\mathcal{H}_{\mathbb{C}}}. \tag{17}$$

For the proof of Lemma 1, cf.,[66] Prop. 2.1 and Lemma 11.6 of[70].

By Lemma 1, if the quadruplet $(\mathcal{H}_{\mathbb{C}}, \phi, A, \Omega)$ satisfies the assumptions of Lemma 1 and in addition the condition of ϕ-bound given in Definition 1 below (historically related to ϕ-bounds introduced in,[56] see also[100]) holds, then the corresponding canonical quantum field theory admits a *moment inequality* for the field operator ϕ (cf. Lemma 4.4 of[94] and Lemma 11.10 of[70]). By using a bound for $|(\Omega, \phi(\xi)^n \Omega)|$, $\xi \in \mathcal{S}(\mathbb{R}^s, \mathbb{R})$, $n \in \mathcal{N}$, derived from the *moment inequality*, we can show that the function $\mathcal{S}(\mathbb{R}^s, \mathbb{R}) \ni \xi \mapsto (\Omega, \exp(i\phi(\xi))\Omega) \in \mathbb{C}$ is a U-functional (cf. Chapter 1). Then applying the *representation theorem* by means of Hida distributions (cf. Chapter 1) the main theorem in this section, Theorem 1 given below, can be proved (cf.,[94] and Section 11.C of[70] for the proof of this theorem).

Definition 1. [ϕ-bound] If there exist constants α, $\beta > 0$, $\gamma \geq 0$ and $p \in \mathbb{N}_0$, so that for all $\xi \in \mathcal{S}(\mathbb{R}^s, \mathbb{R})$, the following inequality holds

$$\pm\phi(\xi) \leq \alpha A + \beta|\xi|^2_{2,p} + \gamma,$$

in the sense of quadratic forms on the domain

$$\mathcal{D}_\xi \equiv \{F(\phi(\xi))\Omega, \ F \in \mathcal{S}(\mathbb{R}, \mathbb{R})\},$$

where $|\xi|^2_{2,p} = \| \left(-\Delta + 1 + |\mathbf{x}|^2\right)^p \xi(\mathbf{x})\|^2_{L^2(\mathbb{R}^s)}$, $\mathbf{x} \in \mathbb{R}^s$ and Δ is the Laplace operator on $\mathbb{R}^s$, then one says that a ϕ-bound holds for $(\mathcal{H}_{\mathbb{C}}, \phi, A, \Omega)$.

Recall that we denote the *Gaussian* white noise process on $\mathbb{R}^d$ by ω (ω is the $\mathcal{S}'(\mathbb{R}^s, \mathbb{R})$-valued centered Gaussian random variable) the probability law of which is μ defined by (1). We shall denote the space of Hida distributions with respect to this *Gaussian* white noise by $(\mathcal{S})'$ (cf. Chapter 1, cf. also (10) in this chapter).

Theorem 1. *Assume that* $(\mathcal{H}_{\mathbb{C}}, \phi, A, \Omega)$ *admits the Araki relation* (16), *and that a ϕ-bound holds for this quadruplet. Then there exists a positive element* $\Xi \in (\mathcal{S})'$ *so that for all* $\xi \in \mathcal{S}(\mathbb{R}^s, \mathbb{R})$, $\omega \in \mathcal{S}'(\mathbb{R}^s, \mathbb{R})$, *the following equality holds:*

$$\left(\Omega, e^{i\phi(\xi)}\Omega\right) = \langle\Xi, \exp(\frac{i}{\sqrt{2}} < \cdot, \xi >)\rangle.$$

Remark 1. By the inclusion (10), the statement of the existence of $\Xi \in (\mathcal{S})'$ given in Theorem 1 can be interpreted by means of a functional analytic expression as the existence of an element of the algebraic direct sum $\bigoplus_{n=0}^{\infty}(\mathcal{S}'(\mathbb{R}^s, \mathbb{R}))^{\hat{\otimes}n}$ that is the S-transform of $\Xi \in (\mathcal{S})'$.

2.2. *Examples*

In this subsection we give some examples of applications of Theorem 1. The following examples show namely that some concrete scalar quantum field models, such as the free field, $P(\phi)_2$, Høegh-Krohn, and Albeverio and Høegh-Krohn models discussed in, e.g.,[4,15–19,56,59,63,100] admit expressions in terms of white noise analysis.

Example 1. [The free field]
i)   Take $\mathcal{H}_{\mathbb{C}}$ = the complex space $L^2(\mathcal{S}'(\mathbb{R}^s, \mathbb{R}), \mu) = (L^2)$, where μ is the white noise measure defined by (1)   (with d replaced by s).
ii)   For $\xi \in \mathcal{S}(\mathbb{R}^s, \mathbb{R})$, denoting the Borel σ-field of $\mathcal{S}'(\mathbb{R}^s, \mathbb{R})$ by $\mathcal{B}$, let $\phi(\xi)$ be the random variable on the probability space $(\mathcal{S}'(\mathbb{R}^s, \mathbb{R}), \mathcal{B}, \mu)$ such that

$$\phi(\xi)(\omega) = \frac{1}{\sqrt{2}} \left\langle \omega(\cdot), (-\Delta + m^2)^{-\frac{1}{4}} \xi \right\rangle, \qquad \omega(\cdot) \in \mathcal{S}'(\mathbb{R}^s, \mathbb{R}),$$

where $m > 0$ is a given *mass* and $(-\Delta + m^2)^{-\frac{1}{4}}$ is a pseudo differential operator built with the Laplace operator Δ on $\mathbb{R}^s$.
iii)   Let $A = d\Gamma\left((-\Delta + m^2)^{\frac{1}{2}}\right)$ on $\mathcal{P}(\mathcal{S}'(\mathbb{R}^s))$ be the operator (called as a $d\Gamma$-operator) such that for

$$F(\omega) \equiv :< \omega(\cdot), f_1 > \cdots < \omega(\cdot), f_n >:, \ f_j \in \mathcal{S}(\mathbb{R}^s, \mathbb{R}), \ j = 1, \ldots, n, \ n \in \mathbb{N},$$

$$\begin{aligned}\left(d\Gamma\left((-\Delta + m^2)^{\frac{1}{2}}\right) F\right)(\omega) = &:< \omega(\cdot), (-\Delta + m^2)^{\frac{1}{2}} f_1 > \cdots \\ &< \omega(\cdot), f_{n-1} >< \omega(\cdot), f_n >: + \cdots \\ &+ :< \omega(\cdot), f_1 > \cdots < \omega(\cdot), f_{n-1} > \\ &< \omega(\cdot), (-\Delta + m^2)^{\frac{1}{2}} f_n >:,\end{aligned}$$

for $F \in \mathcal{P}(\mathcal{S}'(\mathbb{R}^s))$, where $\mathcal{P}(\mathcal{S}'(\mathbb{R}^s))$ is the space of smooth polynomials defined by (2).
Then, it is well known (cf. e.g.,[96,97]) that A is essentially self-adjoint in (L^2). We denote its unique self-adjoint extension by the same notation A.
iv)   Let $\Omega = 1 \in \mathcal{H}_{\mathbb{C}}$.
We easily see that the quadruplet $(\mathcal{H}_{\mathbb{C}}, \phi, A, \Omega)$ defined through i) - iv) satisfies (11) (the CCR) and the ϕ-bound defined by Definition 1 (for the detailed proof, cf. Example 11.12 in section 11 of[70]). Then, by Theorem 2.1, for this $(\mathcal{H}_{\mathbb{C}}, \phi, A, \Omega)$ the relation

$$\left(\Omega, e^{i\,\phi(\xi)}\Omega\right) = \left\langle \Xi, \exp(\frac{i}{\sqrt{2}} < \cdot, \xi >)\right\rangle$$

holds for some $\Xi \in (\mathcal{S})'$, for all $\xi \in \mathcal{S}(\mathbb{R}^s, \mathbb{R})$.

Remark 2. When we interpret s as the space dimension in a relativistic $d = s + 1$ space-time dimensional world, then ϕ is interpreted as the time zero (quantum) free field (operator). The corresponding π (related to ϕ by (11), see also Remark 3 below) is interpreted as the time zero momentum of the free field, A is the Hamiltonian of the (time zero) free field. The corresponding operator $e^{itA}\phi(\xi)e^{-itA}$, defined e.g. on $\mathcal{P}(\mathcal{S}'(\mathbb{R}^s))$ can be seen as an operator in the space $\mathcal{H}_{\mathbb{C}}$. But it can also be identified with Nelson's free field as a random field over $\mathbb{R}^d$, integrated against the test function ξ over $\mathbb{R}^s$. Nelson's free field,[88,99] see below, has as distribution the standard normal distribution with mean zero and covariance $(-\Delta_d + m^2)^{-1}$, with Δ_d the Laplacian in $\mathbb{R}^d$.

Remark 3. 1°) By using D_ξ, the Gâteaux derivative of functions on $\mathcal{S}'(\mathbb{R}^s, \mathbb{R})$ in direction $\xi \in \mathcal{S}(\mathbb{R}^s, \mathbb{R})$, and its adjoint D_ξ^*, also by using the creation operator $a^*(\xi)$ and annihilation operators $a(\xi)$ on the Fock space $\mathrm{Exp}(\mathcal{H}_{\mathbb{C}})$ (cf. (10)), we can express $\phi(\xi)$ in i) as

$$\phi(\xi) = \frac{1}{\sqrt{2}}\left(D^*_{(-\Delta+m^2)^{-\frac{1}{4}}\xi} + D_{(-\Delta+m^2)^{-\frac{1}{4}}\xi}\right)$$

$$= \frac{1}{\sqrt{2}}\left(a^*((-\Delta+m^2)^{-\frac{1}{4}}\xi) + a((-\Delta+m^2)^{-\frac{1}{4}}\xi)\right).$$

Also,

$$\pi(\xi) = \frac{i}{\sqrt{2}}\left(D^*_{(-\Delta+m^2)^{-\frac{1}{4}}\xi} - D_{(-\Delta+m^2)^{-\frac{1}{4}}\xi}\right)$$

$$= \frac{i}{\sqrt{2}}\left(a^*((-\Delta+m^2)^{-\frac{1}{4}}\xi) - a((-\Delta+m^2)^{-\frac{1}{4}}\xi)\right),$$

and

$$A = \sum_{k\in\mathbb{N}} D^*_{(-\Delta+m^2)^{-\frac{1}{4}}\mathbf{e}_k} D_{(-\Delta+m^2)^{-\frac{1}{4}}\mathbf{e}_k},$$

where $\{\mathbf{e}_k\}_{k\in\mathbb{N}}$ is a complete orthonormal system (CONS) of $L^2(\mathbb{R}^s, \mathbb{R})$. For the derivation of the above expressions of $\phi(\xi)$ and $\pi(\xi)$, cf., e.g., Section X.6 of,[97] Section 11.B of[70] and,[6] and for detailed discussions on the directional derivative D_ξ, cf.[35-37,79] and[93].

2°) For the present example, $\Xi \in (\mathcal{S})'$ is the Radon-Nikodym derivative of $\tilde{\mu}$ with respect to μ:

$$\Xi = \frac{d\tilde{\mu}}{d\mu},$$

where $\tilde{\mu}$ is the Gaussian measure on $\mathcal{S}'(\mathbb{R}^s, \mathbb{R})$ such that

$$\int_{\mathcal{S}'(\mathbb{R}^d,\mathbb{R})} e^{i<\omega,\xi>} d\tilde{\mu}(\omega) = \exp\left(-\frac{1}{2}\left\|(-\Delta+m^2)^{-\frac{1}{4}}\xi\right\|^2\right), \qquad \xi \in \mathcal{S}(\mathbb{R}^s, \mathbb{R}).$$

Example 2. $[P(\phi)_2$ model] For the detailed discussion of this example, cf. Example 11.14 in section 11 of[70], and[94]. For the present example we let $s = 1$. We start the explanation from a cutoff model. As in Example 1, we set, for $s = 1$, the following:

i) Take

$$\mathcal{H}_{\mathbb{C}} = \text{the complex space } L^2(\mathcal{S}'(\mathbb{R}, \mathbb{R}), \mu) = (L^2).$$

ii) For $\xi \in \mathcal{S}(\mathbb{R}, \mathbb{R})$, denoting the Borel σ-field of $\mathcal{S}'(\mathbb{R}, \mathbb{R})$ by $\mathcal{B}$, let $\phi(\xi)$ be the random variable on the probability space $(\mathcal{S}'(\mathbb{R}, \mathbb{R}), \mathcal{B}, \mu)$ such that

$$\phi(\xi)(\omega) = \frac{1}{\sqrt{2}} \left\langle \omega(\cdot), (-\frac{d^2}{dx^2} + m^2)^{-\frac{1}{4}} \xi \right\rangle, \qquad \omega(\cdot) \in \mathcal{S}'(\mathbb{R}, \mathbb{R}),$$

where $m > 0$ is a given positive number (called *mass*, cf. Example 1).

iii) Let $A_0 = d\Gamma \left((-\frac{d^2}{dx^2} + m^2)^{\frac{1}{2}} \right)$ on $P(\mathcal{S}'(\mathbb{R}))$ be the $d\Gamma$-operator similarly defined as Example 1-iii) for $s = 1$.

iv) Let $\{\delta_{x,n}\}_{n \in \mathbb{N}} \subset \mathcal{S}'(\mathbb{R}, \mathbb{R})$ be a (weakly) approximation sequence of Dirac measures at $x \in \mathbb{R}$. Let $l > 0$ and set

$$V_{l,n} \equiv \int_{-l}^{l} : P(\phi(\delta_{x,n})) : dx,$$

where P is a given polynomial that satisfies either the following (A) or (B):

(A) $\quad P(u) = \lambda \sum_{k=0}^{2n} a_k u^k, \quad n \in \mathbb{N} \text{ with } \lambda, a_{2n} > 0 \text{ and } \frac{\lambda}{m^2} \text{ small}$

enough.

(B) $\quad P(u) = \lambda \sum_{k=0}^{n} a_k u^{2k} + bu, \quad n \in \mathbb{N} \text{ with } a_n > 0, \quad b \in \mathbb{R}.$

l is called a *"space cutoff"*. For example, in the case where P is given by (B), we precisely have

$$: P(\phi(\xi)) := \sum_{k=0}^{n} a_k : \phi^{2k}(\xi) : + b \phi(\xi),$$

with

$$: \phi^m(\xi) := \sum_{k=0}^{[\frac{m}{2}]} (-1)^k \binom{m}{2k} (2k - 1)!! (\phi(\xi))^{m-2k} \left\| (\frac{d^2}{dx^2} + m^2)^{-\frac{1}{4}} \xi \right\|_{L^2(\mathbb{R})}^{2k},$$

$$\xi \in \mathcal{S}(\mathbb{R}, \mathbb{R}).$$

$(: \phi^m(\cdot) :$ is the m-th Wick power of $\phi(\cdot)$, for this concept see, e.g.,[46,59,67–69,77,78,100]).

v)　　Then, it is known (cf.[63,87,88,100]) that for each fixed $l > 0$, $\lim_{n \to \infty} V_{l,n} = V_l$, $\forall p > 1$ exists in $L^p(\mu)$, and the operator

$$\hat{A}_l \equiv A_0 + V_l \tag{18}$$

is essentially self-adjoint on $\mathcal{P}(\mathcal{S}'(\mathbb{R}))$, on the domain of smooth cylinder functions on $\mathcal{S}'(\mathbb{R})$, and its spectrum is bounded below. By the same symbol $\hat{A}_l$, we shall denote the self-adjoint operator on $L^2(\mu) = (L^2)$, that is the unique self-adjoint extension of $\hat{A}_l$ defined by (18), and then define

$$A_l \equiv \hat{A}_l - \inf \operatorname{spec} \left(\hat{A}_l \right). \tag{19}$$

vi)　　For each fixed $l > 0$, a unique eigenvector Ω_l of A_l to the eigenvalue zero exists, thus

$$\Omega_l = \text{the unique ground state of } A_l.$$

Hence, in particular, $A_l \Omega_l = 0$.

vii)　　It is not hard to see that $\phi(\xi)$ resp., and V_l defined by ii) resp., v) commute and

$$e^{i\phi(\xi)} \Omega_l \in \mathcal{D}(A_0) = \text{the domain of the self-adjoint operator } A_0 \text{ on } (L^2).$$

By this, it is easy to see that, for each fixed $l > 0$, the quadruple $(\mathcal{H}_{\mathbb{C}}, \phi, A_l, \Omega_l)$ defined by i)- vi) satisfies (11) - (17), in particular, satisfies Araki relation.

Moreover, Fröhlich[56] showed that for this model, the following "ϕ-bound" holds:
There exist some real α, β and γ that do not depend on $l > 0$, and such that

$$\pm\phi(\xi) \leq \alpha A_l + \beta \|\|\xi\|\|^2 + \gamma, \qquad \forall \xi \in \mathcal{S}(\mathbb{R}),$$

where

$$\|\|\xi\|\|^2 := \int_{\mathbb{R}} (2 + x^2) |\xi(x)|^2 dx.$$

Since, $\|\|\xi\|\| \leq \sqrt{2} |\xi|_{2,p}$　(cf. Chapter 1), holds, we see the following:

viii)　　For the present example, the ϕ-bound defined by Definition 1 holds.

Consequently, from vii) and viii), by applying Theorem 1, we conclude that for each fixed $l > 0$ there exists $\Xi_l \in (\mathcal{S})'$　(in particular $\Xi_l \in (\mathcal{S})'_+$ (cf. Chapter 1)) so that

$$\left(\Omega_l, e^{i\phi(\xi)} \Omega_l \right) = \langle \Xi_l, \exp(\frac{i}{\sqrt{2}} < \cdot, \xi >) \rangle, \qquad \forall \xi \in \mathcal{S}(\mathbb{R}). \tag{20}$$

This completes the discussion for $P(\phi)_2$ model with a cut off $l > 0$.

In[94] and Example 11.14 in section 11 of[70], by using the fact that the ϕ-bound given by viii) is independent of $l > 0$ and the result that the limit $\lim\limits_{l \to \infty} \Omega_l$ exists in some Hilbert space (cf.[56]), the existence of the limit $\lim\limits_{l \to \infty} \Xi_l \in (\mathcal{S})'$ is proven.

Example 3. [time zero Høegh-Krohn model]   For the detailed discussion of this example, cf. Example 11.14 in section 11 of[70]. As in Example 2 we let $s = 1$. We set the following:

i)   Take

$$\mathcal{H}_{\mathbb{C}} = \text{the complex space } L^2(\mathcal{S}'(\mathbb{R}^s, \mathbb{R}), \mu) = (L^2).$$

ii)   For $\xi \in \mathcal{S}(\mathbb{R}, \mathbb{R})$, denoting the Borel σ-field of $\mathcal{S}'(\mathbb{R}, \mathbb{R})$ by $\mathcal{B}$, let $\phi(\xi)$ be the random variable on the probability space $(\mathcal{S}'(\mathbb{R}, \mathbb{R}), \mathcal{B}, \mu)$ such that

$$\phi(\xi)(\omega) = \frac{1}{\sqrt{2}} \left\langle \omega(\cdot), (-\frac{d^2}{dx^2} + m^2)^{-\frac{1}{4}}\xi \right\rangle, \quad \omega(\cdot) \in \mathcal{S}'(\mathbb{R}, \mathbb{R}), \quad \xi \in \mathcal{S}(\mathbb{R}, \mathbb{R}),$$

where $m > 0$ is a given *mass* (cf. Examples 1, 2).

iii)   Let $A_0 = d\Gamma\left((-\frac{d^2}{dx^2} + m^2)^{\frac{1}{2}}\right)$ on $\mathcal{P}(\mathcal{S}'(\mathbb{R}))$ be the $d\Gamma$-operator similarly defined as in Example 1-iii) for $s = 1$.

iv)   Let $\{\delta_{x,n}\}_{n \in \mathbb{N}} \subset \mathcal{S}'(\mathbb{R}, \mathbb{R})$ be a (weakly) approximation sequence for the Dirac measure at $x \in \mathbb{R}$. Let $V_{l,n}$ be the potential given by the truncated Wick exponential (cf. Chapter 1) such that

$$V_{l,n} \equiv \int_{-l}^{l} : \exp\{\alpha\phi(\delta_{x,n})\} : dx, \quad |\alpha| < \sqrt{2\pi}.$$

v)   Then, $V_{l,n} \geq 0$ for all $l \geq 0$, $n \in \mathbb{N}$, and by[72] the operator defined by

$$A_{l,n} \equiv A_0 + V_{l,n}$$

is shown to be essential self-adjoint on $\mathcal{P}(\mathcal{S}'(\mathbb{R}))$ (even for $|\alpha| \leq \sqrt{4\pi}$ the random field (Nelson's Euclidean field, cf. Remark 2 and Remark 4 below) corresponding to this quantum field model can be considered, cf.,[7,18–22,72,80] cf. also[90]). By the same symbol we denote its unique self-adjoint extension, and we let $\Omega_{l,n}$ be a unique ground state of $A_{l,n}$. Also, it is possible to show that a ϕ-bound holds with constants α, β and γ in Definition 1 that are independent of $l \geq 0$, $n \in \mathbb{N}$. Since, $V_{l,n}$ commutes with $\phi(\xi)$, $A_{l,n}$ satisfies

Araki's relation (16).

vi)   By[72] and[19] there exists a limiting vacuum $\lim\limits_{l\to\infty,n\to\infty} \Omega_{l,n} = \Omega,$
which belongs to some Hilbert space $\mathcal{H}$, and by which $(\Omega, \exp(i\phi(\xi)))_{\mathcal{H}}$
defines a U-functional as proven in [16,17] (cf. Chapter 1).

vii)   Consequently, from v) and vi), by applying Theorem 1, we conclude that for each fixed $l > 0$ there exists $\Xi_l \in (\mathcal{S})'$ (in particular $\Xi_l \in (\mathcal{S})'_+$ (cf. Chapter 1)) so that

$$\left(\Omega_l, e^{i\,\phi(\xi)}\Omega_l\right) = \langle \Xi_l, \exp(\frac{i}{\sqrt{2}} < \cdot, \xi >)\rangle, \qquad \forall \xi \in \mathcal{S}(\mathbb{R}). \qquad (21)$$

A corresponding result holds for $l \to \infty$, see.[16,17]

Remark 4.   The models discussed in Examples 1, 2 and 3 have been shown to lead to models satisfying all axioms of a (scalar) Euclidean quantum field theory on space-time dimension $d = 2$ (and for all $d \geq 2$ in the case of Example 1). (For other examples for $s = 1$ see also[58].) These models are given as moments (called Schwinger functions) of probability measures on $\mathcal{S}'(\mathbb{R}^2)$ (non Gaussian in the cases of Examples 2 and 3, Gaussian in the case of Example 1). These measures can be characterized as positive generalized white noise functionals, as shown in[15-17] (where the $P(\phi)_2$ resp. $(\exp\phi)_2$ models are considered, the $(\sin\phi)_2$ model can also be treated similarly). Analytic continuation in the time variables of the Schwinger functions lead to corresponding Wightman functions (correlation functions for relativistic quantum fields) satisfying all axioms of a (scalar) relativistic quantum field theory on space-time dimension $d = 2$ (and for all $d \geq 2$ in the case of Example 1). These are the relativistic $P(\phi)_2$-field model resp. $(\exp\phi)_2$-field model resp. $(\sin\phi)_2$-field model, see, e.g.,[7,13,18-22,27,28,59,60,72,100]. Only partial results for $d \geq 3$ are known, see e.g.,[30,47,98] and references therein. These models can be seen as canonical realized in $L^2(\mu)$, with μ a probability measure on $\mathcal{S}'(\mathbb{R}^s)$ for all $s \geq 1$ for Example 1 and for $s = 1$ in Examples 2 and 3, even if some uniqueness problems on the determination of the Hamiltonian (and generators of the Poincaré group) are still open in Examples 2 and 3 (see, e.g.,[1-3,5,6,22,23,31,41,82]).

3. Application of *Gaussian* white noise analysis to relativistic quantum field models with indefinite metric

In this section we present some results discussed in[64]. In order to make the notations and notions clear, we adopt here slightly modified ones to these used in[64].

Let d be a given *space-time* dimension. We denote $\mathbf{x} \equiv (t, \vec{x}) \in \mathbb{R} \times \mathbb{R}^{d-1}$.

As in Example 1 in section 2, let μ be the **Gaussian** white noise measure on $\mathcal{S}'(\mathbb{R}^d, \mathbb{R})$, and define as usual

$$(L^2) = \text{the complex Hilbert space } L^2(\mathcal{S}'(\mathbb{R}^d, \mathbb{R}), \mu). \tag{22}$$

We recall the definition of space of Kondratiev distributions $(\mathcal{S})^{-1}$ (cf. Chapter 1). Precisely, let

$$(\mathcal{S})^1 \equiv \bigcap_{p,q \geq 0} (\mathcal{H}_p)_q^1, \tag{23}$$

with

$$(\mathcal{H}_p)_q^1 \equiv \left\{ f \in (L^2) \,\middle|\, f(\omega) = \sum_{n=0}^{\infty} \left\langle : \omega^{\otimes n} :, f^{(n)} \right\rangle \text{ satisfying } \|f\|_{p,q,1}^2 < \infty \right\}, \tag{24}$$

where

$$\|f\|_{p,q,1}^2 \equiv \sum_{n=0}^{\infty} (n!)^2 \, 2^{nq} \left\| (-\Delta + x^2 + 1)^{\otimes n} f^{(n)} \right\|_{L^2(\mathbb{R}^{nd})}^2,$$

Δ being the *Laplace operator* on $\mathbb{R}^d$.

We let $(\mathcal{H}_{-p})_{-q}^{-1}$ be the dual space of $(\mathcal{H}_p)_q^1$ with respect to $L^2(\mu)$, and let $(\mathcal{S})^{-1}$ be the dual space of $(\mathcal{S})^1$ with respect to $L^2(\mu)$. As a consequence, we have the definition

$$(\mathcal{S})^{-1} \equiv \bigcup_{p,q \geq 0} (\mathcal{H}_{-p})_{-q}^{-1}, \tag{25}$$

and the dense inclusions (cf. (10))

$$(\mathcal{S})^1 \subset (L^2) = L^2(\mu) \subset (\mathcal{S})^{-1}. \tag{26}$$

For the Dirac measure $\delta_{t,\vec{x}} \in \mathcal{S}'(\mathbb{R}^d, \mathbb{R})$ at $(t, \vec{x}) \in \mathbb{R} \times \mathbb{R}^{d-1}$, define (cf. Example 2 in section 2)

$$\phi(t, \vec{x}) \equiv \langle \omega, \delta_{t,\vec{x}} \rangle \in (\mathcal{S})^{-1}. \tag{27}$$

Let $H(z)$, $z \in U \subset \mathbb{C}$, be a given *holomorphic* function in U, an open neighborhood of $0 \in \mathbb{C}$, having the representation

$$H(z) = \sum_{k=0}^{\infty} \frac{a_k z^k}{k!}, \qquad z \in U, \tag{28}$$

with

$$a_0 = 0. \tag{29}$$

Define (for notations such as $\diamond$ used below, cf. Chapter 1)

$$H^\diamond \left(\phi(t,\vec{x}) \right) \equiv \sum_{k=0}^{\infty} \frac{a_k}{k!} \, \phi(t,\vec{x})^\diamond = \sum_{k=0}^{\infty} \frac{a_k}{k!} \left\langle : \omega^{\otimes k} :, \delta_{t,\vec{x}}^{\otimes k} \right\rangle. \tag{30}$$

Then, we see that (cf. Theorem III.5, Corollary III.6 of[64])

$$H^\diamond \left(\phi(t,\vec{x}) \right) \in (\mathcal{S})^{-1}, \quad (t,\vec{x}) \in \mathbb{R} \times \mathbb{R}^{d-1}; \quad \int_{\mathbb{R}^d} H^\diamond \left(\phi(t,\vec{x}) \right) \, dt \, d\vec{x} \in (\mathcal{S})^{-1},$$

and

$$\Phi_H \equiv \exp^\diamond \left(- \int_{\mathbb{R}^d} H^\diamond \left(\phi(t,\vec{x}) \right) \, dt \, d\vec{x} \right)$$

$$= \sum_{n=0}^{\infty} \frac{(-1)^n}{n!} : \left(\int_{\mathbb{R}^d} H^\diamond \left(\phi(t,\vec{x}) \right) \, dt \, d\vec{x} \right)^n : \; \in (\mathcal{S})^{-1}. \tag{31}$$

For a given $m_0 > 0$, which is understood as a mass in quantum field theory, and for each $\alpha \in (0, \frac{1}{2}]$, let $\mathcal{G}_\alpha$ be the pseudo differential operator given by

$$\mathcal{G}_\alpha \equiv \left(-\Delta + m_0^2 \right)^{-\alpha}, \tag{32}$$

where Δ is again the *Laplace operator* on $\mathbb{R}^d$. For each fixed and given $\alpha \in (0, \frac{1}{2}]$ and a given Φ_H defined by (31), we define a system of *"Schwinger functions"* $\left\{ S_n^{\Phi_H} \right\}_{n \in \mathbb{N}_0}$ as follows:
$S_0^{\Phi_H} = 1$ and

$$S_n^{\Phi_H}(f_1 \otimes \cdots \otimes f_n) \equiv \left(\Phi_H, \, < \omega, \mathcal{G}_\alpha f_1 > \cdots < \omega, \mathcal{G}_\alpha f_n > \right), \; f_j \in \mathcal{S}(\mathbb{R}^d, \mathbb{R}),$$
$$j = 1, \ldots, n, \, n \in \mathbb{N}, \tag{33}$$

where $(\cdot, \cdot)$ denotes the bilinear dual pairing between $\mathcal{P}' \left(\mathcal{S}'(\mathbb{R}^d, \mathbb{R}) \right)$ and $\mathcal{P} \left(\mathcal{S}'(\mathbb{R}^d, \mathbb{R}) \right)$, in particular in (33), between $(\mathcal{S})^{-1}$ and $(\mathcal{S})^1$ (for the notations cf. (2), and for general notions such as $\mathcal{P}(\mathcal{N}')$ and $\mathcal{P}'(\mathcal{N}')$ cf. Chapter 1).

In the following Theorem 2, we combine the results given in Theorems III.9, III.15, IV.1 and IV.5 of[64] (for the proof cf.[64]). The statement on the analytic continuations of the *"Schwinger functions"* to *"Wightman functions"*, IV.5 of[64], depends on the explicit formulas of Wightman functions derived by[9,10] and[62].

Theorem 2. Let $\left\{ S_n^{\Phi_H} \right\}_{n \in \mathbb{N}_0}$ be a system of *"Schwinger functions"* defined by (33). Then $\left\{ S_n^{\Phi_H} \right\}_{n \in \mathbb{N}_0}$ satisfies all the Osterwalder-Schrader axioms, the OS axioms in short, in constructive quantum field theory (cf., *e.g.,*[59,91,92,100]*) except for the reflection positivity OS3. Namely, it satisfies OS1 (temperedness), OS2 (Euclidean invariance), OS4 (symmetry) and OS5 (cluster property).*

Also, the system of "Schwinger functions" $\left\{S_n^{\Phi_H}\right\}_{n\in\mathbb{N}_0}$ admits an analytic continuation to a system of "Wightman functions" $\left\{W_n^{\Phi_H}\right\}_{n\in\mathbb{N}_0}$ which are explicitly given by (53), (55) and (56) in the next section, with a modification that c_n in (55) is exchanged by the a_n appearing in (28).

Moreover, the system of "Wightman functions" fulfills the **modified Wightman axioms W1-W4** and **modified W5** given in the next section that were proposed by Morchio and Strocchi (cf.[84] ; and cf. next section for a concise explanation on the modified Wightman axioms; for a discussion of axioms for relativistic quantum fields and corresponding Euclidean quantum fields see, e.g.,[85,86,91,92,104–106,108–111] .

As a consequence, the scalar quantum field model with the system of "Schwinger functions" $\left\{S_n^{\Phi_H}\right\}_{n\in\mathbb{N}_0}$ provides an example in the framework of **relativistic quantum field theory with indefinite metric**.

Remark 5. [Gaussian white noise to generalized white noise via T-transform] $1°$) The formulas presented in[64] (cf. Definition III.2 and (11) in Theorem III.9 of[64]), by which differ from (31), (32) and (33). However, the system of Schwinger functions $\left\{S_n^{\Phi_H}\right\}_{n\in\mathbb{N}_0}$ defined through each of the family of formulas are identical.

$2°$) In the next section we consider a system of Schwinger functions given by (49) (cf. also (50) and (51)), which is defined through moment functions corresponding to some **generalized white noise measures**. The system of Schwinger functions $\left\{S_n^{\Phi_H}\right\}_{n\in\mathbb{N}_0}$ discussed in the present subsection in the framework of the **Gaussian white noise** analysis is identical with it. In fact, through the T-transform on the Gaussian white noise functional, it is possible to see a general correspondence between $\left\{S_n^{\Phi_H}\right\}_{n\in\mathbb{N}_0}$ and the moment functions of some generalized white noise measures (cf. Remark III.7 of[64] in detail).

4. Application of *non-Gaussian* white noise to constructive quantum field theory

In this section we introduce mainly the results developed by[10] .

4.1. A general framework of relativistic quantum field theory with indefinite metric

Let us begin this subsection by introducing the axioms **W1** - **W4** for system of Wightman functions $\{W_n\}_{n\in\mathbb{N}_0}$ ($W_0 = 1$ for simplicity) for **scalar** quantum field theory with the space time dimension d.

W1 (Temperedness)   For each $n \in \mathbb{N}$, $W_n \in \mathcal{S}'(\mathbb{R}^{dn}, \mathbb{C})$.

W2 (Poincaré invariance)   There exists a representation $\mathcal{T}$ of the proper orthogonal Lorentz group $\mathcal{L}_+^\uparrow(\mathbb{R}^d)$ acting on $\mathbb{R}^d$, and for any Poincaré transformation $\mathbf{a}+\Lambda$, $\mathbf{a} \in \mathbb{R}^d$, $\Lambda \in \mathcal{L}_+^\uparrow(\mathbb{R}^d)$ and any $n \in \mathbb{N}$ the n-point functions $W_n(\mathbf{x}_1, \cdots, \mathbf{x}_n)$ is invariant under the transformation $\mathbf{a} + \Lambda$:

$$\mathcal{T}(\Lambda)^{\otimes n} W_n(\Lambda^{-1}(\mathbf{x}_1 - \mathbf{a}), \cdots, \Lambda^{-n}(\mathbf{x}_1 - \mathbf{a})).$$

W3 (Spectral condition)   For any $n \in \mathbb{N}$, the Fourier transform $\hat{W}_n(\mathbf{y}_1, \cdots, \mathbf{y}_{n-1})$ is supported in the backward light cones

$$\left\{ (\mathbf{y}_1, \cdots, \mathbf{y}_{n-1} \in \mathbb{R}^{d(n-1)} \,|\, \mathbf{y}_j^2 \geq 0, \ \tau_j < 0, \ 1 \leq j \leq n-1 \right\},$$

where $\mathbf{y}_j = (\tau_j, \vec{y}_j) \in \mathbb{R} \times \mathbb{R}^{d-1}$, and $\mathbf{y}_j^2 \equiv |\tau_j|^2 - |\vec{y}_j|^2$, is the Minkowski length of y_j.

W4 (Locality)   For $n \geq 2$, if $(\mathbf{x}_{j+1} - \mathbf{x}_j)^2 < 0$ for some $j \in \{1, \cdots, n-1\}$, then

$$W_n(\mathbf{x}_1, \cdots, \mathbf{x}_j, \mathbf{x}_{j+1}, \cdots, \mathbf{x}_n) = W_n(\mathbf{x}_1, \cdots, \mathbf{x}_{j+1}, \mathbf{x}_j, \cdots, \mathbf{x}_n).$$

Let $\mathcal{D}_0$ be the Borchers algebra over $\mathcal{S}(\mathbb{R}^d, \mathbb{C})$ such that

$$\mathcal{D}_0 \equiv \left\{ F = (f_0, f_1, \cdots) \,|\, f_0 \in \mathbb{C}, \ f_n \in \mathcal{S}(\mathbb{R}^{nd}, \mathbb{C}), \ n \in \mathbb{N} \right\},$$

and for $F = (f_0, f_1, \cdots)$, $G = (g_0, g_1, \cdots) \in \mathcal{D}_0$ define

$$F + G \equiv (f_0 + g_0, f_1 + g_1, \cdots), \quad F \otimes G \equiv \left(\sum_{j+l=0} f_j \otimes g_l, \ \sum_{j+l=1} f_j \otimes g_l, \cdots \right).$$

Suppose that we are given a system of Wightman functions $\{W_n\}_{n\in\mathbb{N}_0}$ satisfying **W1** - **W4**. Define a functional $\underline{W}$ on $\mathcal{D}_0$ as follows:

$$\underline{W}(F) \equiv \sum_{n=0}^{\infty} W_n(f_n), \qquad F = (f_0, f_1, \dots) \in \mathcal{D}_0, \tag{34}$$

where $W_0(f_0) = f_0$. Then, define a sesquilinear form $\langle \cdot, \cdot \rangle_W$ on $\mathcal{D}_0$ as follows:

$$\langle F, G \rangle_W \equiv \underline{W}(F^* \otimes G), \qquad F, G \in \mathcal{D}_0, \tag{35}$$

with $F^* \equiv (f_0^*, f_1^*, \cdots)$, $f_n^*(\mathbf{x}_1, \cdots, \mathbf{x}_n) \equiv \overline{f_n(\mathbf{x}_n, \cdots, \mathbf{x}_1)}$, $n \in \mathbb{N}_0$ for $F = (f_0, f_1, \dots) \in \mathcal{D}_0$.

Throughout this *subsection*, we assume that W_n satisfies the *hermiticity condition*, which means that W_n is identical with its complex conjugate having an opposite order of variables:

$$W_n(\mathbf{x}_1, \cdots, \mathbf{x}_n) = \overline{W_n(\mathbf{x}_n, \cdots, \mathbf{x}_1)}. \tag{36}$$

Next, define a quotient space $\mathcal{D}$ such that

$$\mathcal{D} \equiv \mathcal{D}_0 / \mathcal{N}_W \qquad \text{with} \qquad \mathcal{N}_W \equiv \{F \in \mathcal{D}_0 \,|\, \langle F, G \rangle = 0, \, \forall G \in \mathcal{D}_0\}. \tag{37}$$

Definition 2. [Majorant Hilbert space and metric operator] Under the condition that we are given a system of Wightman functions $\{W_n\}_{n \in \mathbb{N}_0}$ which satisfies **W1** - **W4** and the hermiticity condition (36), and the sesquilinear form $\langle \cdot, \cdot \rangle_W$ and the space $\mathcal{D}$ defined by (34), (35) and (37), suppose that there exists a Hilbert inner product $(\cdot, \cdot)$ on $\mathcal{D}$ such that

$$|\langle F, G \rangle_W| \le (F, F)^{\frac{1}{2}} (G, G)^{\frac{1}{2}}, \qquad F, G \in \mathcal{D}. \tag{38}$$

The **majorant Hilbert space** $\mathcal{H}$ is defined as the closure of $\mathcal{D}$ with respect to the norm corresponding to the inner product $(\cdot, \cdot)$ in (38) :

$$\mathcal{H} \equiv \overline{\mathcal{D}}^{(\cdot, \cdot)} \quad \text{with the inner product} \quad (\cdot, \cdot).$$

By (38), there exists a bounded self-adjoint operator T on $\mathcal{H}$ such that

$$\langle F, G \rangle_W = (F, TG), \qquad \forall F, G \in \mathcal{H}.$$

The operator T is called the **metric operator**.

On the dense domain $\mathcal{D} \subset \mathcal{H}$, for each $\xi \in \mathcal{S}(\mathbb{R}^d, \mathbb{C})$ by setting $F_\xi = (0, \xi, 0, \cdots) \in \mathcal{D}_0$, we can define a *field operator* $\phi(\xi)$ as follows:

$$(\phi(\xi))(G) \equiv F_\xi \otimes G + \mathcal{N}_W, \qquad G \in \mathcal{D}.$$

Then, for $f_n = \xi_1 \otimes \cdots \otimes \xi_n$, $\xi_j \in \mathcal{S}(\mathbb{R}^d, \mathbb{C})$, $j = 1, \dots, n$, $n \in \mathbb{N}$, we have

$$W_n(f_n) = (\Omega, T\phi(\xi_1) \cdots \phi(\xi_n)\Omega) = \langle \Omega, \phi(\xi_1) \cdots \phi(\xi_n)\Omega \rangle_W, \tag{39}$$

where $\Omega \equiv (1, 0, \cdots) + \mathcal{N}_W$.

By (39) and the hermiticity condition (36), the field operator $\phi(\xi)$ discussed above is *T-symmetric* for the *metric operator* T in the sense that for $\xi \in \mathcal{S}(\mathbb{R}^d, \mathbb{R})$:

$$T \phi(\bar{\xi})^* T^{-1} = \phi(\xi). \tag{40}$$

[84] shows that for a given system of Wightman functions $\{W_n\}_{n\in\mathbb{N}_0}$ a necessary and sufficient condition under which there exists a majorant Hilbert space (topology) is the following:

Modified W5 (Hilbert space structure condition)   *There exists a sequence* $\{p_n\}_{n\in\mathbb{N}}$ *such that*

$$p_n \: : \: \mathcal{S}(\mathbb{R}^d, \mathbb{C}) \longrightarrow [0, \infty), \qquad n \in \mathbb{N}$$

are Hilbert seminorms and

$$|W_{m+n}(\varphi^* \otimes \eta)| \leq p_m(\varphi)p_n(\eta), \ \forall \varphi \in \mathcal{S}(\mathbb{R}^{dm}, \mathbb{C}), \forall \eta \in \mathcal{S}(\mathbb{R}^{dn}, \mathbb{C}), \ m, n \in \mathbb{N}. \tag{41}$$

Axioms **W1-W4** together with **Modified W5** are called *modified Wightman axioms*.

4.2. Non-Gaussian white noise and relativistic quantum field models with indefinite metric

Firstly, we define a generalized white noise measure P on

$$\left(\mathcal{S}'(\mathbb{R}^d, \mathbb{R}), \mathcal{B}(\mathcal{S}'(\mathbb{R}^d, \mathbb{R})) \right),$$

that includes Gaussian white noise as a particular case, where $\mathcal{B}(\mathcal{S}'(\mathbb{R}^d, \mathbb{R}))$ is the Borel σ-field of $\mathcal{S}'(\mathbb{R}^d, \mathbb{R})$.

By the Bochner-Minlos theorem (cf. Chapter 1), we can define a unique probability measure P on $\left(\mathcal{S}'(\mathbb{R}^d, \mathbb{R}), \mathcal{B}(\mathcal{S}'(\mathbb{R}^d, \mathbb{R})) \right)$ such that its Fourier transform satisfies

$$\int_{\mathcal{S}'(\mathbb{R}^d, \mathbb{R})} e^{i<\xi, \omega>} \, dP(\omega) \: = \: \exp\left\{ \int_{\mathbb{R}^d} \psi(\xi(x)) dx \right\}, \qquad \xi \in \mathcal{S}(\mathbb{R}^d, \mathbb{R}), \tag{42}$$

where

$$\psi(t) \equiv i \, at - \frac{1}{2}\sigma^2 t^2 + \int_{\mathbb{R}\backslash\{0\}} \left(e^{ist} - 1 - \frac{i \, st}{1 + s^2} \right) dM(s), \quad t \in \mathbb{R} \tag{43}$$

with $a, \sigma \in \mathbb{R}$ are some given number and M is a given non-decreasing function satisfying

$$\int_{\mathbb{R}\backslash\{0\}} \min(1, \, s^2) \, dM(s) < \infty.$$

As in section 4.1, for some given $\alpha \in (0, \frac{1}{2}]$ and $m > 0$, let us define a pseudo differential operator G_α such that

$$G_\alpha \equiv \left(-\Delta + m^2 \right)^{-\alpha}, \tag{44}$$

where Δ denotes the Laplace operator on $\mathbb{R}^d$. If we denote its kernel again by $G_\alpha(\cdot)$, we have

$$G_\alpha(x) \equiv (2\pi)^{-d} \int_{\mathbb{R}^d} \frac{e^{ikx}}{(|k|^2 + m^2)^\alpha} \, dk, \qquad x \in \mathbb{R}, \tag{45}$$

then, we see that

$$(G_\alpha \xi)(x) = \int_{\mathbb{R}^d} G_\alpha(x-y)\,\xi(y)\,dy = (G_\alpha * \xi)(x), \qquad \xi \in \mathcal{S}(\mathbb{R}^d, \mathbb{R}), \tag{46}$$

where $G_\alpha * \xi$ stand for the convolution.

By (42) and (46), we can define a *convoluted generalized white noise measure* P_{G_α} on $\left(\mathcal{S}'(\mathbb{R}^d, \mathbb{R}), \mathcal{B}(\mathcal{S}'(\mathbb{R}^d, \mathbb{R}))\right)$, that is characterized by the following:

$$\int_{\mathcal{S}'(\mathbb{R}^d, \mathbb{R})} e^{i<\xi, \omega>} \, dP_{G_\alpha}(\omega) = \exp\left\{ \int_{\mathbb{R}^d} \psi\left(\xi\left(G_\alpha(x)\right)\right) dx \right\}, \qquad \xi \in \mathcal{S}(\mathbb{R}^d, \mathbb{R}). \tag{47}$$

For each $\xi \in \mathcal{S}(\mathbb{R}^d, \mathbb{R})$, define a *real valued* random variable on the probability space $\left(\mathcal{S}'(\mathbb{R}^d, \mathbb{R}), \mathcal{B}(\mathcal{S}'(\mathbb{R}^d, \mathbb{R})), P_{G_\alpha}\right)$, by the dualization such that

$$_{\mathcal{S}(\mathbb{R}^d, \mathbb{R})}\langle \xi, \omega \rangle_{\mathcal{S}'(\mathbb{R}^d, \mathbb{R})}, \tag{48}$$

and denote it simply by $\langle \xi, \omega \rangle$, that is taken as a *field operator* $\phi(\xi)$ of a *Euclidean field* (cf. previous sections in the present chapter) that corresponds to the probability space (a *random field*). Then by making use of these random variables through their *expectations* we define *Schwinger functions* S_n, $n \in \mathbb{N}_0$ on $\left(\mathcal{S}(\mathbb{R}^d, \mathbb{R})\right)^{\otimes n}$ as follows:

$$S_n(\xi_1, \cdots, \xi_n) \equiv \int_{\mathcal{S}'(\mathbb{R}^d, \mathbb{R})} \prod_{j=1}^{n} \langle \xi_j, \omega \rangle \, dP_{G_\alpha}(\omega), \quad \xi_j \in \mathcal{S}(\mathbb{R}^d, \mathbb{R}),$$

$$j = 1, \ldots, n, \ n \in \mathbb{N}, \tag{49}$$

with $S_0 = 1$. For the Schwinger functions S_n, we have the following expression in terms of *truncated Schwinger functions* S_k^T, $k \in \mathbb{N}$:

$$S_n(\xi_1, \cdots, \xi_n) = \sum_{I \in \mathcal{P}^n} \prod_{\{j_1, \cdots, j_k\} \in I} S_k^T(\xi_{j_1}, \cdots, \xi_{j_k}), \qquad n \in \mathbb{N}, \tag{50}$$

where $\mathcal{P}^n$ stands for the collection of all partitions I of $\{1, \cdots, n\}$ into sums of disjoint subsets, and for each $k \in \mathbb{N}$, by (47) S_k^T is explicitly given by

$$S_k^T(\xi_1, \cdots, \xi_k) = c_k \int_{\mathbb{R}^{dk}} G_\alpha^{(k)}(x_1, \cdots, x_k) \prod_{j=1}^{k} \xi_j(x_j) \prod_{j=1}^{k} dx_j,$$

$$\xi_1, \cdots, \xi_k \in \mathcal{S}(\mathbb{R}^d, \mathbb{R}), \tag{51}$$

with

$$c_1 = a + \int_{\mathbb{R}\setminus\{0\}} \frac{s^2}{1+s^2} dM(s), \quad c_2 = \sigma^2 + \int_{\mathbb{R}\setminus\{0\}} s^2 \, dM(s),$$
$$c_k = \int_{\mathbb{R}\setminus\{0\}} s^k \, dM(s), \ k \geq 3,$$

$$G_\alpha^{(k)}(x_1, \cdots, x_k) \equiv \int_{\mathbb{R}^d} \prod_{j=1}^k G_\alpha(x - x_j) \, dx, \qquad k \in \mathbb{N}. \tag{52}$$

Now, the Wightman functions W_n, $n \in \mathcal{N}_0$, are defined as functions whose *Laplace transforms* are the Schwinger functions. Namely, for each $n \in \mathbb{N}$, W_n is given by (cf. (50))

$$W_n(\xi_1, \cdots, \xi_n) = \sum_{I \in \mathcal{P}^n} \prod_{\{j_1, \cdots, j_k\} \in I} W_k^T(\xi_{j_1}, \cdots, \xi_{j_k}),$$
$$\text{with } \xi_j \in \mathcal{S}(\mathbb{R}^d, \mathbb{C}), \ j = 1, \cdots, n, \quad n \in \mathbb{N}, \tag{53}$$

where the *truncated Wightman functions* W_n^T, $n \in \mathbb{N}$, whose Fourier transforms are denoted by $\hat{W}_n^T$, $n \in \mathbb{N}$, are such that

$$S_n^T(\mathbf{x}_1, \cdots, \mathbf{x}_n) = (2\pi)^{-\frac{dn}{2}} \int_{\mathbb{R}^{dn}} e^{-\sum_{l=1}^n \tau_l t_l + i \vec{y}_l \vec{x}_l} \hat{W}_n^T(\mathbf{y}_1, \cdots, \mathbf{y}_l) \prod_{l=1}^n d\mathbf{y}_l, \tag{54}$$

for $t_1 < \cdots < t_n$, with the notations such that $\mathbf{x}_l = (t_l, \vec{x}_l), \mathbf{y}_l = (\tau_l, \vec{y}_l) \in \mathbb{R} \times \mathbb{R}^{d-1}$ (cf. **W1-W4**). Then, $W_1^T \equiv 0$ for simplicity, W_2 is given as c_2 times the two point function of the relativistic free field with mass m for $\alpha = \frac{1}{2}$ (cf. Example 1 in section 2), and for $n \geq 3$ with $\alpha \in (0, \frac{1}{2}]$ or $n = 2$ with $\alpha \in (0, \frac{1}{2})$ by[9,10] and[62] the following explicit formula holds:

$$\hat{W}_n^T(\mathbf{y}_1, \cdots, \mathbf{y}_n)$$
$$= c_n 2^{n-1} (2\pi)^d \left\{ \sum_{j=1}^n \prod_{l=1}^{j-1} \mu_\alpha^-(\mathbf{y}_l) \mu_\alpha(\mathbf{y}_j) \prod_{l=j+1}^n \mu_\alpha^+(\mathbf{y}_l) \right\} \delta\left(\sum_{l=1}^n \mathbf{y}_l\right), \tag{55}$$

with

$$\mu_\alpha^+(\mathbf{y}) \equiv (2\pi)^{-\frac{d}{2}} (\sin \pi \alpha) \frac{1_{\{\mathbf{y}^2 > m^2, \, \tau > 0\}}(\mathbf{y})}{(\mathbf{y}^2 - m^2)^\alpha},$$

$$\mu_\alpha^-(\mathbf{y}) \equiv (2\pi)^{-\frac{d}{2}} (\sin \pi \alpha) \frac{1_{\{\mathbf{y}^2 > m^2, \, \tau < 0\}}(\mathbf{y})}{(\mathbf{y}^2 - m^2)^\alpha},$$

$$\mu_\alpha(\mathbf{y}) \equiv (2\pi)^{-\frac{d}{2}} \left\{ (\cos \pi \alpha) 1_{\{\mathbf{y}^2 > m^2\}}(\mathbf{y}) + 1_{\{\mathbf{y}^2 < m^2\}}(\mathbf{y}) \right\} \frac{1}{|\mathbf{y}^2 - m^2|^\alpha},$$

where $\mathbf{y} \equiv (\tau, \vec{y}) \in \mathbb{R} \times \mathbb{R}^{d-1}$, $\mathbf{y}^2 \equiv \tau^2 - |\vec{y}|^2$ and $1_{\{\dots\}}(\mathbf{y})$ stands for the indicator function.

The following is a main result of this subsection, which is a part of Theorems 7 and 8 of[10].

Theorem 3. *Let $\{W_n\}_{n \in \mathbb{N}}$ be the sequence of Wightman functions defined by (53) with (52) and (55). It satisfies axioms* **W1-W4, Modified W5,** *the hermiticity condition (36) and the cluster property given by (56) below.*

In particular there exists a Hilbert space $(\mathcal{H}, (\cdot, \cdot))$, a continuous self-adjoint metric operator T on $\mathcal{H}$ fulfilling $T^2 = 1$ and local T-symmetric (cf. (40)) field operator $\phi(\xi)$ defined on a common dense domain $\mathcal{D} \subset \mathcal{H}$ for $\xi \in \mathcal{S}(\mathbb{R}^d, \mathbb{C})$ such that (39) holds.

Definition 3. A sequence of Wightman functions $\{W_n\}_{n \in \mathbb{N}}$ satisfies the cluster property if for any m, $n \in \mathbb{N}$ and any space-like $\mathbf{a} \in \mathbb{R}$ (i.e., $\mathbf{a}^2 < 0$ in Minkowski metric (cf. **W3**))

$$\lim_{\lambda \to \infty} W_{m+n}(\xi_1, \cdots, \xi_m, T_{\lambda \mathbf{a}} \xi_{m+1}, \cdots, T_{\lambda \mathbf{a}} \xi_{m+n})$$
$$= W_m(\xi_1, \cdots, \xi_m) W_n(\xi_{m+1}, \cdots, \xi_{m+n}), \tag{56}$$

for $\xi_1, \cdots, \xi_{m+n} \in \mathcal{S}(\mathbb{R}^d, \mathbb{C})$, where $T_{\lambda \mathbf{a}}$ denotes the representation of translation to the direction $\lambda \mathbf{a}$ on $\mathcal{S}(\mathbb{R}^{dn}, \mathbb{C}^n)$.

Remark 6.
1°) (48)-(51) gives the explicit structure of $S_n \in (\mathcal{S}'(\mathbb{R}^d, \mathbb{R}))^{\hat{\otimes}n}$, the notation being explained in the introduction of this chapter (cf. (10)). For further discussions of these models and related ones, see[11,12,42,43]. In[44], some "system of Schwinger functions" for a modified ϕ_4^4-model, lying outside of the framework of $(\mathcal{H}_{-p})_{-q}^{-\beta}$ but which can be realized in the framework of the algebraic direct product space $\bigoplus_{n=0}^{\infty} (\mathcal{S}'(\mathbb{R}^d, \mathbb{R}))^{\hat{\otimes}n}$, is considered.

2°) [9] discusses not only the **scalar field** models introduced in this subsection, but also **vector field** models, and a result corresponding to Theorem 3 is derived for the vector field models as well. See also the subsequent publications[48,49].

3°) The question whether relativistic quantum field models of the form of those discussed above have a non trivial scattering has been solved positively, at least in a sub class of models, see[8,50].

4°) The question whether it is possible to associate to these non trivial relativistic quantum fields, which are first realized in an indefinite metric relativistic Hilbert space, some "natural" positive metric Hilbert space has also been intensively discussed. Some asymptotic such spaces have been found in[10]. The question whether the Wightman relativistic correlation functions ("Wightman functions") of these models satisfy the axiomatic positivity condition leading to a positive metric Hilbert state space for the quantum fields has been recently clarified. This is related to the question whether the corresponding Euclidean quantum fields satisfy the reflection positivity property, see[8,50].

5°)   To the measures and generalized white noise functionals discussed above one can associate some SPDEs driven by space-time Gaussian white noise (equations of the stochastic quantization type, going back to Parisi and Wu), see, e.g.[1–3,31,35–37,42,52–54,95], and references therein. For related equations discussed by white noise calculus see the pioneer work[71,75,76] and, e.g.,[1,31,38].

6°)   Another important application of white noise calculus to quantum fields concerns the construction of certain topological quantum fields. The first model which has been constructed is the abelian Chern-Simons model. This is a particular case of the non-abelian Chern-Simons functional described by a heuristic "complex measure" of the form

$$\mu(d\gamma) = \text{``}Z^{-1}e^{\frac{i}{\kappa}S(\gamma)}\,d\gamma\text{''}, \tag{57}$$

where γ runs in the space of connections 1-forms of a 3-dimensional (Riemannian) manifold M with values in the Lie algebra of a compact Lie group $SU(n)$ (the abelian case is for $n = 1$),

$$S(\gamma) := \int_M \gamma \wedge d\gamma + \frac{2}{3} \int \gamma \wedge \gamma \wedge \gamma$$

is the Chern-Simons functional (the cubic term vanishes identically in the abelian case). κ is a real-valued parameter.

Interesting quantities, both from the physical and topological point of view are "integrals" of the form

$$\int f(\gamma)\mu(d\gamma), \tag{58}$$

where f belongs to the space of finite products of holonomy operators $\text{Hol}_{C_i}(\gamma)$ of the form

$$f(\gamma) = \prod_{i=1}^{n} \text{Hol}_{C_i}(\gamma),$$

where C_i are links, $n \in \mathbb{N}$. It turns out that in the abelian case and for $M = \mathbb{R}^3$ integrals of the above form can be given a meaning as white noise functionals[81] (for a related construction using Fresnel integrals in the sense of[24,32,33], see[39]). In the non abelian case for $M = \mathbb{R}^3$ the construction has been carried through again using white noise calculus in[40]. For concrete computation of Chern-Simons path integrals (58) for $M = \mathbb{R}^3$ (see[39,81], for the abelian case, and[14] in the general case). They yield knot invariants. Extension to the case of more general underlying manifolds, of great interest because of topological applications, see[65] and references therein. See also[34] for asymptotics in κ of a regularized version of (58). For references on functional integration see in addition, e.g.,[2,107].

References

1. Albeverio, S., Along paths inspired by Ludwig Streit: Stochastic equations for quantum fields and related systems, Stochastic and Infinite Dimensional Analysis, Eds. C. Bernido, M. Carpio-Bernido, M. Grothaus, T. Kuna, J. L. daSilva, M. J. Oliveira (*Proceedings of the conference in honor of Prof. L. Streit, in Univ. Bielefeld, 2014*), Birkhäuser (2016)
2. Albeverio, S., *Wiener and Feynman-Path integrals and their applications,* AMS, Proceedings of Symposia in Applied Mathematics **50** (1997), 163-194.
3. Albeverio, S., *Theory of Dirichlet forms and applications,* In: P. Bernard (ed.) Lectures on probability theory and statistics (Saint-Flour, 2000), In Lecture Notes in Math. **1816** (2003), 1-106, Springer, Berlin.
4. Albeverio, S., Fenstad, J.E., Høegh-Krohn, R., Lindstrøm, T., *Nonstandard methods in stochastic analysis and mathematical physics,* Pure and Applied Mathematics **122** (1986), 1-106, Academic Press Inc., Orlando, FL. Reprint: Dover Publications, Mineola, NY (2009).
5. Albeverio, S., Ferrario, B., *Some methods of infinite dimensional analysis in hydrodynamics: an introduction,* SPDE in hydrodynamic: recent progress and prospects, Lecture Notes in Math. **1942** (2008), 1-50, Springer, Berlin.
6. Albeverio, S., Ferrario, B., Yoshida, M.W., On the essential self-adjointness of Wick powers of relativistic fields and of fields unitary equivalent to random fields, *Acta Applicande Mathematicae* **80** (2004), 309-334.
7. Albeverio, S., Gallavotti, G., Høegh-Krohn, R., Some results for the exponential interaction in two or more dimensions, *Comm. Math. Phys.* **70** (1979), 187-192.
8. Albeverio, S., Gottschalk, H., Quantum fields obtained from convoluted white noise never have positive metric, *Lett. Math. Phys.* **106** (2015), 575-581.
9. Albeverio, S., Gottschalk, H., Wu, J.-L., Convoluted generalized white

noise, Schwinger functions and their analytic continuation to Wightman functions, *Rev. Math. Phys.* **8** (1996), 763-817.

10. Albeverio, S., Gottschalk, H., Wu, J.-L., Models of local relativistic quantum fields with indefinite metric (in all dimensions), *Comm. Math. Phys.* **184** (1997), 509-531.

11. Albeverio, S., Gottschalk, Yoshida, M.W., Representing Euclidean quantum fields as scaling limit of particle systems, *J. Statist. Phys.* **108** (2002), 361-369.

12. Albeverio, S., Gottschalk, Yoshida, M.W., System of classical particle in the grand canonical ensamble, scaling limits and quantum field theory, *Rev. Math. Phys.* **17** (2005), 175-226.

13. Albeverio, S., Haba, Z., Russo, F., A two-space dimensional semilinear heat equation perturbed by (Gaussian) white space, *Probab. Theory Related Fields* **121** (2001), 319-366.

14. Albeverio, S., Hahn, A, Sengupta, A., Chern-Simons theory, Hida distributions and state models, *IDAOP* **6** (2003), 65-81 (Special Issues on Diff. Geom. And Stoch. Anal. I).

15. Albeverio, S., Hida, T., Potthoff, J., Streit, L., *The vacuum of the Høegh-Krohn model as a generalized white noise functional,* Phys. Lett. **B 217**, (1989), 511.

16. Albeverio, S., Hida, T., Potthoff, J., Röckner, M., Streit, L., Dirichlet forms in terms of white noise analysis I-Construction and QFT examples, *Rev. Math. Phys.* **1** (1990), 291-312.

17. Albeverio, S., Hida, T., Potthoff, J., Röckner, M., Streit, L., Dirichlet forms in terms of white noise analysis II-Closability and Diffusion Processes,, *Rev. Math. Phys.* **1** (1990), 313-323.

18. Albeverio, S., Høegh-Krohn, R., Uniqueness of the physical vacuum and the Wightman functions in infinite volume limit for some non polynomial interactions, *Comm. Math. Phys.* **30** (1973), 171-200.

19. Albeverio, S., Høegh-Krohn, R., The Wightman axioms and the mass gap for strong interactions of exponential type in two-dimensional space-time. *J. Funct. Anal.* **16** (1974), 39-82.

20. Albeverio, S., Høegh-Krohn, R., *Quasi invariant measures, symmetric diffusion processes and quantum fields,* Editions du CNRS, Proceedings of the international Colloquium on Mathematical Models of Quantum Field Theory, Colloques Internationaux du Centre National de la Recherche Scientifique, No. **248** (1976), 11-59.

21. Albeverio, S., Høegh-Krohn, R., Dirichlet forms and diffusion processes on rigged Hilbert spaces, *Zeitschrift für Wahrscheinlichkeitstheorie und verw. Gebiete* **40** (1977), 1-57.

22. Albeverio, S., Høegh-Krohn, R., Hunt processes and analytic potential theory on rigged Hilbert spaces. *Ann. Inst. H. Poincar Sect. B (N.S.)* **13** (1977), no. 3, 269291.

23. Albeverio, S., Høegh-Krohn, R., *Canonical relativistic quantum fields in two space-time dimensions,* Preprint Series, Inst. Maths., Univ. Oslo, No. **23** (1975).

24. Albeverio, S., Høegh-Krohn, R., Mazzucchi, S., *Mathematical Theory of Feynman Path Integrals An Introduction*, 2nd Edition, Springer (2008).
25. Albeverio, S., Høegh-Krohn, R., Streit, L., Energy forms, Hamiltonians, and distorted Brownian paths, *J. Math. Phys.* **18** (1977), 907-917.
26. Albeverio, S., Jost, J., Paycha, S., Scarlatti, S., *A mathematical introduction to string theory*, Cambridge Univ. Press (1997).
27. Albeverio, S., Kawabi, H., Röckner, M., Strong uniqueness for both Dirichlet operators and stochastic dynamics to Gibbs measures on a path space with exponential interactions, *J. Functional Analysis* **262** (2012), 602-638.
28. Albeverio, S., Kawabi, H., Mihalache, S., Röckner, M., *Dirichlet form approach to stochastic quantization under exponential interaction in finite volume*, in preparation.
29. Albeverio, S., Kondratiev, Y., Kozitsky, Y., Röckner, M., *Statistical Mechanics of Quantum Lattice Systems: A Path Integral Approach*, EMS Tracts Math. **8** (2009), European Mathematical Society.
30. Albeverio, S., Liang, S., Zegarliński, B., Remarks on the integration by parts formula for the ϕ_3^4-quantum field model, *Infinite Dimensional Analysis, Quantum Probability and Related Topics.* **9** (2006), 149-154.
31. Albeverio, S., Zhi Ming Ma, Röckner, M., Quasi regular forms and the stochastic quantization problem, *Festschrift Masatoshi Fukushima, 2758, Interdiscip. Math. Sci., 17*, World Sci. Publ., Hackensack, NJ, 2015.
32. Albeverio, S., Mazzucchi, S., A survey on mathematical Feynman path integrals: construction, asymptotics, applications, *Quantum field theory*, 49-66, Birkhäuser, Basel, 2009.
33. Albeverio, S., Mazzucchi, S., Infinite dimensional oscillatory integrals as projective systems of functionals. *J. Math. Soc. Japan* **67** (2015), no. 4, 12951316.
34. Albeverio, S., Mitoma, I., Asymptotic expansion of perturbative Chern-Simons theory via Wiener space. *Bull. Sci. Math.* **133** (2009), no. 3, 272-314.
35. Albeverio, S., Röckner, M., Classical Dirichlet forms on topological vector spaces- the construction of the associated diffusion processes, *Probab. Theory Related Fields* **83** (1989), 405-434.
36. Albeverio, S., Röckner, M., Classical Dirichlet forms on topological vector spaces-closability and a Cameron-Martin formula, *J. Functional Analysis* **88** (1990), 395-43.
37. Albeverio, S., Röckner, M., Stochastic differential equations in infinite dimensions: solution via Dirichlet forms, *Probab. Theory Related Fields* **89** (1991), 347-386.
38. Albeverio, S., Rüdiger B., Wu, J.-L., *Analytic and probabilistic aspects of Lévy processes and fields in quantum theory*, pp. 187-224 in Lévy processes, Theory and Application, Edts. O. Barndorff-Nielsen, T. Mikoski, S. Resnick, (2001), Birkhäuser, Besel.
39. Albeverio, S., Schäfer, J., Abelian Chern-Simons theory and linking numbers via oscillatory integrals. *J. Math. Phys.* **36** (1995), no. 5, 2157-2169.

40. Albeverio, S., Sengupta, A., A mathematical construction of the non-abelian Chern-Simons functional integral. *Comm. Math. Phys.* **186** (1997), no. 3, 563-579.

41. Albeverio, S., Sengupta, A., *From classical to quantum fields - A mathematical approach,* book in preparation.

42. Albeverio, S., Yoshida, M. W., $H - C^1$ maps and elliptic SPDEs with polynomial and exponential perturbations of Nelson's Euclidean free field, *J. Functional Analysis* **196** (2002), 265-322.

43. Albeverio, S., Yoshida, M. W., Hida distribution construction of non-Gaussian reflection positive generalized random fields, *Infinite Dimensional Analysis, Quantum Probability and Related Topics* **12** (2009), 21-49.

44. Albeverio, S., Yoshida, M. W., Hida distribution construction of indefinite metric $(\phi^p)_d$ $(d \geq 4)$ quantum field theory, *Lecture Note Series, Institute for Mathematical Sciences, National University of Singapore (NUS), (Proceedings of the conference on white noise analysis held in NUS, March 2014).*

45. Araki, H., Hamiltonian formulation and the canonical commutation relations in quantum field theory, *J. Math. Phys.,* **1** (1960), 492-504.

46. Baez, J.C., Segal, I.E., Zhou, Z., *Introduction to Algebraic and Constructive Quantum Field Theory,* Princeton Univ. Press 1992

47. Battle, G. *Wavelets and renormalization.* Series in Approximations and Decompositions, 10, World Scientific Publishing Co., Inc., River Edge, NJ, 1999.

48. Becker, C., Wilson loops in two-dimensional space-time regarded as white noise, *J. Funct. Anal.* 134 (1995), 321-349.

49. Becker, C., Gielerak, R., Lugiewicz, P., Covariant SPDEs and quantum field structures. *J. Phys. A* 31 (1998), no. 1, 231-258.

50. Borchers, H.-J., *Algebraic aspects of Wightman field theory,* in R.N. Sen and C. Weil (eds.), Statistical mechanics and Field Theory, Haifa Lectures 1971; New York: Halstedt Press, 1972

51. Borasi, I., *Ph. D. Thesis, Univ. of Bonn, in preparation.*

52. Da Prato, G., Debussche, A., Strong solution to stochastic quantizations, *Ann. Probab.* **31** (4) (2003), 1900-1916.

53. Da Prato, G., Röckner, M., Singular dissipative stochastic equation in Hilbert spaces, *Probab. Theory Related Fields* **124 (2)**, (2002), 261-303.

54. Da Prato, G., Tubaro, L., Self-adjointness of some infinite dimensional elliptic operators and applications to stochastic quantization, *Probab. Theory Related Fields* **118** (1) (2003), 131-145.

55. Epstein, H., On the Borchers class of a free field, *Nuovo Cimento* **27** (1963), 886-893.

56. Fröhlich, J., Schwinger functions and their generating functionals I., *Helv. Phys. Acta.* **47** (1974), 265-306, II.

57. Fukushima, M., *Dirichlet forms and Markov processes,* North-Holland Mathematical Library, **23**, *North-Holland Publishing Co.,* Amsterdam-New York, 1980.

58. Gielerak, R., Tugiewicz, P., $4D$ local quantum field theory models from covariant stochastic partial differential equations I. Generalities, *Rev. Math. Phys.* **13** (2001), 335-408.

59. Glimm, J., Jaffe, A., *Quantum Physics, 2nd ed.*, Springer, Berlin, 1986.

60. Glimm, J., Jaffe, A., Spencer, T., The Wightman axioms and particle structure in $P(\varphi)_2$ quantum field model, *Ann. of Math. (2)* **100** (1974), 585-632.

61. Glimm, J., Jaffe, A., Spencer, T., Phase transitions for φ_2^4 quantum fields, *Comm. Math. Phys.* **45** (1975), 203-216.

62. Gottschalk, H., Die Momente gefalteten Gauss-Poissonschen Weißen Rauschens als Schwingerfunktionen. *Diplomarbeit, Bochum 1995.*

63. Guerra, F., Rosen, L., Simon, B., The $P(\phi)_2$ Euclidean quantum field theory as classical statistical mechanics. I, II, *Ann. of Math.* **101** (1975) 111-189.

64. Grothaus, M., Streit, L., Construction of relativistic quantum fields in the framework of white noise analysis, *J. Math. Phys.* **40** (1999), 5387-5405.

65. Hahn, A., Torus knots and the Chern-Simons path integral:a rigorous treatment,arXiv:1508.03804v5 math phys 27 Jan 2016.

66. Herbst, I., On canonical field theories, *J. Math. Phys.* **17** (1976), 1210-1221.

67. Hida, T., Generalized multiple Wiener integrals, *Proc. Japan Acad. Ser. A Math. Sci.* **54** (1978), 55–58.

68. Hida, T., *Brownian motion*, Springer-Verlag, New York Heidelberg Berlin 1980.

69. Hida, T., Hitsuda, M., *Gaussian processes*, Translated from the 1976 Japanese original by the authors. Translations of Mathematical Monographs, 120. American Mathematical Society, Providence, RI, 1993.

70. Hida, T., Kuo, H.-K., Potthoff, J., Streit, L., *White Noise: An Infinite Dimensional Calculus*, Kluwer Academic Publishers, Dordrecht, 1993.

71. Hida, T., Streit,L, On quantum theory in terms of white noise, *Nagoya Math. J.* **68** (1977), 21-34.

72. Høegh-Krohn, R., A general class of quantum fields without cut-off in two space-time dimensions, *Comm. Math. Phys.* **21** (1971), 244-251.

73. Holden, H., Øksendal, B., Uboe, J., Zhang, T., *Stochastic Partial Differential Equations- A modeling, white noise functional approach,*, 2nd edition, Springer, 2010.

74. Ikeda, N., Watanabe, S., *Stochastic differential equations and diffusion processes*, second edition, North-Holland, 1989.

75. Iwata, K., An infinite dimensional stochastic differential equation with state space $C(\mathbb{R})$, *Probab. Theory Related Fields* **74** (1987), 141-159.

76. Jona-Lasinio, G., Mitter, P., K., On the stochastic quantization of field theory, *Comm. Math. Phys.* **101** (1985), 409-436.

77. Klein, A., Renormalized products of the generalized free field and its derivatives, *Pac. J. Math.* **45** (1973) 275-292.

78. Kondratiev,Y.G., Leukert,P., Streit,L., Wick calculus in Gaussian analysis, *Acta Appl. Math.* **44** (1996) 269-294.

79. Kusuoka, S., Dirichlet forms and diffusion processes on Banach space, *J. Fac. Sci., Univ. Tokyo, Sect. IA* **29** (1982), 79-95.
80. Kusuoka, S., *Høegh-Krohn's model of quantum fields and the absolute continuity of measures*, pp. 405-421 in S. Albeverio, J.E. Fenstad, H. Holden, T. Lindstrøm, eds., Ideas and Methods in Mathematical Analysis, Stochastics, and Applications, Cambridge Univ. Press (1992). (1982), 79-95.
81. Leukert, P., Schäfer, J., A rigorous construction of abelian Chern-Simons path integrals using white noise analysis, *Rev. Math. Phys.* **8** (1996), 445-456.
82. Ma, Z. M., Röckner, M., *Introduction to the theory of (nonsymmetric) Dirichlet forms*, Universitext, Springer-Verlag, Berlin, 1992.
83. Mizohata, S., Functional-integral solution for the Schrödinger equation with polynomial potential: a white noise approach, *IDAQP*, **14** (2011) 675-688.
84. Morchio, G., Strocchi, F., Infrared singulalities, vacuum structure and pure phase in local quantum field theory, *Ann. Inst. H. Poincaré* **A33** (1980) 251-282.
85. Nagamachi, S., Mugibayashi, N., Hyperfunction quantum field theory, *Comm. Math. Phys.* **46** (1976), 119-134.
86. Nagamachi, S., Mugibayashi, N., Hyperfunction quantum field theory. II. Euclidean Green's functions, *Comm. Math. Phys.* **49** (1976), 257-275.
87. Nelson, E., Construction of quantum fields from Markoff fields, *J. Functional Analysis* **12** (1973), 97-112.
88. Nelson, E., The free Markov field, *J. Functional Analysis* **12** (1973), 221-227.
89. Nualart, D., *The Malliavin calculus and related topics*, Springer-Verlag, New York/Heidelberg/Berlin, 1995.
90. Okabe,Y., On a stationally Gaussian process with T-positivity and its associated Langevin equation and S-matrix, *J. Fac. Sci. Univ. Tokyo Sect. IA Math.* **26** (1979) 115-165.
91. Osterwalder,K.,Schrader, R., Axioms for Euclidean Green's functions I, *Comm. Math. Phys.* **31** (1973), 83-112.
92. Osterwalder,K.,Schrader, R., Axioms for Euclidean Green's functions II, *Comm. Math. Phys.* **42** (1975), 281-305.
93. Potthoff, J., Röckner, M., On the contraction property of energy forms on infinite-dimensional space. *J. Functional Analysis* **92** (1975), 155-165.
94. Potthoff, J., Streit, L., Invariant states on Random and Quantum Fields: ϕ-bounds and White Noise Analysis, *J. Functional Analysis* **111** (1993), 265-311.
95. Prévôt, C., Röckner, M., *A concise course on stochastic partial differential equations*, Lecture Notes in Mathematics, **1905** Springer, Berlin, 2007.
96. Reed, M., Simon, B., *Methods of modern mathematical physics. I. Functional analysis*, Academic Press, 1975.
97. Reed, M., Simon, B., *Methods of modern mathematical physics. II. Fourier analysis, self-adjointness*, Academic Press, 1975.

98. Rivasseau, V., Constructive field theory and applications: Perspectives and open problems, *J. Math. Phys.*, **41**, (2000), 3764-3775.

99. Röckner, M., Traces of harmonic functions and a new path space for the free quantum field, *J. Functional Analysis* **79** (1988), 211-249.

100. Simon, B., *The $P(\Phi)_2$ Euclidean (Quantum) Field Theory*, Princeton Univ. Press, Princeton, NJ., 1974.

101. Simon, B., Borel summability of the ground-state energy in spacially cutoff $(\varphi^4)_2$, *Phys. Rev. letters* **25** (1970), 1583-1586.

102. Si Si, An aspect of quadratic Hida distribution in realization of duality between Gaussian and Poisson noise, *Infinite Dimensional Analysis, Quantum Probability and Related Topics* **11** (2008), 109-118.

103. Si Si, Introduction to Hida distribution, World Scientific Pub., 2011.

104. Smirnov, A.G., Soloviev, M.A., Spectral properties of Wick power series of a free field with an indefinite metric, *Theoret. and Math. Phys.* **125** (2000), 1349-1362.

105. Streater, R.F., Outline of axiomatic relativistic quantum field theory, *Reports on Progress in Physics* **38** (1975), 771-846.

106. Streater R.F., Wightman A.S., *PCT, Spin and Statistics, and all that*, Princeton Univ. Press 1964.

107. Streit, L., An introduction to theory of integration over function spaces, *Acta Phys. Austriaca, Suppl.* **II** (1965), 2-21.

108. Wightman, A.S., *Recent achievements of axiomatic field theory*, Sem. on Theor. Phys. Trieste 1962 Int. At. En. Ag., 11-58 Vienna (1963)

109. Wightman, A.S., *Introduction to new aspects of relativistic dynamics of quantum fields*, Cargèse Lect. Theor. Phys, Ed. M. Lévy, 171-291, Gordon and Breach New York (1967)

110. Wightman, A.S., *Troubles in the external field problem for invariant wave equations*, Lectures from the Coral Gables Conference on fundamental interactions at High energy. At Miami Univ., Edt. by Mario Dal Cin, et al., Gordon and Breach 1971.

111. Zinoviev, Yu.M., Equivalence of Euclidean and Wightman field theories, *Comm. Math. Phys.* **174** (1995), 1-27.

Chapter 3

How to Use White Noise Analysis to Make Feynman Integrals Mathematically Rigorous

Wolfgang Bock

Technische Universität Kaiserslautern
Fachbereich Mathematik
AG Technomathematik
Postfach 3049
67653 Kaiserslautern, Germany,
bock@mathematik.uni-kl.de

In this chapter we summarize applications of White Noise Analysis within the framework of Feynman Path Integrals. With more than 30 years of intensive study this field can be seen as one of the most important applications. Of course this chapter cannot give an exhaustive summary of all approaches to different path integrals in White Noise Analysis. Here we give a slice through different attemtps to path integrals in the White Noise framework.

1. Introduction

As proposed by Feynman[14,16], quantum mechanical transition amplitudes may be thought of as a kind of averaging over fluctuating paths, with oscillatory weight functions given in terms of the classical action

$$S(x) = \int_t -0^t L(x(\tau), \dot{x}(\tau), \tau) \, d\tau.$$

The ideas go back on the previous work of Norbert Wiener and the succeeding work of Paul Dirac in 1933[11]. The Lagrangian (hence the action as the time integral of the Lagrangian) is given by the difference of the kinetic energy and the potential, e.g.

$$L(x(t), \dot{x}(t), t) = -\frac{1}{2} m(\dot{x}(t))^2 - V(x(t), \dot{x}(t), t).$$

Informally, the Feynman path integral is then expressed as

$$K(t, y|t_0, y_0) = N \int \exp\left(\frac{i}{\hbar} S(x)\right) \prod_{0 < \tau < t} dx(\tau). \tag{1}$$

Here $\hbar$ is Planck's constant and the integral is thought of as being over all paths with $x(t_0) = y_0 \in \mathbb{R}^d$ and $x(t) = y \in \mathbb{R}^d$. The quantum mechanical propagator $K(t, y|t_0, y_0)$ represents the transition amplitude for a particle to be found at position y at time t given that the particle was at position y_0 at an earlier time t_0. The propagator $K(t, y|t_0, y_0)$ is the integral kernel of the unitary operator $U_t = \exp(it\hat{H})$, where $\hat{H}$ is the Hamilton operator. The operator U_t defines the semi-group of the Schrödinger equation, i.e.

$$i\hbar\partial_t \Big(U(t, t_0)\Psi\Big) = H(t)\Big(U(t, t_0)\Psi\Big), \quad U(t_0, t_0)\Psi = \Psi.$$

Although the path integral first appeared to give an alternative approach to quantum mechanics it developed to a very useful tool in many different areas, among them are finance, statistics, polymer science and of course quantum theory. For many examples and a good overview of applications and path integral methods we refer to the book of Kleinert[35] and the recent monograph of Klauder[29]. A good introduction to mathematical methods to treat path integrals is given in[2].

As useful as the theory of Feynman path integrals is, so meaningless is the expression in (1) from the mathematical point of view. This is due to the fact that the integral is thought of being performed over continuous paths with respect to a flat measure. However such a measure does not exist in infinite dimensions, furthermore the normalization constant in front of the integral turns out to be infinity.

There are many attempts to give a mathematically rigorous meaning to the Feynman path integral and enlarging the treatable potential classes. Notable among them are the prodistributions of DeWitt-Morette[10], the oscillatory integrals of Albeverio and Høegh-Krohn[2] the apporach by Smolyanov[45] and the coherent state approach by Klauder[32]. Of course we cannot give a comprehensive survey of all attempts and refer the interested reader to[2] and the references therein. For a historical overview of continuous time regularization of path integrals we refer to the survey of Klauder[34].

Here we will use techniques from White Noise Analysis to give a mathematical rigor to not only the integral but the integrand itself as a suitable generalized function. The T-transform of the integrand is providing us with the moment generating functional of the Feynman measure, compare also to[46] and has - as we will see - also an interpretation as a propagator for a time dependent potential itself. The Feynman integral is given by the generalized expectation of the integrand, i.e. the T-transform in the Schwartz test function 0. The use of White Noise techniques also serves to study quantum mechanical properties such as the canonical commutation relations in a version of Feynman and Hibbs[17]. So not only the mathematical rigor can be established but it can also be guaranteed that the objects have also a correct physical interpretation. There are two separate approaches to the Feynman integrand in White Noise Analysis. On the one hand there is the approach of Hida and Streit. Here the aim is to give a meaning to the product of Donsker's delta function, a normalized exponential and a potential term. On the other hand there is the scaling approach, which goes back to Cameron[6] and Doss[12]. Here the idea is to scale the Feynman-Kac integrand, i.e. the integrand in the Euclidean case to a complex time. The potentials treatable in this class are beyond perturbation theory.

The Hida-Streit approach was introduced in 1983[26], in 1991 it could be shown that the Feynman integrand for the potential free case and the harmonic oscillator exists as a Hida distribution[8]. The class of potentials was in the following years expanded to potentials being

- (signed) finite measures which are "small" at infinity (Khandekar-Streit class)[31,41],
- Fourier transforms of measures (Albeverio-Høegh-Krohn class)[50],
- Laplace transforms of measures[37],
- and combinations of them[9].

In the scaling approach there are different potential classes established such as the Doss potentials[23] and the Mazzucchi potentials[42]. Moreover scaling techniques have been used for the integrand for electrons in random media[24].

The chapter is organized as follows: first we will give some brief summary about results from White Noise Analysis which are needed in the following but not yet mentioned in Chaper 1. Then we will follow the path of Hida and Streit by constructing the integrand for the Feynman path integral in the framework of White Noise analysis. Afterwards we sketch the development of potential classes which can be treated. Moreover we will introduce the approach to Hamiltonian path integral - i.e. Feynman integrals in phase space - as well as the scaling approach. The last section is dedicated to the canonical commutation relations in the sense of Feynman and Hibbs.

2. Preliminaries

Here we review a special class of Hida distributions which are defined by their T-transform, see e.g.[22,25,26]. Proofs and more details can be found in[4]. Let $\mathcal{B}$ be the set of all continuous bilinear mappings $B : S_d(\mathbb{R}) \times S_d(\mathbb{R}) \to \mathbb{C}$. Then the functions

$$S_d(\mathbb{R}) \ni \mathbf{f} \mapsto \exp\left(-\frac{1}{2}B(\mathbf{f}, \mathbf{f})\right) \in \mathbb{C}$$

for all $B \in \mathcal{B}$ are U-functionals. Therefore, by using the Characterization Theorem of Hida distributions in Chapter 1, the inverse T-transform of these functions

$$\Phi_B := T^{-1}\exp\left(-\frac{1}{2}B\right)$$

are elements of $(S)'$.

Definition 1. The set of **generalized Gauss kernels** is defined by

$$GGK := \{\Phi_B, \ B \in \mathcal{B}\}.$$

Example 1.[22] We consider a symmetric trace class operator $\mathbf{K}$ on $L_d^2(\mathbb{R})$ such that $-\frac{1}{2} < \mathbf{K} \le 0$, then

$$\int_{S_d'(\mathbb{R})} \exp\left(-\langle\boldsymbol{\omega}, \mathbf{K}\boldsymbol{\omega}\rangle\right) d\mu(\boldsymbol{\omega}) = (\det(\mathbf{Id} + 2\mathbf{K}))^{-\frac{1}{2}} < \infty.$$

For the definition of $\langle\cdot, \mathbf{K}\cdot\rangle$ see the remark below. Here $\mathbf{Id}$ denotes the identity operator on the Hilbert space $L_d^2(\mathbb{R})$, and $\det(\mathbf{A})$ of a symmetric trace class operator $\mathbf{A}$ on $L_d^2(\mathbb{R})$ denotes the infinite product of its eigenvalues, if it exists. In the present situation we have $\det(\mathbf{Id} + 2\mathbf{K}) \neq 0$. Therefore we obtain that the exponential $g = \exp(-\frac{1}{2}\langle\cdot, \mathbf{K}\cdot\rangle)$ is square-integrable and its T-transform is given by

$$Tg(\mathbf{f}) = (\det(\mathbf{Id} + \mathbf{K}))^{-\frac{1}{2}} \exp\left(-\frac{1}{2}(\mathbf{f}, (\mathbf{Id} + \mathbf{K})^{-1}\mathbf{f})\right), \quad \mathbf{f} \in S_d(\mathbb{R}).$$

Therefore $(\det(\mathbf{Id} + \mathbf{K}))^{\frac{1}{2}} g$ is a generalized Gauss kernel.

Remark 1.

i) Since a trace class operator is compact, see e.g.[44], we have that $\mathbf{K}$ in the above example is diagonalizable, i.e.

$$\mathbf{Kf} = \sum_{k=1}^{\infty} k_n(\mathbf{f}, \mathbf{e}_n)\mathbf{e}_n, \quad \mathbf{f} \in L_d^2(\mathbb{R}),$$

where $(e_n)_{n\in\mathbb{N}}$ denotes an eigenbasis of the corresponding eigenvalues $(k_n)_{n\in\mathbb{N}}$ with $k_n \in (-\frac{1}{2}, 0]$, for all $n \in \mathbb{N}$. Since K is compact, we have that $\lim_{n\to\infty} k_n = 0$ and since $\mathbf{K}$ is trace class we also have $\sum_{n=1}^{\infty}(e_n, -\mathbf{K}e_n) < \infty$. We define for $\omega \in S'_d(\mathbb{R})$

$$-\langle\omega, \mathbf{K}\omega\rangle := \lim_{N\to\infty} \sum_{n=1}^{N}\langle\omega, e_n\rangle(-k_n)\langle e_n, \omega\rangle.$$

Then as a limit of measurable functions $\omega \mapsto -\langle\omega, \mathbf{K}\omega\rangle$ is measurable and hence

$$\int_{S'_d(\mathbb{R})} \exp(-\langle\omega, \mathbf{K}\omega\rangle)\, d\mu(\omega) \in [0, \infty].$$

The explicit formula for the T-transform and expectation then follow by a straightforward calculation with help of the above limit procedure.

ii) In the following, if we apply operators or bilinear forms defined on $L_d^2(\mathbb{R})$ to generalized functions from $S'_d(\mathbb{R})$, we are always having in mind the interpretation as in Ref. 16.

Definition 2. [4] Let $\mathbf{K} : L_d^2(\mathbb{R})_{\mathbb{C}} \to L_d^2(\mathbb{R})_{\mathbb{C}}$ be linear and continuous such that:

(i) $\mathbf{Id} + \mathbf{K}$ is injective.
(ii) There exists $p \in \mathbb{N}_0$ such that $(\mathbf{Id} + \mathbf{K})(L_d^2(\mathbb{R})_{\mathbb{C}}) \subset H_{p,\mathbb{C}}$ is dense.
(iii) There exist $q \in \mathbb{N}_0$ such that $(\mathbf{Id} + \mathbf{K})^{-1} : H_{p,\mathbb{C}} \to H_{-q,\mathbb{C}}$ is continuous with p as in (ii).

Then we define the normalized exponential

$$\mathrm{Nexp}(-\frac{1}{2}\langle\cdot, \mathbf{K}\cdot\rangle) \tag{2}$$

by

$$T(\mathrm{Nexp}(-\frac{1}{2}\langle\cdot, \mathbf{K}\cdot\rangle))(\mathbf{f}) := \exp(-\frac{1}{2}\langle\mathbf{f}, (\mathbf{Id} + \mathbf{K})^{-1}\mathbf{f}\rangle), \quad \mathbf{f} \in S_d(\mathbb{R}).$$

Remark 2. The "normalization" of the exponential in the above definition can be regarded as a division of a divergent factor. In an informal way one can write

$$T(\mathrm{Nexp}(-\frac{1}{2}\langle\cdot, \mathbf{K}\cdot\rangle))(\mathbf{f}) = \frac{T(\exp(-\frac{1}{2}\langle\cdot, \mathbf{K}\cdot\rangle))(\mathbf{f})}{T(\exp(-\frac{1}{2}\langle\cdot, \mathbf{K}\cdot\rangle))(0)}$$

$$= \frac{T(\exp(-\frac{1}{2}\langle\cdot, \mathbf{K}\cdot\rangle))(\mathbf{f})}{\sqrt{\det(\mathbf{Id} + \mathbf{K})}}, \quad \mathbf{f} \in S_d(\mathbb{R}),$$

i.e. if the determinant in Example 1 above is not defined, we can still define the normalized exponential by the T-transform without the diverging prefactor. The assumptions in the above definition then guarantee the existence of the generalized Gauss kernel in (2).

Example 2. For sufficiently "nice" operators $\mathbf{K}$ and $\mathbf{L}$ on $L_d^2(\mathbb{R})_{\mathbb{C}}$ we can define the product

$$\mathrm{Nexp}\big(-\frac{1}{2}\langle\cdot,\mathbf{K}\cdot\rangle\big)\cdot\exp\big(-\frac{1}{2}\langle\cdot,\mathbf{L}\cdot\rangle\big)$$

of two square-integrable functions. Its T-transform is then given by

$$T\Big(\mathrm{Nexp}(-\frac{1}{2}\langle\cdot,\mathbf{K}\cdot\rangle)\cdot\exp(-\frac{1}{2}\langle\cdot,\mathbf{L}\cdot\rangle)\Big)(\mathbf{f})$$

$$=\sqrt{\frac{1}{\det(\mathbf{Id}+\mathbf{L}(\mathbf{Id}+\mathbf{K})^{-1})}}\,\exp(-\frac{1}{2}\langle\mathbf{f},(\mathbf{Id}+\mathbf{K}+\mathbf{L})^{-1}\mathbf{f}\rangle),\quad\mathbf{f}\in S_d(\mathbb{R}),$$

in the case the right-hand side indeed is a U-functional.

Definition 3. Let $\mathbf{K}:L_d^2(\mathbb{R})_{\mathbb{C}}\to L_d^2(\mathbb{R})_{\mathbb{C}}$ be as in Definition 2, i.e.

$$\mathrm{Nexp}(-\frac{1}{2}\langle\cdot,\mathbf{K}\cdot\rangle)$$

exists. Furthermore let $\mathbf{L}:L_d^2(\mathbb{R})_{\mathbb{C}}\to L_d^2(\mathbb{R})_{\mathbb{C}}$ be trace class. Then we define

$$\mathrm{Nexp}(-\frac{1}{2}\langle\cdot,\mathbf{K}\cdot\rangle)\cdot\exp(-\frac{1}{2}\langle\cdot,\mathbf{L}\cdot\rangle)$$

via its T-transform, whenever

$$T\Big(\mathrm{Nexp}(-\frac{1}{2}\langle\cdot,\mathbf{K}\cdot\rangle)\cdot\exp(-\frac{1}{2}\langle\cdot,\mathbf{L}\cdot\rangle)\Big)(\mathbf{f})$$

$$=\sqrt{\frac{1}{\det(\mathbf{Id}+\mathbf{L}(\mathbf{Id}+\mathbf{K})^{-1})}}\,\exp(-\frac{1}{2}\langle\mathbf{f},(\mathbf{Id}+\mathbf{K}+\mathbf{L})^{-1}\mathbf{f}\rangle),\quad\mathbf{f}\in S_d(\mathbb{R}),$$

is a U-functional.

In the case $\mathbf{g}\in S_d(\mathbb{R})$, $c\in\mathbb{C}$ the product between the Hida distribution Φ and the Hida test function $\exp(i\langle\mathbf{g},.\rangle+c)$ can be defined because (S) is a continuous algebra under point-wise multiplication. The next definition is an extension of this product.

Definition 4. The point-wise product of a Hida distribution $\Phi\in(S)'$ with an exponential of a linear term, i.e.

$$\Phi\cdot\exp(i\langle\mathbf{g},\cdot\rangle+c),\quad\mathbf{g}\in L_d^2(\mathbb{R})_{\mathbb{C}},c\in\mathbb{C},$$

is defined by

$$T(\Phi \cdot \exp(i\langle \mathbf{g}, \cdot \rangle + c))(\mathbf{f}) := T\Phi(\mathbf{f} + \mathbf{g})\exp(c), \quad \mathbf{f} \in S_d(\mathbb{R}),$$

if $T\Phi$ has a continuous extension to $L_d^2(\mathbb{R})_{\mathbb{C}}$ and the term on the right-hand side is a U-functional in $\mathbf{f} \in S_d(\mathbb{R})$.

Definition 5. Let $D \subset \mathbb{R}$ with $0 \in \overline{D}$. Under the assumption that $T\Phi$ has a continuous extension to $L_d^2(\mathbb{R})_{\mathbb{C}}$, $\boldsymbol{\eta} \in L_d^2(\mathbb{R})_{\mathbb{C}}$, $y \in \mathbb{R}$, $\lambda \in \gamma_\alpha :=$ $\{\exp(-i\alpha)s|\, s \in \mathbb{R}\}$ and that the integrand

$$\gamma_\alpha \ni \lambda \mapsto \exp(-i\lambda y)T\Phi(\mathbf{f} + \lambda\boldsymbol{\eta}) \in \mathbb{C}$$

fulfills the conditions of Corollary 2 in Chapter 1 for all $\alpha \in D$. Then one can define the product

$$\Phi \cdot \delta_0(\langle \boldsymbol{\eta}, \cdot \rangle - y),$$

by

$$T(\Phi \cdot \delta_0(\langle \boldsymbol{\eta}, \cdot \rangle - y))(\mathbf{f}) := \lim_{\alpha \to 0} \int_{\gamma_\alpha} \exp(-i\lambda y)T\Phi(\mathbf{f} + \lambda\boldsymbol{\eta})\, d\lambda.$$

Of course under the assumption that the right-hand side converges in the sense of Corollary 1 from Chapter 1, see e.g.[22].

This definition is motivated by the definition of Donsker's delta in Chapter 1.

Lemma 1.[4] *Let* $\mathbf{L}$ *be a* $d \times d$ *block operator matrix on* $L_d^2(\mathbb{R})_{\mathbb{C}}$ *acting component-wise such that all entries are bounded operators on* $L^2(\mathbb{R})_{\mathbb{C}}$. *Let* $\mathbf{K}$ *be a* $d \times d$ *block operator matrix on* $L_d^2(\mathbb{R})_{\mathbb{C}}$, *such that* $\mathbf{Id} + \mathbf{K}$ *and* $\mathbf{N} = \mathbf{Id} + \mathbf{K} + \mathbf{L}$ *are bounded with bounded inverse. Furthermore assume that* $\det(\mathbf{Id} + \mathbf{L}(\mathbf{Id} + \mathbf{K})^{-1})$ *exists and is different from zero (this is e.g. the case if* $\mathbf{L}$ *is trace class and -1 in the resolvent set of* $\mathbf{L}(\mathbf{Id} + \mathbf{K})^{-1})$. *Let* $M_{\mathbf{N}^{-1}}$ *be the matrix given by an orthogonal system* $(\boldsymbol{\eta}_k)_{k=1,\dots J}$ *of non-zero functions from* $L_d^2(\mathbb{R})$, $J \in \mathbb{N}$, *under the bilinear form* $(\cdot, \mathbf{N}^{-1}\cdot)$, *i.e.* $(M_{\mathbf{N}^{-1}})_{i,j} = (\boldsymbol{\eta}_i, \mathbf{N}^{-1}\boldsymbol{\eta}_j)$, $1 \leq i, j \leq J$. *Under the assumption that either*

$$\Re(M_{\mathbf{N}^{-1}}) > 0 \quad or \quad \Re(M_{\mathbf{N}^{-1}}) = 0 \; and \; \Im(M_{\mathbf{N}^{-1}}) \neq 0,$$

where $M_{\mathbf{N}^{-1}} = \Re(M_{\mathbf{N}^{-1}}) + i\Im(M_{\mathbf{N}^{-1}})$ *with real matrices* $\Re(M_{\mathbf{N}^{-1}})$ *and* $\Im(M_{\mathbf{N}^{-1}})$, *then*

$$\Phi_{\mathbf{K},\mathbf{L}} := \mathrm{Nexp}\left(-\frac{1}{2}\langle \cdot, \mathbf{K}\cdot \rangle\right) \cdot \exp\left(-\frac{1}{2}\langle \cdot, \mathbf{L}\cdot \rangle\right) \cdot \exp(i\langle \cdot, \mathbf{g} \rangle) \cdot \prod_{i=1}^{J} \delta_0(\langle \cdot, \boldsymbol{\eta}_k \rangle - y_k),$$

for $\mathbf{g} \in L^2_d(\mathbb{R}, \mathbb{C})$, $t > 0$, $y_k \in \mathbb{R}$, $k = 1 \ldots, J$, *exists as a Hida distribution.*
Moreover for $\mathbf{f} \in S_d(\mathbb{R})$

$$T\Phi_{\mathbf{K},\mathbf{L}}(\mathbf{f}) = \frac{1}{\sqrt{(2\pi)^J \det((M_{\mathbf{N}^{-1}}))}} \sqrt{\frac{1}{\det(\mathbf{Id} + \mathbf{L}(\mathbf{Id} + \mathbf{K})^{-1})}}$$

$$\times \exp\left(-\frac{1}{2}((\mathbf{f} + \mathbf{g}), \mathbf{N}^{-1}(\mathbf{f} + \mathbf{g}))\right) \exp\left(-\frac{1}{2}(u, (M_{\mathbf{N}^{-1}})^{-1}u)\right), \quad (3)$$

where

$$u = \left((iy_1 + (\boldsymbol{\eta}_1, \mathbf{N}^{-1}(\mathbf{f} + \mathbf{g}))), \ldots, (iy_J + (\boldsymbol{\eta}_J, \mathbf{N}^{-1}(\mathbf{f} + \mathbf{g})))\right).$$

3. The Approach of Hida and Streit

3.1. *Constructing the Free Integrand*

In this section we will follow the path of[8,26] to construct the free integrand I_0 as a Hida distribution. Therefore we will use the notion of generalized Gauss kernels, see Chapter I.

First let us again consider the formula of Feynman yielding the propagator

$$K(x, t; x_0, t_0) = \mathrm{N} \int_{x(0)=x_0, x(t)=x} \exp(\frac{i}{\hbar} S(x(T), \dot{x}(T))) \mathcal{D}x,$$

as mentioned before the integrand consists of an oscillatory phase factor based on the classical action, the flat, i.e. translation invariant measure on the infinite dimensional paths space and the paths are thought to start in x_0 at time t_0 and end in x at time t.

These ingredients we have to insert in the ansatz for the integrand.

Modeling the path:

   The paths in the Feynman integral are thought to be continuous. From the White Noise setting it is thus natural to model them by Brownian fluctuation. Moreover one has to take into account that the physical dimensions are that of a space variable. Hida and Streit proposed the following ansatz:

$$x(t) = x_0 + \sqrt{\frac{\hbar}{m}}\langle \cdot, \mathbb{1}_{[t_0,t)}\rangle,$$

see also[25].

Though for sake of simplicity, if not stated otherwise we set in the following $m = \hbar = 1$.

Kinetic energy and flat measure:

We cannot expect to obatin a flat non-trivial measure on the whole infinite dimensional space. However we can use the (Gaussian) white noise measure μ and simulate a locally flat measure by compensating the Gaussian fall-off. Note that we are interested in the propagation on the time interval $[t_0, t]$, thus the flattening of the measure in this region is enough for our observations.

To compensate the Gaussian fall-off on this interval one needs the informal factor

$$\omega \mapsto \exp(\frac{1}{2}\langle \omega, P_{[t_0,t]}\omega\rangle), \quad \omega \in S'(\mathbb{R}),$$

where $P_{[t_0,t)}$ denotes the orthogonal projection to the time interval $[t_0, t)$. For $\omega \in L^2(\mathbb{R})$ this would read:

$$\exp(\frac{1}{2}\langle \omega, P_{[t_0,t]}\omega\rangle) = \exp(\frac{1}{2}\int_{t_0}^{t} \omega^2(s)\,ds).$$

Note that we have a dual pairing of two tempered distributions, which we have to define in a rigorous way.

An analogous consideration is valid for the classical action of I_0. Note that we consider the free integrand, i.e. $V \equiv 0$. Thus the only contribution to the classical action is given by the kinetic energy

$$\frac{1}{2}m\dot{x}(t)^2.$$

Thus with

$$\dot{x}(\omega, t) = \langle \omega, \delta_t \rangle = \omega(t),$$

we obtain informally

$$\exp\left(\frac{i}{\hbar}S(\dot{x}(T))\right) = \exp\left(i\langle \cdot, P_{[t_0,t]}\cdot\rangle\right).$$

We can model the kinetic energy term and the compensation of the Gaussian fall-off with the Hida distribution

$$J_{\sqrt{i}} = \text{Nexp}\left(\frac{i+1}{2}\langle \cdot, P_{[t_0,t]}\cdot\rangle\right) \in (S)',$$

which is defined as a generalized Gauss kernel as before.

Pinning the initial and final point:

To pin the initial and final point, we have to make sure that the paths starts in x_0 (which is given by definition) and ends in x at time point t. For this it is natural to use a Delta function, which just gives a contribution if the path at final time t hits y, i.e. we have

$$\omega \mapsto \delta(x(t) - y) = \delta(x_0 - y + \langle \omega, \mathbb{1}_{[t_0,t]}\rangle \in (S)',$$

which is the well known Donskers' delta function introduced in Chapter 1.

Informally the free Feynman integrand I_0 reads

$$I_0 = \text{Nexp}\left(\frac{i+1}{2}\langle \cdot, P_{[t_0,t]}\cdot\rangle\right) \cdot \delta(x_0 - y + \langle \cdot, \mathbb{1}_{[t_0,t]}\rangle,$$

which is a pointwise product of two Hida distributions, which has to be suitably defined.

Indeed, since the two objects are well-known one can proceed as follows: We use $\text{Nexp}\left(\frac{i+1}{2}\langle \cdot, P_{[t_0,t]}\cdot\rangle\right)\mu$ as complex (Hida) measure and approximate Donskers' delta function with the help of the Fourier representation of the Delta function and apply the integral corollary from the Characterization Theorem. We then obtain the following theorem, see also e.g.[8,25].

Theorem 1 ([8]). *The free integrand*

$$I_0 = \text{Nexp}\left(\frac{i+1}{2}\langle \cdot, P_{[t_0,t]}\cdot\rangle\right) \cdot \delta(x_0 - y + \langle \cdot, \mathbb{1}_{[t_0,t]}\rangle,$$

exists as a Hida distribution. Its T-transform in $\xi \in S(\mathbb{R})$ is given by

$$T(I_0)(\xi) = \frac{1}{\sqrt{2\pi i(t - t_0)}}\exp\left(-\frac{1}{2}\langle \xi, (P_{[t_0,t)^c} + iP_{[t_0,t)})\xi\rangle\right)$$

$$\times \exp\left(\frac{i}{2(t - t_0)}(y - x_0 - \langle, \xi, P_{[t_0,t)}\mathbb{1}_{[t_0,t)})^2\rangle\right). \quad (4)$$

Remark 3.

- Note that the generalized expectation of the free integrand, i.e. its T-transform in $\xi = 0$ yields

$$\mathbb{E}(I_0) = T(I_0)(0) = \frac{1}{\sqrt{2\pi i(t - t_0)}}\exp\left(\frac{i}{2(t - t_0)}(y - x_0)^2\right),$$

which is the propagator for the Schrödinger equation of a free particle.

- Note that I_0 itself can be interpreted as a generalized Gauss kernel itself.

- The approximation based on the Fourier transform is just one particulat approximation scheme. The same result can be obtained by using a heat kernel approximation of the delta function.

The T-transform itself has an interesting interpretation as a time dependent potential. Consider again $T(I_0)(\xi)$, then an informal integration by parts yields, compare e.g.[25,49,50]

$$T(I_0)(\xi) = \exp\left(-\frac{1}{2}|\xi^2_{[t_0,t]^c} + i\xi(t)x - i\xi(t_0)x_0)\right)$$

$$\times \mathbb{E}\left(I_0 \exp(-i\int_{t_0}^t \dot{\xi}(r)x(r)\,dr)\right). \quad (5)$$

Then the term $\exp(-i\int_{t_0}^t \dot{\xi}(r)x(r)\,dr)$ would arise from a time dependent potential

$$V_{\dot{\xi}}(x,t) = \dot{\xi}(t)x, \quad x \in \mathbb{R}, g \in S(\mathbb{R}), 0 < t_0 < t < \infty. \quad (6)$$

In particular

$$T(I_0)(\xi) = \exp(-\frac{1}{2}|\xi_{[t_0,t]^c}| + i\xi(t)x - i\xi(t_0)x_0)K_0^{\dot{\xi}}(x,t;x_0,t_0), \quad (7)$$

where

$$K_0^{\dot{\xi}}(x,t;x_0,t_0) := \frac{1}{\sqrt{2\pi i(t-t_0}} \exp(i(\xi(t_0)x_0 - \xi(t)x)$$

$$\times \exp\left(-\frac{i}{2}|\xi_{[t_0,t)}|^2 + \frac{i}{2(t-t_0)}\left(\int_{t_0}^t \xi(r)\,dr + x - x_0\right)^2\right) \quad (8)$$

denotes the Greens function to the potential $V_{\dot{\xi}}$, i.e. it solves the Schrödinger equation

$$\left(i\frac{\partial}{\partial t} - \frac{1}{2}\Delta - \dot{\xi}(t)x\right)K_0^{\dot{\xi}}(x,t;x_0,t_0) = 0 \quad (9)$$

$$\text{with } \lim_{t \to t_0} K_0^{\dot{\xi}}(x,t;x_0,t_0) = \delta(x - x_0).$$

3.2. *Quadratic Actions*

The assumptions of Lemma 1 allow for a wide class of quadratic functionals. This enables us to apply it to a wider class of Feynman integrals with quadratic actions. An action is called quadratic if it is a polynomial in space and velocity of at most second order. These potentials are of special interest in semi-classical (WKB) approximation, see e.g.[7,47].

The Feynman integrand for the harmonic oscillator
First we start with the Feynman integrand for the harmonic oscillator.
The Lagrangian is then given by

$$L(x, \dot{x}) = \frac{1}{2}\dot{x} + cx^2, \quad c \in \mathbb{R}. \tag{10}$$

The ansatz for the normalized exponential thus stays the same as in the
case of the free integrand.

The free integrand is then multiplied with an exponential of quadratic
type

$$\exp(-\frac{1}{2}\langle \cdot, icA \cdot \rangle),$$

where

$$A f(s) = i \mathbb{1}_{[0,t)}(s) \int_s^t \int_0^\tau f(r) \, dr \, d\tau, f \in L^2(\mathbb{R}, \mathbb{C}), s \in \mathbb{R},$$

we refer to[22] for properties of the operator as the trace class property,
invertibility and spectrum.
We have for $f, g \in L^2(\mathbb{R})_\mathbb{C}$

$$\langle f, Ag \rangle = \int_0^t \int_0^\tau f(s) \, ds \cdot \int_0^\tau g(s) \, ds \, d\tau.$$

Hence the operator is used to implement the integral over the squared
Brownian motion.

Thus the integrand is given by

$$I_{HO} = \text{Nexp}\left(\frac{i+1}{2}\langle \cdot, P_{[0,t]} \cdot \rangle\right) \cdot \delta(x_0 - y + \langle \cdot, \mathbb{1}_{[0,t]} \rangle) \cdot \exp\left(-\frac{1}{2}\langle \cdot, icA \cdot \rangle\right). \tag{11}$$

To apply Lemma 1 we choose

$$K = -(i+1)P_{[0,t]}, \quad and \quad L = icA,$$

hence we have

$$N = Id + K + L = -iP_{[0,t]}(Id - cA) + P_{[0,t]^c}.$$

For determining the inverse of N we use the decomposition of $L_2^2(\mathbb{R})_\mathbb{C}$ into
the orthogonal subspaces $L_2^2([0,t])_\mathbb{C}$ and $L_2^2([0,t]^c)_\mathbb{C}$. The operator N leaves
both spaces invariant and on $L_2^2([0,t)^c)$ it is already the identity. Therefore

we need just an inversion of N on $L_2^2([0,t))$. By formal calculation we obtain

$$N^{-1} = iP_{[0,t]}(Id - cA)^{-1} + P_{[0,t]^c},$$

if $(Id - cA)^{-1}$ exists, i.e. $(Id - cA)$ is bijective on $L_2^2([0,t))$. The operator $cAf(s) = 1\!\!1_{[0,t)}(s)k\int_s^t \int_0^\tau f(r)\, dr\, d\tau$, $f \in L_2^2([0,t))_{\mathbb{C}}, s \in [0,t)$, diagonalizes and the eigenvalues l_n different from zero have the form:

$$l_n = c\left(\frac{t}{(n - \frac{1}{2})\pi}\right)^2, \quad n \in \mathbb{N}.$$

Thus $(Id - cA)^{-1}$ exists if $l_n \neq 1$ for all $n \in \mathbb{N}$. For $0 < t < \pi/(2\sqrt{c})$ this is true. The corresponding normalized eigenvectors to l_n are

$$[0,t] \ni s \mapsto e_n(s) = \sqrt{\frac{2}{t}} \cos\left(\frac{s}{t}\left(n - \frac{1}{2}\right)\pi\right), \quad s \in [0,t] \quad n \in \mathbb{N}.$$

Hence by using formula (1) on p. 421 of[20] we obtain:

$$\frac{1}{\det(Id + L(Id + K)^{-1})} = \left(\prod_{n=1}^{\infty}(1 - c)\left(\frac{t}{(n - \frac{1}{2})\pi}\right)^2\right)^{-1} = \frac{1}{\cos(\sqrt{k}t)}$$

and

$$(1\!\!1_{[0,t)}, (P_{[0,t]^c} + i(P_{[0,t]} - cA)^{-1})\, 1\!\!1_{[0,t)}) = i\sum_{n=1}^{\infty}(1 - l_n)^{-1}(1\!\!1_{[0,t)}, e_n)^2$$

$$= i\sum_{n=1}^{\infty}\frac{1}{1 - c\left(\frac{t}{((n-\frac{1}{2})\pi}\right)^2}\frac{2t}{(n - \frac{1}{2})\pi)^2}$$

$$= 2it\sum_{n=1}^{\infty}\frac{1}{((n - \frac{1}{2})\pi)^2 - ct^2}$$

$$= \frac{i}{\sqrt{c}}8\sqrt{k}t\sum_{n=1}^{\infty}\frac{1}{((2n - 1)\pi)^2 - 4ct^2}$$

$$= \frac{i}{\sqrt{k}}\tan(\sqrt{c}t) = i\frac{\tan(\sqrt{k}t)}{\sqrt{c}},$$

by using formula (1) on p. 421 of[20]. Hence we have for the T-transform in $\xi \in S(\mathbb{R})$

$$T(I_{HO})(\xi)$$

$$= T\left(\text{Nexp}(\frac{i+1}{2}\langle\cdot, P_{[0,t]}\cdot\rangle) \cdot \delta(x_0 - y + \langle\cdot, 1\!\!1_{[0,t]}\rangle) \cdot \exp\left(-\frac{1}{2}\langle\cdot, icA\cdot\rangle\right)\right)(\xi)$$

$$= \sqrt{\left(\frac{\sqrt{k}}{2\pi i \sin(\sqrt{c}t)}\right)}\exp\left(\frac{1}{2}\frac{\sqrt{k}}{i\tan(\sqrt{c}t)}(iy + \langle\xi, 1\!\!1_{[0,t]}\rangle)^2\right)$$

$$\times \exp\left(-\frac{1}{2}\langle\xi, N^{-1}\xi\rangle\right). \quad (12)$$

Summarizing we have the following theorem:

Theorem 2. *Let $y \in \mathbb{R}$, $0 < t < \frac{\pi}{2\sqrt{c}}$, then the Feynman integrand for the harmonic oscillator I_{HO} exists as a Hida distribution and its generating functional is given by (12). Moreover its generalized expectation*

$$\mathbb{E}(I_{HO}) = T(I_{HO})(0) = \sqrt{\left(\frac{\sqrt{k}}{2\pi i \sin(\sqrt{k}t)}\right)} \exp\left(i\frac{\sqrt{k}}{2\tan(\sqrt{k}t)}y^2\right)$$

is the Green's function to the Schrödinger equation for the harmonic oscillator, compare e.g. with[30].

The Feynman integrand for the charged particle in a constant magnetic field The Feynman integrand for the charged particle in a constant magnetic field was treated in the White Noise framework in[15,21] and[5]. Here we have - in contrast to the harmonic oscillator - also a part of the potential which is dependent on the velocity of the particle. However Lemma 1 also applies here.

In classical physics a charged particle moving through a magnetic field $\mathbf{H} = (0, 0, H_3)$ has the Lagrangian

$$L(x_1, x_2, x_3, \dot{x_1}, \dot{x_2}, \dot{x_3}) = \frac{1}{2}m(\dot{x_1}^2 + \dot{x_2}^2 + \dot{x_3}^2) + \frac{qH_3}{c}(x_1\dot{x_2} - \dot{x_1}x_2),$$

where m is the mass of the particle. We denote the constant in front of the potential term by $k := \frac{qH_3}{c}$. Moreover we use the notation $\mathbf{x} = (x_1, x_2, x_3)$ and $\dot{\mathbf{x}} = (\dot{x_1}, \dot{x_2}, \dot{x_3})$, for the vectors of the space and velocity variables respectively. We see that, besides the dependence on the spatial coordinates, the potential term depends explicitly on the velocities.

Since the above three dimensional system can be separated to the free motion parallel to the magnetic field vector and a motion in the plane orthogonal to the magnetic field vector, we restrict ourselves to the two-dimensional system.

In the following we realize rigorously the ansatz

$$I_{\text{mag}} = \text{Nexp}\left(\frac{i}{\hbar}\int_0^t \frac{\dot{\mathbf{x}}(\tau)^2}{2m}d\tau + \frac{1}{2}\int_{t_0}^t \dot{\mathbf{x}}(\tau)^2\right)$$

$$\times \exp\left(-\frac{ik}{\hbar}\int_0^t (x_1(\tau)\dot{x_2}(\tau) - \dot{x_1}(\tau)x_2(\tau)) \, d\tau\right) \cdot \delta_0(\mathbf{x}(t) - \mathbf{y}), \quad (13)$$

for the Feynman integrand of a charged particle in a constant magnetic field, with the help of Lemma 1. In (13) the path x is realized by a two-dimensional Brownian motion starting in 0 at time $t_0 = 0$. Then the first term in (13) can be written as an exponential of quadratic type and gives a generalized Gauss kernel, see Definition 1. Indeed with $\hbar = m = 1$,

$$\text{Nexp}\left(i\int_0^t \frac{\dot{\mathbf{x}}(\tau)^2}{2}d\tau + \frac{1}{2}\int_0^t \dot{\mathbf{x}}(\tau)^2\right) = \text{Nexp}\left(-\frac{1}{2}\langle(\omega_1, \omega_2), \mathbf{K}(\omega_1, \omega_2)\rangle\right),$$

(14)

with $\mathbf{K} := -(i+1)\mathbf{P}_{[0,t)} := -(i+1)\begin{pmatrix} P_{[0,t)} & 0 \\ 0 & P_{[0,t)} \end{pmatrix}$, where $P_{[0,t)}$ denotes the orthogonal projection as before.

First we write also the potential term in (13) in a quadratic way.

Proposition 1. *The operator matrix*

$$\mathbf{L} = \mathbf{P}_{[0,t)}\begin{pmatrix} 0 & ik(A - A^*) \\ ik(A^* - A) & 0 \end{pmatrix}\mathbf{P}_{[0,t)},$$

(15)

fulfills

$$\frac{1}{2}\langle\mathbf{f}, \mathbf{Lf}\rangle = -ik\int_0^t \left(\int_0^\tau f_1(s)\,ds\,f_2(\tau) - f_1(\tau)\int_0^\tau f_2(s)\,ds\right)d\tau, \quad 0 \leq t < \infty,$$

where $\mathbf{f} = (f_1, f_2) \in L_2^2(\mathbb{R})$ *and operator* A *is defined by*

$$Af(\tau) = \mathbb{1}_{[0,t)}(\tau)\int_{[0,\tau)} f(s)\,ds, \quad f \in L^2(\mathbb{R}), \tau \in \mathbb{R}.$$

A^* *denotes its adjoint w.r.t. the bilinear dual pairing* $\langle\cdot, \cdot\rangle$*. Moreover* $\mathbf{L}$ *is symmetric w.r.t. the dual pairing.*

Theorem 3 (Feynman integrand for a charged particle in a magnetic field). *Let* $0 \leq 0 < t < \infty$ *with*

$$\frac{2kt}{\pi} \notin \mathbb{Z}.$$

Then the Feynman integrand I_{mag} *for a charged particle in a constant magnetic field exists as a Hida Distribution. Moreover the integrand can be written as*

$$I_{\text{mag}} = \text{Nexp}\left(-\frac{1}{2}\langle\omega, \mathbf{K}\omega\rangle\right) \cdot \exp\left(-\frac{1}{2}\langle\omega, \mathbf{L}\omega\rangle\right) \cdot \delta_0(\mathbf{B}_t - \mathbf{y}),$$

where $\mathbf{y} = (y_1, y_2)^T \in \mathbb{R}^2$ *and the operator* $\mathbf{K}$ *and* $\mathbf{L}$ *as in Proposition 1. Its* T-*transform in* $\varphi \in S_2(\mathbb{R})$ *is given by*

$$TI_{mag}(\varphi) = \frac{k}{2\pi i} \frac{1}{\cos((kt)} \exp\left(-\frac{1}{2}\left(\varphi, \mathbf{N}^{-1}\varphi\right)\right)$$

$$\times \exp\left(-\frac{ik}{2} \cot(kt)\left(\left(iy_1 + \frac{1}{2}\left(\mathbb{1}_{[0,t)}, (\mathbf{N}^{-1}\varphi)_1\right) + \frac{1}{2}\left(\varphi, \mathbf{f}\right)\right)^2\right.\right.$$

$$\left.\left. + \left(iy_2 + \frac{1}{2}\left(\mathbb{1}_{[0,t)}, (\mathbf{N}^{-1}\varphi)_2\right) + \frac{1}{2}\left(\varphi, \mathbf{g}\right)\right)^2\right)\right),$$

for all $\varphi \in S_2(\mathbb{R})$. *Here* $\mathbf{f} = (\mathbb{1}_{[0,t)}f_1, \mathbb{1}_{[0,t)}f_2)^T$, $\mathbf{g} = (\mathbb{1}_{[0,t)}g_1, \mathbb{1}_{[0,t)}g_2)^T$ *with*

$$f_1(s) = i\cos(2kt) + i\frac{\sin(2kt)}{\cos(2kt) + 1}\sin(2ks) = g_2(s), \quad s \in [0, t)$$

$$f_2(s) = i\frac{\sin(2kt)}{\cos(2kt) + 1}\cos(2ks) - i\sin(2ks) = -g_1(s), \quad s \in [0, t).$$

The generalized expectation (T-*transform in* $\varphi = 0$*) gives*

$$TI_{mag}(0) = \frac{k}{2\pi i} \frac{1}{\cos(kt)} \exp\left(\frac{ik}{2}\cot(kt)\left(y_1^2 + y_2^2\right)\right), \tag{16}$$

which coincides with the Green's function for a charged particle in a magnetic field see e.g.[21,30].

4. Potential Classes

In order to pass from quadratic actions to more general cases, one has to give a rigorous definition of the expression

$$I_V = I_0 \exp\left(-i\int_{t_0}^t V(\mathbf{x}(\tau)\,d\tau\right).$$

There are several potential classes, which can be seen as perturbative approaches. Among these are the Khandekar-Streit class, which was used in[31] and generalized in[41] and the Albeverio-Høegh-Krohn class, which is one of the first potential classes in rigorous approaches to Feynman integrals see e.g.[2,19,40].

4.1. *Khandekar-Streit Potentials*

Khandekar and Streit used perturbative methods in the case $d = 1$ and V a finite signed Borel measure with compact support.[31] This construction was generalized in[41] to time-dependent potentials with Gaussian fall-off in the space variable, instead of compact support. If one just considers an exponential fall-off, it could be shown that the integrand is at least a Kondratiev distribution.

We will follow the path of[41] here.

Therefore let $D = [T_0, T] \supset \Delta = [t_0, t)$ and ν a finite signed Borel measure on $D \times \mathbb{R}$ with marginal measures

$$\nu_x(A) = \nu(A \times D), \quad A \in \mathcal{B}(\mathbb{R}),$$
$$\nu_t(B) = \nu(\mathbb{R} \times B), \quad B \in \mathcal{B}(D).$$

Now we assume ν fulfills the following two assumptions

1. $\exists \rho > 0, \forall r > \rho : |\nu|_x(\{x : |x| > r\}) < e^{-\beta r^2}, \quad \beta > 0$ (17)

2. $|\nu|_t$ has an L^∞-density with essential bound C_ν.   (18)

Let us now consider the construction by heuristically pretending that the measure ν can be treated like an ordinary function. Consider the power series expansion

$$\exp(-i \int_{t_0}^t V(x(\tau), \tau)\, d\tau) = \sum_{n=0}^\infty (-i)^n \int_{\Lambda_n} d^n t \prod_{i=1}^n \int dx_i V(x_i, t_i) \delta(x(t_i) - x_i),$$

where

$$\Lambda_n = \{(t_1, \ldots, t_n) | t_0 < t_1 < \cdots < t_n < t\}.$$

Based on this definition in[41] the following theorem was shown.

Theorem 4. *The Feynman integrand*

$$I = I_0 + \sum_{n=1}^\infty (-i)^n \int_{\mathbb{R}^n} \int_{\Lambda_n} I_0 \prod_{j=1}^n \delta(x(t_j) - x_j) \prod_{i=1}^n v(dx_i, dt_i)$$

exists as a Hida distribution in case V obeys 1. and 2.

Sketch of Proof: Using the integral Corollary 2 from Chapter 1 one obtains:

$$I_n = \int_{\mathbb{R}^n} \int_{\Lambda_n} I_0 \prod_{j=1}^n \delta(x(t_j) - x_j) \prod_{i=1}^n \nu(dx_i, dt_i) \in (S)'.$$

We have for $q > 2$, $\frac{1}{p} + \frac{1}{q} = 1$, $0 < \gamma < \frac{\beta}{\gamma}$:

$$\left| T\left(I_0 \prod_{j=1}^{n} \delta(x(t_j) - x_j) \right)(z\xi) \right|$$

$$\leq \left(\prod_{j=1}^{n} \frac{1}{\sqrt{2\pi|t_i - t_{i-1}|}} \right) \exp(\gamma(\sup_{0 \leq i \leq n+1} |x_i|)^2) \exp\left(\frac{1}{2}(1 + \frac{2}{\gamma} + |\Delta|)(|z|^2|\xi|_s^2) \right),$$

such that $\sup_{t \in \Delta} |\xi(t)| \leq |\xi|_s$

Then by (17) we have $e^{\gamma x^2} \in L^q(\mathbb{R} \times \Delta, |\nu|)$. Moreover let

$$Q = \left(\int_{\mathbb{R}} e^{\gamma q x^2} |\nu_x|(dx) \right)^{\frac{1}{q}}.$$

Then

$$\left(\int_{\mathbb{R}^n} \int_{Delta_n} \exp(\gamma q \sup_{0 \leq i \leq n+1} |x_i|)^2) \prod_{i=1}^{n} |\nu|(dx_i, dt_i) \right)^{\frac{1}{q}} < Q^n e^{\gamma|x|^2} e^{\gamma|x_0|^2}.$$

Now we consider the time integration. With

$$\int_{\Delta} \prod_{j=1}^{n+1} \frac{1}{(2\pi|t_j - t_{j-1}|)^u} = \left(\frac{\Gamma(1-u)}{(2\pi)^u} \right)^{n+1} \frac{|t - t_0|^{n(1-u)-u}}{\Gamma((n+1)(1-u))} d^n t, \quad u < 1,$$

we obtain

$$\left(\int_{\mathbb{R}^n} \int_{\Delta_n} \left(\prod_{j=1}^{n+1} \frac{1}{(2\pi|t_j - t_{j-1}|)^u} \right)^p \prod_{i=1}^{n} |\nu|(dx_i, dt_i) \right)^{\frac{1}{p}}$$

$$\leq |\nu_t|_\infty^{\frac{n}{p}} \left(\frac{\Gamma(\frac{2-p}{2})^{\frac{n+1}{p}}}{(2\pi)^{\frac{n+1}{2}}} \right) \frac{|\Delta|^{\frac{n}{p} - \frac{1}{2}(n+1)}}{\Gamma((n+1)(\frac{2-p}{2}))^{\frac{1}{p}}}. \quad (19)$$

Then

$$\left| \int_{\mathbb{R}^n} \int_{\Lambda_n} T(I_0 \prod_{j=1}^{n} \delta(x(t_j) - x_j))(z\xi) \prod_{i=1}^{n} \nu(dx_i, dt_i) \right|$$

$$\leq C_n \exp(\frac{1}{2}(1 + \frac{2}{\gamma} + |\Delta|)|z|^2|\xi|_s^2),$$

with

$$C_n(x, \Delta) = e^{\gamma|x_0|^2} e^{\gamma|x|^2} Q^n |\nu_t|^{\frac{n}{p}} \left(\frac{\Gamma(\frac{2-p}{2})^{\frac{n+1}{p}}}{(2\pi)^{\frac{n+1}{2}}} \right) \frac{|\Delta|^{\frac{n}{p} - \frac{1}{2}(n+1)}}{\Gamma((n+1)(\frac{2-p}{2}))^{\frac{1}{p}}}.$$

Hence I_n exists as Hida distribution. Moreover, since C_n is rapidly decreasing in n, we have

$$I = \sum_{n=0}^{\infty} I_n \in (S)',$$

by Chapter 1, Corollary 1. □

Remark 4. The class of Khandekar-Streit potentials involves Dirac type potentials. These potentials are of special interest in systems with boundaries.

4.2. *Albeverio-Høegh-Krohn Potentials*

In this subsection we consider another class of potentials for which we will represent the Green's function of the Schrödinger equation by a path integral in a rigorous manner. Potentials of this kind already appeared in mathematically rigorous works on Feynman integrals in the 60s and 70s of the last century, see e.g.[19,40]. Albeverio and Høegh gave a very elegant construction of a path integral by using so-called Fresnel integrals. For this reason many authors named the potential class Albeverio Høegh-Krohn class.

The treatment of this potential class in the framework of White Noise Analysis, see e.g.[50] differs from the earlier mathematical studies in two core points

(1) In[19] the initial wave functions had to fulfill smoothness conditions. In the framework of White Noise one is able to use Delta functions as initial conditions and thus reflects Feynman's original idea of treating propagators by path integrals.
(2) The integrand itself can be characterized as an object of mathematical rigor in a suitable White Noise distribution space. Its expectation yields the propagator.

Definition 6. Let m denote a bounded complex measure on the Borel sets of $\mathbb{R}^d$, $d \geq 1$. A complex valued function V on $\mathbb{R}^d$ is called Fresnel integrable if

$$V(x) = \int_{\mathbb{R}^d} e^{i\alpha \cdot x}\, d^d m(\alpha). \tag{20}$$

Since the bounded complex Borel measures form an algebra under convolution, Fresnel integrable functions also form an algebra $\mathcal{F}(\mathbb{R}^d)$ under pointwise multiplication. We call $\mathcal{F}(\mathbb{R}^d)$ the Albeverio Høegh-Krohn class. A potential $V \in \mathcal{F}(\mathbb{R}^d)$ is called admissile, if there exists a $\varepsilon > 0$, such that

$$\int_{\mathbb{R}^d} e^{\varepsilon|\alpha|} \, d^d|m|(\alpha) < \infty. \tag{21}$$

Additionally we assume for some $\varepsilon, \delta > 0$ that we have

$$\int_{\mathbb{R}^d} e^{\varepsilon|\alpha|^{1+\delta}} \, d^d|m|(\alpha) < \infty. \tag{22}$$

Since all moments of admissible potentials of the corresponding measure m exist we have that these potentials are infinitely often differentiable. In particular admissible potentials are analytic in the open ε-ball. In fact formula (20) extends to $x \in \mathbb{C}^d$ with $\|\mathrm{Im}(x)\| < \varepsilon$, i.e. an admissible potential is regular on a strip containing the real axis, see e.g.[50] .

Example 3. Consider the 2-dimensional potential

$$V(x) = (\cos(x_1)\sin(x_2))^{\beta},$$

where $\beta \in 2\mathbb{Z}$, which is of interest in the theory of antidot lattices, compare[18] . The potential generates a chaotic behavior and its Hamiltonian has a fractal spectrum. However all conditions of Definition 6 are fulfilled, since the measure m corresponding to V is a linear combination of delta functions and thus has compact support.

4.2.1. *Construction of the integrand*

To construct the integrand, again we have to give a rigorous meaning to the poniwise product of the free integrand and the exponential of the potential. In contrast to the construction in the case of Khandekar-Streit-potentials we do not expand the potentials in terms of delta functions. This is due to the fact, that delta functions cause problems w.r.t. time integration in higher dimensions. Therefore here we expand the potential with the help of its Fourier decomposition. Then - as in the Khandekar-Streit-case - one can show that the expansion of the exponential leads to a convergent series in the space of Kondratiev distributions. In fact one has

Theorem 5.[50] *Let V be an admissible potential in the Albeverio-Høegh-Krohn class, i.e. there exists a bounded complex Borel measure m satisfying (21). Then*

$$I \equiv I_0 + \sum_{n=1}^{\infty}(-i)^n \frac{1}{n!}\int_{[t_0,t]^n} d^n\tau \int_{\mathbb{R}^{dn}} \prod_{j=1}^{n} d^d m(\alpha_j) I_0 \cdot \prod_{j=1}^{n} e^{i\boldsymbol{\alpha}_j \mathbf{x}(\tau_j)},$$

exists as a generalized White Noise functional, i.e. $I \in (S)^{-1}$.

Remark 5. It is well known, that $\mathbb{E}(I)$ is the solution of the time dependent Schrödinger equation, since it coincides with the series representation developed in[2]. An explicit proof can be found in[19].

4.3. *Westerkamp-Kuna class*

Kuna, Westerkamp and Streit[37] and also de Faria, Oliveira and Streit[9] investigated another potenital class based on Laplace transforms of measures. The potentials of the Westerkamp-Kuna class are defined as follows

Definition 7. Let m be a complex measure on the Borel sets on $\mathbb{R}^d$, $d \geq 1$ fulfilling the condition

$$\int_{\mathbb{R}^d} e^{C|\alpha|} \, \mathrm{d}|m|(\alpha) < \infty, \quad \forall C > 0. \tag{23}$$

Then the potential V on $\mathbb{R}^d$ given by

$$V(x) = \int_{\mathbb{R}^d} e^{\alpha \cdot x} \, \mathrm{d}m(\alpha),$$

is said to be in the Westerkamp-Kuna class.

The conditions in the definition ensure that the measure m is finite. One can show by dominated convergence that the potentials are restrictions of entire functions to the real line. They can grow strongly over all boundaries at infinity however they are locally bounded and free of singularities.

Example 4.

(1) A compactly supported measure fulfills (23).
(2) For a Dirac measure δ_a, $a > 0$ the potential is $V(x) = e^{ax}$. As a consequence all polynomials of exponentials are in the Westerkamp-Kuna class.
(3) The Morse potential $V(x) := g(e^{-2ax} - 2\gamma e^{-ax})$, $g, a, x \in \mathbb{R}$ and $\gamma > 0$ fulfills (23),[37].

In order to give a rigorous meaning of the pointwise multiplication

$$I = I_0 \cdot \exp\left(-i \int_{t_0}^t V(x(\tau)) \, d\tau\right),$$

one informally, as before, expands the exponential and obtains

$$I = \sum_{n=0}^{\infty} \frac{(-i)^n}{n!} \int_{[t_0, t]^n} \int_{\mathbb{R}^{dn}} I_0 \cdot \prod_{j=1}^{n} e^{\alpha_j \cdot x(\tau_j)} \prod_{j=1}^{n} \mathrm{d}m(\alpha_j) \mathrm{d}^n \tau.$$

Indeed Kuna, Westerkamp and Streit could show the following:

Theorem 6. *Let V in the Westerkamp-Kuna class. Then*

$$I := \sum_{n=0}^{\infty} \frac{(-i)^n}{n!} \int_{[t_0,t]^n} \int_{\mathbb{R}^{dn}} I_0 \cdot \prod_{j=1}^{n} e^{\alpha_j \cdot x(\tau_j)} \prod_{j=1}^{n} dm(\alpha_j) d^n \tau,$$

converges as series in the strong topology of $(S_d)^{-1}$. The integrals exist as Bochner intergrals and one has

$$T(I)(\xi) = \sum_{n=0}^{\infty} \frac{(-i)^n}{n!} \int_{[t_0,t]^n} \int_{\mathbb{R}^{dn}} T\left(I_0 \cdot \prod_{j=1}^{n} e^{\alpha_j \cdot x(\tau_j)}\right)(\xi) \prod_{j=1}^{n} dm(\alpha_j) d^n \tau,$$

for all $\xi \in S_d(\mathbb{R})_{\mathbb{C}}$ with $2^q |\xi|_p < 1$, for some $p, q \in \mathbb{N}_0$.

Remark 6. If we take a look at the proof of the above theorem one has a quite suprising result for the dependence of the coupling constant g. We obtain for the T-transform

$$TI(\xi) = \sum_{n=0}^{\infty} \frac{(-ig)^n}{n!} \int_{[t_0,t]^n} \int_{\mathbb{R}^{dn}} T\left(I_0 \cdot \prod_{j=1}^{n} e^{\alpha_j \cdot x(\tau_j)}\right)(\xi) \prod_{j=1}^{n} dm(\alpha_j) d^n \tau,$$

which is indeed a perturbation series in g. In the proof of the above theorem one can find the following bound[37] ,

$$|TI(\xi)| \leq (2\pi(t-t_0))^{-\frac{d}{2}} \exp(|\xi|^2 + \frac{|x-x_0|}{\sqrt{t-t_0}}|\xi|)$$

$$\times \exp\left(|g|(t-t_0) \int_{\mathbb{R}^d} \exp\left(|\alpha|(|x-x_0| + |x_0| + 2\sqrt{t-t_0}|\xi|)\right) dm(|\alpha|)\right).$$

$$(24)$$

This shows that the T-transform of the integrand is entire in the coupling constant. This is a remarkable fact, since the Hamilton operators, even if they are essentially self-adjoint, for $g > 0$ are not essentially self-adjoint for $g < 0$ in general. So eigenvectors and eigenvalues will not be analytic in the coulping constant.

Remark 7. Additionally to the potentials mentioned above, one can also consider a time dependent version of the Westerkamp-Kuna class. For this we would like to refer to[37] and[9] and the proofs therein. Moreover in[37] the Morse potential is treated as a special case and it is shown that the corresponding propagator solves the Schrödinger equation. In[9] the authors also investigated a combination of the Khandekar-Streit and the Westerkamp-Kuna class.

5. Hamiltonian Path Integrals

Although the first aim of Feynman was to develop path integrals based on a Lagrangian, they also can be used for various systems which have a law of least action, see e.g.[14]. Since classical quantum mechanics is based on a Hamiltonian formulation rather than a Lagrangian one, it is worthwhile to take a closer look at the so-called Hamiltonian path integral, which means the Feynman integral in phase space. This has therefore many advantages:

- At first the semi-classical limit of quantum mechanics is more natural in a Hamiltonian setting, i.e. the phase space is more natural in classical mechanics than the configuration space, see also[2,32] and the references therein.
- In[13] the authors state, that potentials which are time-dependent or velocity dependent should be treated with the Hamiltonian path integral.
- The idea of canonical transformations due to a Hamiltonian system can be done easier in a Hamiltonian setting.
- Also momentum space propagators can be investigated.

Feynman gave a heuristic formulation of the phase space Feynman Integral in[16]

$$K(t, y|0, y_0) = N \int_{x(0)=y_0, x(t)=y} \int \exp\left(\frac{i}{\hbar} S(x)\right) \prod_{0<\tau<t} \frac{dp(\tau)}{(2\pi)^d} dx(\tau). \quad (25)$$

Here the action (and hence the dynamic) is expressed by a canonical (Hamiltonian) system of generalized space variables and their corresponding conjugate momentums. The canonical variables can be found by a Legendre-transformation, see e.g.[48]. The Hamiltonian action:

$$S(x, p, t) = \int_0^t p(\tau) \cdot \dot{x}(\tau) - H(x(\tau), p(\tau), \tau) d\tau,$$

where H is the Hamilton function and given by the sum of the kinetic energy and the potential.

$$H(x, p, t) = \frac{1}{2m} p^2 + V(x, p, t).$$

The phase space is therefore $2n$-dimensional if we have n so called degrees of freedom. Furthermore the variables are independent.

By the Heisenberg uncertainty principle the momentum variables are fulfilling no boundary conditions, because the space variables are fixed.

Note that both integrals, the Feynman integral as well as the Hamiltonian path integral are thought to be integrals w.r.t. a flat, i.e. translation invariant measure on the infinite dimensional path space. Such a measure does not exist, hence the integral at first - as it stands - is not a mathematical rigorous object. The normalization constant in both integrals turns out to be infinity. Nevertheless there is no doubt that it has a physical meaning.

There are many attempts to give a meaning to the Hamiltonian path integral as a mathematical rigorous object. Among these are analytic continuation of probabilistic integrals via coherent states[32,33] and infinite dimensional distributions e.g.[10]. Most recently also an approach using time-slicing was developed by Naoto Kumano-Go[38] and also by Albeverio et al. using Fresnel integrals[1,2]. As a guide to the literature on many attempts to formulate these ideas we point out the list in[2].

In the following for the two settings and objects we give a meaning as distributions of White Noise Analysis. First we consider the ansatz for the configuration space Hamiltonian path integrand then for one in momentum space. The propagators are related to each other by Fourier transform. Hence we can check if the propagators are fulfilling the right physics.

Hamiltonian Path Integral in coordinate space representation
First we introduce the space trajectories as Brownian motion starting in x_0.

$$x(\tau) = x_0 + \sqrt{\frac{\hbar}{m}} B(\tau), \quad 0 \leq \tau \leq t.$$

Furthermore the momentum variable is modeled by white noise, i.e.

$$p(\tau) = \sqrt{\hbar m} \omega_p(\tau), \quad 0 \leq \tau \leq t.$$

This is a meaningful definition, since a path has always start and end points which a noise does not have. Moreover since we have that if the initial and end conditions are fully known, the momentum is completely uncertain, which means has variance infinity.

The white noise process is intrinsically fulfilling the no boundary condition property and has as well infinite variance. Furthermore one can think of (for a potential just depending on the space variable) the momentum to be $p = m\dot{x}$, which in our approach would correspond to a noise in terms of derivative of the Brownian path.

The model for the space path can be found in[25] to model the momentum path we take a closer look to the physical dimensions of $x(\tau)$.

$y(s)$ has as a space variable the dimension of a length, i.e. also $\sqrt{\frac{\hbar}{m}}B(\tau)$ has to have the dimension of a length. We have

$$[\sqrt{\frac{\hbar}{m}}] = \sqrt{\frac{Js}{kg}} = \sqrt{\frac{kgm^2}{skg}} = \frac{m}{\sqrt{s}}.$$

Thus since the norm of the Brownian motion gives again a $\sqrt{t}$ which has the dimension $\sqrt{s}$, we have that $x(\tau)$ has the dimension of a length.

Considering the momentum variable we have to obtain that the dimension is the dimension of a momentum. We have

$$[\sqrt{\hbar m}] = \sqrt{Nmskg} = \sqrt{\frac{kg^2m^2s}{s^2}} = \frac{kgm}{\sqrt{s}},$$

hence ω_p has the dimension $\frac{1}{\sqrt{s}}$, such that $p(\tau)$ has the dimension of a momentum.

A definition which goes in the same direction using the momentum as a kind of derivative of the path can also be found in[2] and[1]. Here the authors modeled the path space as the space of absolutely continuous functions and the momentum to be in $L^2(\mathbb{R})$. Then we propose the following informal ansatz for the Feynman integrand in phase space with respect to the Gaussian measure μ,

$$I_V = N\exp\left(\frac{i}{\hbar}\int_{t_0}^t p(\tau)\dot{x}(\tau) - \frac{p(\tau)^2}{2m}d\tau + \frac{1}{2}\int_{t_0}^t \dot{x}(\tau)^2 + p(\tau)^2 d\tau\right) \quad (26)$$

$$\times \exp\left(-\frac{i}{\hbar}\int_{t_0}^t V(x(\tau), p(\tau), \tau)\, d\tau\right) \cdot \delta(x(t) - y).$$

In this expression the sum of the first and the third integral is the action $S(x, p)$, and the Donsker's delta function serves to pin trajectories to y at time t. The second integral is introduced to simulate the Lebesgue integral by compensation of the fall-off of the Gaussian measure in the time interval (t_0, t).

Hamiltonian path integral in momentum space

If we know the initial and the end momentums it is clear by Heisenberg's uncertainty principle that we have no certain information about the corresponding space variables. This means we model the momentum trajectories as a Brownian fluctuation starting in the initial momentum p_0.

$$p(\tau) = p_0 + \frac{\sqrt{\hbar m}}{t - t_0}B(\tau), \quad 0 \le \tau \le t. \quad (27)$$

Furthermore the space variable is modeled by white noise, i.e.

$$x(\tau) = \sqrt{\frac{\hbar}{m}} \cdot (t - t_0)\omega_x(\tau), \quad 0 \le \tau \le t. \tag{28}$$

The Hamiltonian path integral for the momentum space propagator is formally given by, see e.g.[35]

$$K(p', t', p_0, t_0) = N \int_{p(t_0)=p_0, p(t)=p'} \exp(\frac{i}{\hbar} \int_{t_0}^{t} -q(s)\dot{p}(s) - H(p, q) \, ds) \, DpDq. \tag{29}$$

This path integral can be obtained by a Fourier transform of the coordinate space path integral in both variables, see e.g.[34]. Then we propose the following informal ansatz for the Feynman integrand in phase space with respect to the Gaussian measure μ,

$$I_V = N \exp\left(\frac{i}{\hbar} \int_{t_0}^{t} -x(\tau)\dot{p}(\tau) - \frac{p(\tau)^2}{2m} d\tau + \frac{1}{2} \int_{t_0}^{t} \omega_x(\tau)^2 + \omega_p(\tau)^2 d\tau\right) \tag{30}$$

$$\times \exp\left(-\frac{i}{\hbar} \int_{t_0}^{t} V(x(\tau), p(\tau), \tau) \, d\tau\right) \cdot \delta(p(t) - p').$$

5.1. *Example: The free integrand in momentum space representation*

It is well known, see e.g.[35] that the momentum space propagator for a free particle is given in form of a Dirac Delta function. We want to show therefore at least that we can find an expression which converges to this propagator. As above we consider first the action

$$S = \int_0^t -x(\tau)\dot{p}(\tau) - \frac{1}{2m}p^2(\tau)d\tau$$

then we have with (27) and (28)

$$S = \int_0^t -\sqrt{\frac{\hbar}{m}}\sqrt{\hbar m}\omega_p(\tau)\omega_x(\tau) - \frac{1}{2m}(p_0 + \frac{\sqrt{\hbar m}}{t}\langle\omega_p, \mathbb{1}_{[0,\tau)}\rangle)^2 \, d\tau$$

$$= -\hbar \int_0^t \omega_p(\tau)\omega_x(\tau)d\tau - \frac{p_0^2}{2m}t + \langle p_0\sqrt{\frac{\hbar}{mt^2}}(s - t), \omega_p(s), \mathbb{1}_{[0,t)}(s)\rangle$$

$$-\frac{1}{2}\int_0^t \frac{\hbar}{t^2}\langle\cdot, \omega_p, \mathbb{1}_{[0,\tau)}\rangle^2 \, d\tau$$

Then we can write (30) in the following form using $m = \hbar = 1$:

$$N\exp\left(-\frac{1}{2}\langle(\omega_x, \omega_p), K_{mom}(\omega_x, \omega_p)\rangle\right)$$

$$\times \exp(\langle p_0\frac{1}{t}(\cdot - t)\mathbb{1}_{[0,t)}(\cdot), \omega_p\rangle)\delta(p' - p_0 - \langle\frac{1}{t}\mathbb{1}_{[0,t)}, \omega_p\rangle)$$

where the operator matrix K on $L_2^2(\mathbb{R})_{\mathbb{C}}$ can be written as

$$K_{mom} = \begin{pmatrix} -P_{[0,t)} & iP_{[0,t)} \\ iP_{[0,t)} & -P_{[0,t)} + \frac{i}{t^2}A \end{pmatrix}. \tag{31}$$

Here

$$A f(s) = \mathbb{1}_{[0,t)}(s) \int_s^t \int_0^\tau f(r)\, dr\, d\tau, f \in L^2(\mathbb{R},\mathbb{C}), s \in \mathbb{R},$$

we refer to[22] for properties of the operator as the trace class property, invertibility and spectrum.
We have for $f, g \in L^2(\mathbb{R})_{\mathbb{C}}$

$$\langle f, Ag \rangle = \int_0^t \int_0^\tau f(s)\, ds \cdot \int_0^\tau g(s)\, ds\, d\tau.$$

Hence the operator is used to implement the integral over the squared Brownian motion. The last term pins the momentum variable to p' at t. Note that the space variable is not pinned.
Our aim is to apply Lemma 1 with K as above and $\mathbf{g} = (0, \frac{p_0}{t}(s-t)\mathbb{1}_{[0,t)}(s))$, $L = 0$ and as $\boldsymbol{\eta} = (0, \frac{1}{t}\mathbb{1}_{[0,t)})$. The inverse of $(Id + K)$ is given by

$$N^{-1} = (Id + K_{mom})^{-1} = \begin{pmatrix} P_{[0,t)^c} & 0 \\ 0 & P_{[0,t)^c} \end{pmatrix} + i \begin{pmatrix} \frac{1}{t^2}A & -P_{[0,t)} \\ -P_{[0,t)} & 0 \end{pmatrix}, \tag{32}$$

hence $(\boldsymbol{\eta}, N^{-1}\boldsymbol{\eta}) = 0$.
To apply Lemma 1 we use a small perturbation of the matrix N^{-1}.
Let $\epsilon > 0$ then we define

$$N_\epsilon^{-1} = \begin{pmatrix} P_{[0,t)^c} & 0 \\ 0 & P_{[0,t)^c} \end{pmatrix} + \begin{pmatrix} \frac{i}{t^2}A & -i\mathbb{1}_{[0,t)} \\ -i\mathbb{1}_{[0,t)} & +\epsilon \end{pmatrix},$$

and obtain $(\boldsymbol{\eta}, N_\epsilon^{-1}\boldsymbol{\eta}) = \frac{\epsilon}{t}$ and Lemma 1 can be applied. Therefore the assumptions of Lemma 1 are fulfilled. Thus we define the regularized free momentum integrand by its T-transform in $(f_x, f_p) \in S_2(\mathbb{R})$

$$T(I_{0mom,\epsilon})(f_x, f_p) = \frac{1}{\sqrt{2\pi\frac{\epsilon}{t}}} \exp(-\frac{ip_0^2}{2}t)$$

$$\times \exp\left(-\frac{1}{2} \left\langle \begin{pmatrix} f_x \\ f_p + \frac{p_0}{t}(\cdot - t)\mathbb{1}_{[0,t)} \end{pmatrix}, N_\epsilon^{-1} \begin{pmatrix} f_x \\ f_p + \frac{p_0}{t}(\cdot - t)\mathbb{1}_{[0,t)} \end{pmatrix} \right\rangle \right)$$

$$\times \exp\left(\frac{1}{2\frac{\epsilon}{t}} \left(i(p' - p_0) + \left\langle \begin{pmatrix} f_x \\ f_p + (\cdot - t)\mathbb{1}_{[0,t)} \end{pmatrix}, N_\epsilon^{-1} \begin{pmatrix} 0 \\ \mathbb{1}_{[0,t)} \end{pmatrix} \right\rangle \right)^2 \right).$$

Hence its generalized expectation

$$\mathbb{E}(I_{0,mom,\epsilon}) = T I_{0,mom,\epsilon}(0)$$

$$= \frac{\sqrt{t}}{\sqrt{2\pi\epsilon}} \exp(-\frac{1}{2} \left\langle \left(\begin{matrix} 0 \\ 0 + \frac{p_0}{t}(\cdot - t)\mathbb{1}_{[0,t)} \end{matrix} \right), N_\epsilon^{-1} \left(\begin{matrix} 0 \\ \frac{p_0}{t}(\cdot - t)\mathbb{1}_{[0,t)} \end{matrix} \right) \right\rangle)$$

$$\times \exp\left(\frac{t}{2\epsilon} \left(i(p' - p_0) + \left\langle \left(\begin{matrix} 0 \\ 0 + \frac{p_0}{t}(\cdot - t)\mathbb{1}_{[0,t)} \end{matrix} \right), N_\epsilon^{-1} \left(\begin{matrix} 0 \\ \frac{1}{t}\mathbb{1}_{[0,t)} \end{matrix} \right) \right\rangle \right)^2 \right)$$

$$\cdot \exp(-\frac{ip_0^2}{2}t) = \frac{\sqrt{t}}{\sqrt{2\pi\epsilon}} \exp(-\frac{t}{2\epsilon}(p' - p_0)^2) \exp(-\frac{ip_0^2}{2}t)$$

$$\times \exp\left(-\frac{\epsilon}{2t^2}p_0^2 \int_0^t (s - t)^2 \, ds \right) \exp\left(\frac{p_0^2\epsilon}{2t^3} \left(\int_0^t (s - t) \, ds \right)^2 \right)$$

$$\cdot \exp(i\frac{p_0}{t}(p' - p_0) \int_0^t (s - t) \, ds).$$

In the limit $\epsilon \to 0$ we obtain:

$$\lim_{\epsilon \to 0} \mathbb{E}(I_{0mom,\epsilon}) = \delta(p' - p_0) \cdot \exp(-\frac{ip_0^2}{2}t) \cdot \exp\left(\frac{ip_0}{t}(p' - p_0) \int_0^t (s - t) \, ds \right)$$

$$= \delta(p' - p_0) \cdot \exp(-\frac{ip_0^2}{2}t),$$

Since the last term vanishes note that the Delta function just gives values if $p' = p_0$. The generalized expectation is up to a factor 2π exactly the propagator of the free particle in momentum space, see[35]. Note that the Delta function serves to conserve the momentum of the free particle. If there is no potential the momentum must be the same as the initial momentum since the space is free of any force.

6. Feynman integrals via complex scaling

First we summarize the complex scaling method which is originated from Cameron and then developed by Doss. Let $\lambda > 0$. Under some regularity properties on the initial function ψ_0 and the potential v, it is well-known that the solution of the heat equation

$$\begin{cases} \lambda\frac{\partial}{\partial t}\psi(x,t) = \frac{\lambda^2}{2m}\Delta\psi(x,t) + v(x)\psi(x,t) \\ \psi(x,0) = \psi_0(x), \quad (x,t) \in \mathbb{R}^d \times [0,\infty), \end{cases} \tag{33}$$

is given by the famous Feynman-Kac formula

$$\psi(x,t) = \mathbb{E}_W \left(\psi_0 \left(x + \sqrt{\frac{\lambda}{m}}B_t \right) \exp\left(\frac{1}{\lambda} \int_0^t v \left(x + \sqrt{\frac{\lambda}{m}}B_s \right) \, ds \right) \right),$$

where $(B_t)_{t \geq 0}$ is a d-dimensional Brownian motion and $\mathbb{E}_W$ denotes the expectation with respect to the Wiener measure on the Wiener space. Moreover for $t \in [0, \infty)$ and $x, x_0 \in \mathbb{R}^d$, the heat kernel $K_v : \mathbb{R}^d \times \mathbb{R}^d \times [0, \infty) \to \mathbb{R}$ is given by

$$K_v(x, t; x_0, 0) = \frac{1}{\sqrt{2\pi t}} \exp\left(-\frac{m}{2\lambda t}(x - x_0)^2\right)$$

$$\times \mathbb{E}_W \left(\exp\left(\frac{1}{\lambda} \int_0^t v\left(\frac{s}{t}(x - x_0) + \sqrt{\frac{\lambda}{m}}\left(B_s - \frac{s}{t}B_t\right)\right) ds\right)\right).$$

We notice that in the potential energy part we are dealing with a Brownian bridge starting at time 0 in x_0 and ending at time t in x. For more information concerning Feynman-Kac formula we refer to [28, Section 4.4]. Informally, by replacing λ in the heat equation (33) by $i\hbar$ we obtain the Schrödinger equation. H. Doss in 1980 has proved that under some hypotheses of analyticity and integrability on ψ_0 and v the unique solution of the Schrödinger eqution can be represented explicitly in a Feynman-Kac type formula. Below we describe the main result in Doss' approach, for details and proofs see[12] . Let O be a nonempty open set in $\mathbb{R}^d$ and D be an open set in $\mathbb{C}^d$ defined by

$$D := \left\{ \left(x_1 + \sqrt{i}y_1, \ldots, x_d + \sqrt{i}y_d\right) : (x_1, \ldots, x_d) \in O, (y_1, \ldots, y_d) \in \mathbb{R}^d \right\}.$$

Doss built the following two assumptions:

(1). There exist two holomorphic functions $\tilde{\psi}_0$ and $\tilde{v}$ from D to $\mathbb{C}$ such that $\tilde{\psi}_0\big|_O = \psi_0$ and $\tilde{v}\big|_O = v$. Let $\mathcal{H}(D) := \{f : D \to \mathbb{C} : f \text{ is holomorphic}\}$ and $L : \mathcal{H}(D) \to \mathcal{H}(D)$ given by

$$Lf(x) := \frac{1}{2}i\hbar \sum_{j=1}^d \frac{\partial^2}{\partial x_j^2} f(x) + \frac{1}{i\hbar}\tilde{v}(x)f(x),$$

for $f \in \mathcal{H}(D)$ and $x = (x_1, \ldots, x_d) \in D$. Furthermore, let us define functions F, G, H_1, R_1, H_2, and R_2 from $D \times [0, \infty) \times \Omega$ to $\mathbb{C}$ as follows:

$$G(x, t, \omega) := \tilde{\psi}_0\left(x + \sqrt{i\hbar}B_t(\omega)\right),$$

$$F(x, t, \omega) := \frac{1}{i\hbar} \int_0^t \tilde{v}\left(x + \sqrt{i\hbar}B_s(\omega)\right) ds,$$

$$H_1(x, t, \omega) := G(x, t, \omega) \exp\left(F(x, t, \omega)\right),$$

$$R_1(x, t, \omega) := \sup_{s \in [0,t]} |H_1(x, s, \omega)|,$$

$$H_2(x, t, \omega) := L\tilde{\psi}_0\left(x + \sqrt{i\hbar}B_t(\omega)\right) \exp\left(F(x, t, \omega)\right), \quad \text{and}$$

$$R_2(x, t, \omega) := \sup_{s \in [0,t]} |H_2(x, s, \omega)|.$$

(2). For all $(x,t) \in D \times [0, \infty)$ the random variables $R_1(x, t, \cdot)$ and $R_2(x, t, \cdot)$ are Wiener integrable and the mapping $D \ni x \mapsto \psi(x,t) := \mathbb{E}\left(H_1(x,t)\right)$ is holomorphic.

Theorem 7. *Under the assumptions (1) and (2), there exists a unique strong solution $\psi = (\psi(x,t))_{(x,t)\in\mathcal{O}\times[0,\infty)}$ of the Schrödinger equation which satisfies the following conditions: $\psi = (\psi(x,t))_{(x,t)\in\mathcal{O}\times[0,\infty)}$ can be extended to a C^1-function on $D \times [0, \infty)$ which is analytic in the space variable $x \in D$. Moreover, the following representation holds:*

$$\psi(x,t) = \mathbb{E}\left(\tilde{\psi}_0\left(x + \sqrt{i\hbar}B_t(\omega)\right)\right) \exp\left(\frac{1}{i\hbar} \int_0^t \tilde{v}\left(x + \sqrt{i\hbar}B_s(\omega)\right)\, ds\right),$$
(34)

for $(x,t) \in D \times [0, \infty)$.

Proof. See[12] . $\square$

Note that the assumptions on ψ_0 and v are sufficient to allow the application of Fubini's theorem and Morera's theorem in order to show that the solution (34) is holomorphic on D.

Some examples of elements from the Doss class are polynomial potentials, the harmonic oscillator and non-perturbative accessible potentials such as $V : D \to \mathbb{C}$ given by

$$V(x) = \frac{a}{|(x,b)_{euc} - c|^n},$$

where $n \in \mathbb{N}$, $a \in \mathbb{C}$, $b \in \mathbb{R}^d$, and $c \in \mathbb{R}$. For more information see e.g.[49] .

In the following, since we are interested in solutions to the Schrödinger equation we focus on complex scaled heat equations, i.e. for $z \in \mathbb{C}$, we consider

$$\begin{cases} \frac{\partial}{\partial t}\psi(t,x) = -z^2\frac{1}{2}\Delta\psi(t,x) + \frac{1}{z^2}V(x)\psi(t,x) \\ \psi(0,x) = f(x), \quad 0 \le t \le T < \infty, x \in \mathcal{O} \subset \mathbb{R}^d, \end{cases}$$
(35)

for suitable functions f and suitable time-independent potentials V.
In the configuration space, this has been done in[50] and[23,49] in phase space there is an approach to be found in[3] . Scaling approaches for potentials which are not from the Doss class are treated in[42] . The Doss scaling approach has several advantages:

- Treatable potentials are beyond perturbation theory such as

$$V(x) = (-1)^{n+1}a_{4n+2}x^{4n+2} + \sum_{j=1}^{4n+1} a_j x^j, \quad x \in \mathbb{R}, n \in \mathbb{N},$$

with $a_{4n+2} > 0, a_j \in \mathbb{C}$.

- Due to a Wick formula we have a convenient structure (i.e. "Brownian motion is replaced by a Brownian bridge")
- The kinetic energy " '$\sigma_z \delta$' " and the potential can be treated separately.
- The Wick product of two Hida distributions always exists as Hida distribution, thus one does not need to justify the well-definedness of the pointwise product.

6.1. *Preliminaries about scaling operators*

First note, that every test function $\varphi \in (S)$ can be extended to $S_d(\mathbb{R})'_\mathbb{C}$, see e.g.[39] . Thus the following definition makes sense.

Definition 8. Let φ be the continuous version of an element of (S). Then for $0 \neq z \in \mathbb{C}$ we define the scaling operator σ_z by

$$(\sigma_z \varphi)(\omega) = \varphi(z\omega), \quad \omega \in S'_d(\mathbb{R}).$$

Proposition 2.

(i) For all $0 \neq z \in \mathbb{C}$ we have $\sigma_z \in L((S),(S))$,
(ii) for $\varphi, \psi \in (S)$ we have

$$\sigma_z(\varphi \cdot \psi) = (\sigma_z \varphi)(\sigma_z \psi).$$

More precisely we have, compare to[49] and[50] the following proposition.

Proposition 3. *Let $\varphi \in (S)$, $z \in \mathbb{C}$, then*

$$\sigma_z \varphi = \sum_{n=0}^{\infty} \langle \varphi_z^{(n)}, : \cdot^n : \rangle,$$

with kernels

$$\varphi_z^{(n)} = z^n \sum_{k=0}^{\infty} \frac{(n+2k)!}{k!n!} \left(\frac{z^2 - 1}{2} \right)^k \cdot tr^k \varphi^{(n+2k)}.$$

Definition 9. Since σ_z is a continuous mapping from (S) to itself we can define its dual operator $\sigma_z^\dagger : (S)' \to (S)'$ by

$$\langle\!\langle \varphi, \sigma_z^\dagger \Phi \rangle\!\rangle = \langle\!\langle \sigma_z \varphi, \Phi \rangle\!\rangle,$$

for $\Phi \in (S)'$ and $\varphi \in (S)$.

The following proposition can be found in[50] and[49] .

Proposition 4. *Let $\Phi \in (S)^*$, $\varphi, \psi \in (S)$ and $z \in \mathbb{C}$ then we have*

(i)

$$\sigma_z^\dagger \Phi = J_z \diamond \Gamma_z \Phi,$$

where Γ_z is defined by

$$S(\Gamma_z \Phi)(\xi) = S(\Phi)(z\xi), \quad \xi \in S_d(\mathbb{R}),$$

and $J_z = \text{Nexp}(-\frac{1}{2}z^2\langle\cdot,\cdot\rangle)$. In particular we have

$$\sigma_z^\dagger \mathbb{1} = J_z.$$

(ii) $J_z\varphi = \sigma_z^\dagger(\sigma_z\varphi)$.

We want to generalize the notion of scaling to bounded operators. More precisely we investigate for which kind of linear mappings $B \in L(S(\mathbb{R})', S(\mathbb{R})')$ there exists some operator $\sigma_B : (\mathcal{N}) \to (\mathcal{N})$ such that

$$\Phi_{(BB^*)} \cdot \varphi := \sigma_B^\dagger \sigma_B \varphi.$$

Further we state a generalization of the Wick formula to Gauss kernels. We start with the definition of σ_B.

Definition 10. Let $B \in L(S_d(\mathbb{R})_\mathbb{C}, S_d'(\mathbb{R})_\mathbb{C})$. By tr_B we denote the element in $S_d'(\mathbb{R})_\mathbb{C} \otimes S_d'(\mathbb{R})_\mathbb{C}$, which is defined by

$$\forall \xi, \eta \in S_d(\mathbb{R})_\mathbb{C} : \ tr_B(\xi \otimes \eta) := \langle \xi, B\eta \rangle.$$

Note that tr_B is not symmetric. Further there exists a $q \in \mathbb{Z}$ such that $tr_B \in \mathcal{H}_{q,\mathbb{C}} \otimes \mathcal{H}_{q,\mathbb{C}}$.

Proposition 5. Let $B \in L(\mathcal{H}_\mathbb{C}, \mathcal{H}_\mathbb{C})$ be a Hilbert-Schmidt operator. Then $tr_B \in \mathcal{H}_\mathbb{C}\otimes\mathcal{H}_\mathbb{C}$. Further for each orthonormal basis $(e_j)_{j\in\mathbb{N}}$ of $\mathcal{H}_\mathbb{C}$ it follows:

$$tr_B = \sum_{i=0}^\infty Be_i \otimes e_i.$$

Proposition 6. In the case $S(\mathbb{R}) = S(\mathbb{R})$ and $B \in L(S_\mathbb{C}(\mathbb{R}), S_\mathbb{C}'(\mathbb{R}))$ we have

$$tr_B = \sum_{i=0}^\infty Bh_i \otimes h_i.$$

Definition 11. For $B \in L(S_d'(\mathbb{R})_\mathbb{C}, S_d'(\mathbb{R})_\mathbb{C})$ we define $\sigma_B\varphi$, $\varphi \in (S)$, via its chaos decomposition, which is given by

$$\sigma_B\varphi = \sum_{n=0}^\infty \left\langle \varphi_B^{(n)}, :\cdot^{\otimes n}: \right\rangle, \tag{36}$$

with kernels

$$\varphi_B^{(n)} = \sum_{k=0}^{\infty} \frac{(n+2k)!}{k!n!} \left(-\frac{1}{2}\right)^k (B^*)^{\otimes n} (\text{tr}_{(Id-BB^*)}^k \varphi^{(n+2k)}).$$

Here, B^* means the dual operator of B with respect to $\langle \cdot, \cdot \rangle$. Further for $A \in L(S_d(\mathbb{R})_{\mathbb{C}}, S_d'(\mathbb{R})_{\mathbb{C}})$, the expression $\text{tr}_A^k \varphi^{(n+2k)}$ is defined by

$$\text{tr}_A^k \varphi^{(n+2k)} := \left\langle \text{tr}_A^{\otimes k}, \varphi^{(n+2k)} \right\rangle \in \mathcal{N}^{\hat{\otimes} n},$$

where the generalized trace kernel tr_A is defined in (10).

Proposition 7. *Let* $B \in L(S_d(\mathbb{R})_{\mathbb{C}}, S_d(\mathbb{R})_{\mathbb{C}})$ *and* $n \in \mathbb{N}$, $n > 0$. *Then for all* $p \in \mathbb{N}$ *there exists a* $K > 0$ *and* $q_1, q_2 \in \mathbb{N}$, *with* $p < q_1 < q_2$ *such that for all* $\theta \in S_d(\mathbb{R})_{\mathbb{C}}^{\otimes n}$ *we have*

$$\left| B^{\otimes n} \theta \right|_p \leq \left(K \left\| B \right\|_{q_2, q_1} \left\| i_{q_1, p} \right\|_{HS} \right)^n \cdot \left| \theta \right|_{q_2}.$$

Next we show the continuity of the generalized scaling operator.

Proposition 8. *Let* $B : S_d'(\mathbb{R})_{\mathbb{C}} \to S_d'(\mathbb{R})_{\mathbb{C}}$ *a bounded operator. Then* $\varphi \mapsto \sigma_B \varphi$ *is continuous from* (S) *into itself.*

Proposition 9. *Let* $\varphi \in (S)$ *given by its continuous version. Then it holds*

$$\sigma_B \varphi(\omega) = \varphi(B\omega),$$

if $B \in L(S_d'(\mathbb{R}), S_d'(\mathbb{R}))$, $\omega \in S_d'(\mathbb{R})$.

This can be proved directly by an explicit calculation on the set of Wick exponentials, a density argument and a verifying of pointwise convergence, compare[43].

In the same manner the following statement is proved.

Proposition 10. *Let* $B : S(\mathbb{R})' \to S(\mathbb{R})'$ *be a bounded operator. For* $\varphi, \psi \in (S)$ *the following equation holds*

$$\sigma_B(\varphi\psi) = (\sigma_B\varphi)(\sigma_B\psi).$$

Since we consider a continuous mapping from (S) into itself one can define the dual scaling operator with respect to $\langle \cdot, \cdot \rangle$, $\sigma_B^{\dagger} : (S)' \to (S)'$ by

$$\left\langle\!\left\langle \sigma_B^{\dagger} \Phi, \psi \right\rangle\!\right\rangle = \left\langle\!\left\langle \Phi, \sigma_B \psi \right\rangle\!\right\rangle.$$

The Wick formula as stated in[23,49] for Donsker's delta function can be extended to Generalized Gauss kernels.

Proposition 11. *[Generalized Wick formula] Let* $\Phi \in (S)'$, $\varphi, \psi \in (S)$ *and* $B \in L(S_d'(\mathbb{R}), S_d'(\mathbb{R}))$, *then we have*

(i)

$$\sigma_B^\dagger = \Phi_{BB^*} \diamond \Gamma_B^* \Phi,$$

where Γ_B^* is defined by

$$S(\Gamma_B^* \Phi)(\xi) = S(\Phi)(B^* \xi), \quad \xi \in S_d(\mathbb{R}).$$

In particular we have

$$\sigma_B^\dagger \mathbb{1} = \Phi_{BB^*}.$$

(ii) $\Phi_{BB^*} \cdot \varphi = \sigma_B^\dagger(\sigma_B \varphi).$
(iii) $\Phi_{BB^*} \cdot \varphi = \Phi_{BB^*} \diamond (\Gamma_{B^*} \circ \sigma_B(\varphi)).$

6.2. *Complex Scaling of Donsker's delta*

It has been shown that the kinetic enery combined with the compensation of the Gaussian fall-off of the white noise measure yields a well defined Hida distribution. In a more general way one can define

$$J_z := \text{Nexp}\left(\frac{1}{2}(1 - z^{-2}) \int_{t_0}^t \omega^2(r)\, dr\right),$$

for $-\infty < t_0 < t < \infty$ and $z \in \mathbb{C} \setminus \{0\}$. Moreover one has for $\psi, \phi \in (S)$:

$$J_z \phi = \sigma_{z,t_0,t}^\dagger(\sigma_{z,t_0,t}\phi) \text{ and} \tag{37}$$

$$J_z \phi \psi = \sigma_{z,t_0,t}^\dagger(\sigma_{z,t_0,t}\phi \sigma_{z,t_0,t}\psi). \tag{38}$$

To obtain the Feynman integrand however, we would need to scale Donsker's delta function, i.e. we have to generalize the complex scaling to Hida distributions. Lascheck, Leukert, Streit and Westerkamp proved a series of theorems concerning Donsker's delta function and its complex scaling in[41]. We will here collect the results and refer the reader to the proofs in this work.

Consider again the S-transform of Donsker's delta in $\xi \in S(\mathbb{R})$

$$S(\delta(\langle \cdot, h \rangle - a))(\xi) = \frac{1}{\sqrt{2\pi \langle h, h \rangle}} \exp\left(-\frac{1}{2\langle h, h \rangle}(\langle \xi, h \rangle - a)^2\right),$$

$$h \in L^2(\mathbb{R}),\ a \in \mathbb{R}.$$

It is obvious that the expression is analytic in the parameter a. Hence we can extend it to complex a and obtain still a U-functional. The same holds for the complexification of h, where we exclude the case that (h, h) is a real, negative number. We obtain

Theorem 8. Let $a \in \mathbb{C}$ and $h \in L_{\mathbb{C}}^2(\mathbb{R})$ with $(h, h) \notin (-\infty, 0]$. Then $\delta(\langle \cdot, h \rangle - a) \in (S)'$.

The following theorem shows that products of complex scaled Donsker's delta functions are again - under suiatble conditions - Hida distributions. We will use the theorem in particular to define Riemann approximations for the Feynman integrand. A proof can be found in[23,41].

Theorem 9. *The n-fold product of Donsker's delta functions is defined by*

$$\Phi := \prod_{j=1}^{n} \sigma_z \delta(\langle \cdot, h_j \rangle - a_j) := \frac{1}{(2\pi)^n} \prod_{j=1}^{n} \int_{\gamma} \exp(i\lambda_j (z\langle \cdot, h_j \rangle - a_j)) d\lambda_j$$

as a Bochner integral, where $\gamma = \{e^{i\alpha} x \,|\, x \in \mathbb{R}\}$ *and* $z \in \mathbb{C}$ *such that* $|\arg(z)| < \frac{\pi}{4} + \alpha$, $|\alpha| < \frac{\pi}{4}$. *Further* $(h_j)_j$ *is a linear independent system in* $L^2(\mathbb{R})$ *and* $a_j \in \mathbb{C}$, $j = 1, ..., n$. *Then* $\Phi \in (S)'$ *with*

$$S\Phi(\xi) = \frac{1}{\sqrt{(2\pi z^2)^n \det(M)}} \exp\left(-\frac{1}{2} \left((h, \xi) - \frac{1}{z} a \right) M^{-1} \left((h, \xi) - \frac{1}{z} a \right) \right),$$

where $M := (h_j, h_k)_{k,l=1,..,n}$ *is the Gramian of* $(h_j)_j$ *and* $(h, \xi) := ((h_1, \xi), ..., (h_n, \xi))$ *and* $a = (a_1, ..., a_n)$.

6.3. *The complex scaled Feynman-Kac kernel*

As we discussed before, a solution to the heat equation is given via the Feynman-Kac formula for suitable potentials V. For nice potentials the heat kernel can be described via an expectation dealing with rather a Brownian bridge than a Brownian motion, compare e.g. 27. We will follow the path of[23] here to show that under suitable conditions on the potential

$$\exp\left(\frac{1}{z^2} \int_0^t V(x + z(\langle \cdot, \mathbb{1}_{[0,t)} \rangle)) \right) \sigma_z \delta(\langle \cdot, \mathbb{1}_{[0,t)} \rangle + x_0 - x)$$

$$= \exp\left(\frac{1}{z^2} \int_0^t V(x_0 + \frac{s - t_0}{t - t_0}(x - x_0) + z(\langle \cdot, \mathbb{1}_{[0,s)} \rangle)) - \frac{s - t_0}{t - t_0} \langle \cdot, \mathbb{1}_{[0,t)} \rangle \right)$$

$$\diamond \sigma_z \delta(\langle \cdot, \mathbb{1}_{[0,t)} \rangle + x_0 - x). \quad (39)$$

Furthermore the generalized expectation will solve the complex scaled heat eqution (35).

In order to construct the complex scaled Feynman-Kac kernel we will use an approximation scheme with the help of finitely based functions. For this however we need some more assumptions on the potential.

Let $\mathcal{O} \subset \mathbb{R}$ such that $\mathbb{R} \setminus \mathcal{O}$ is a set of Lebesgue measure zero. For $z \in \mathbb{C} \setminus 0$ we define

$$\mathcal{D}_z := \{ x + zy \,|\, x \in \mathcal{O}, y \in \mathbb{R} \}.$$

Assumption 1. Let $z \in S_0 := \{z \in \mathbb{C} \mid \arg(z) \in (-\frac{\pi}{4}, \frac{\pi}{4}\}$ and $0 < T < |infty$. We assume that the potential $V : \mathcal{D}_z \to \mathbb{C}$ is analytic and we have the bounds

$$\left|\exp(\frac{1}{z^2}V(y)\right| \le A\exp(\varepsilon y^2), \quad \left|\exp(\frac{1}{z^2}V(x_0 + zy)\right| \le B(x_0)\exp(\varepsilon y^2),$$

where $0 < A < \infty$, $B : \mathcal{O} \to \mathbb{R}$ is a locally bounded function and $0 < \varepsilon < \frac{1}{4T}$.

Example 5.

(1) $V(x) = -\exp(p(x))$, where p is a polynomial for $z = 1$ fulfilling the assumption.
(2) The before mentioned Doss potentials are admissible for $z = \sqrt{i}$.

The next step is to give a mathematical reasonable definition of the integral. Therefore one writes the integral as a Riemann sum and takes the product with Donsker's delta function. One obtains,

Proposition 12. [23] *The product of a complex scaled Donsker's delta with the Riemann approximation in (39) of the intergal is a Hida distribution, i.e.*

$$\Phi_n = \exp\left(\frac{1}{z^2}\frac{t}{n}\sum_{k=1}^{n-1} V(x_0 + z\langle\cdot, \mathbb{1}_{[0,t_k)}\rangle)\right) \sigma_z\delta(x_0 - x + \langle\cdot, \mathbb{1}_{[0,t)}\rangle) \in (S)'.$$

The next theorem is giving the relation between the Brownian motion and the Brownian bridge in this case. This representation has the advantage that the product of the potential part and the Donsker's delta funtions is now a Wick product, which is always defined between Hida distributions. This means we omit the definition of a pointwise product of two distributions which caused a lot of trouble before.

Theorem 10.[23] *Let* $h_k = \mathbb{1}_{[0,t_k)} - \frac{k}{n}\mathbb{1}_{[0,t)}$, $1 \le k \le n - 1$. *Moreover let* $x \in \mathcal{O}$ *such that* $x_0 + \frac{s}{t}(x - x_0)$ *and* Φ_n *as in Proposition 12. Then for*

$$\Psi_n := \exp\left(\frac{1}{z^2}\frac{t}{n}\sum_{k=1}^{n-1} V(x_0 + \frac{k}{n}(x - x_0) + z\langle\cdot, h_k\rangle)\right) \in (L^2), \quad n \in \mathbb{N},$$

one has

$$\Phi_n = \Psi_n \diamond \sigma_z\delta(x_0 - x + \langle\cdot, \mathbb{1}_{[0,t)}\rangle),$$

which is obviously a Hida distribution.

Now one can show that both sequences $(\Phi_n)_n$ and $(\Psi_n)_n$ converge in $(S)'$. In particular one obtains, compare[23]:

Theorem 11. *Let* $0 \leq t \leq T < \infty$, $t_k = \frac{k}{n}t$, $k = 1, ..., n \in \mathbb{N}$. *Let* $x_0, x \in \mathcal{O}$ *such that* $x_0 + \frac{s}{t}(x - x_0) \in \mathcal{O}$, *fal all* $s \in [0, t)$ *and* Φ_n *as before. Then* $(\Phi_n)_n$ *converges in* $(S)'$ *to a Hida distribution* Φ. *The S-transform of* Φ *is given by*

$$S(\Phi)(\xi) = S(\sigma_z \delta(x_0 - x + \langle \cdot, \mathbb{1}_{[0,t)} \rangle)))(\xi)$$
$$\times S(\exp\left(\frac{1}{z^2} \int_0^t V(x_0 - \frac{s}{t}(x - x_0) + z\langle \cdot, h_s \rangle)\right),$$

for all $\xi \in S(\mathbb{R})$ *and* $h_s = \mathbb{1}_{[0,s)} - \frac{s}{t}\mathbb{1}_{[0,t)}$.

6.4. *Construction of the generalized scaled heat kernel in phase space*

Within this section we consider the case of one degree of freedom, i.e. the underlying space is the space $S_2'(\mathbb{R})$. In the Euclidean configuration space a solution to the heat equation is given by the Feynman-Kac formula with its corresponding heat kernel. In White Noise Analysis one constructs the integral kernel by inserting Donsker's delta function to pin the final point $x \in \mathbb{R}$ and taking the expectation, i.e.,

$$K_V(x, t, x_0, t_0) = \mathbb{E}\left(\exp(\int_{t_0}^t V(x_0 + \langle \mathbb{1}_{[t_0,r)}, \cdot \rangle)\,dr)\delta(x_0 + \langle \mathbb{1}_{[t_0,t)}, \cdot \rangle - x)\right),$$

where the integrand is a suitable distribution in White Noise Analysis (e.g. a Hida distribution).

We will construct in this section by a suitable generalized scaling the Hamiltonian Path Integral as an expectation based on the formula above.

First we construct the scaling operator we need.

Proposition 13. *Let* $N^{-1} = \begin{pmatrix} P_{[0,t)^c} & 0 \\ 0 & P_{[0,t)^c} \end{pmatrix} + i\begin{pmatrix} P_{[0,t)} & P_{[0,t)} \\ P_{[0,t)} & 0 \end{pmatrix}$ *as in the case of the free Hamiltonian integrand. Let* R *be a symmetric operator (w.r.t. the dual pairing) with* $R^2 = N^{-1}$. *Indeed we have:*

$$R = \begin{pmatrix} P_{[0,t)^c} & 0 \\ 0 & P_{[0,t)^c} \end{pmatrix} + \frac{\sqrt{i}}{1 + (\frac{\sqrt{5}+1}{2})^2}U^T\begin{pmatrix} \frac{1+\sqrt{5}}{2}P_{[0,t)} & 0 \\ 0 & \frac{1-\sqrt{5}}{2}P_{[0,t)} \end{pmatrix}U,$$

with

$$U = \begin{pmatrix} -\frac{\sqrt{5}+1}{2} & 1 \\ -1 & -\frac{\sqrt{5}+1}{2} \end{pmatrix}.$$

Then under the assumption that $\sigma_R \delta(\langle(\mathbb{1}_{[0,t)}, 0), \cdot\rangle) = \delta(\langle R(\mathbb{1}_{[0,t)}, 0), \cdot\rangle)$
$\in (S)'$, *we have*

$$I_0 = \sigma_R^\dagger \sigma_R \delta(\langle(\mathbb{1}_{[0,t)}, 0), \cdot\rangle).$$

Consequently the Hamiltonian path integrand for an arbitrary space dependent potential V, can be informally written as

$$I_V = \text{Nexp}\left(-\frac{1}{2}\langle\cdot, K\cdot\rangle\right) \exp\left(-i \int_0^t V(x_0 + \langle(\mathbb{1}_{[0,r)}, 0), \cdot\rangle)\, dr\right)$$

$$\times \delta(x_0 + \langle(\mathbb{1}_{[t_0,t)}, 0), \cdot\rangle - x)$$

$$= \sigma_R^\dagger\left(\sigma_R\left(\exp\left(-i \int_0^t V(x_0 + \langle(\mathbb{1}_{[0,r)}, 0), \cdot\rangle)\, dr\right)\right)\right.$$

$$\left. \times \sigma_R \delta(x_0 + \langle(\mathbb{1}_{[t_0,t)}, 0), \cdot\rangle - x)\right), \quad (40)$$

for $x, x_0 \in \mathbb{R}$ and $0 < t_0 < t < \infty$.

In the following we present some ideas to give a mathematical meaning to the expression in (40). First we consider a quadratic potential, i.e. we consider

$$\exp(-\frac{1}{2}\langle\cdot L\cdot\rangle)\delta(\langle(\mathbb{1}_{[t_0,t)}, 0), \cdot\rangle - x).$$

Definition 12. For L fulfilling the assumption of Lemma 1 and $\delta(\langle(\mathbb{1}_{[t_0,t)}, 0), \cdot\rangle - x)$ we define

$$\sigma_R\left(\exp(-\frac{1}{2}\langle\cdot L\cdot\rangle)\delta(\langle(\mathbb{1}_{[t_0,t)}, 0), \cdot\rangle - x)\right) := \exp(-\frac{1}{2}\langle\cdot RLR\cdot\rangle)$$

$$\times \delta(\langle R(\mathbb{1}_{[t_0,t)}, 0), \cdot\rangle - x).$$

We now take a look at the T-transform of this expression. We have

$$T(\sigma_R^\dagger \sigma_R\left(\exp(-\frac{1}{2}\langle\cdot L\cdot\rangle)\delta(\langle(\mathbb{1}_{[t_0,t)}, 0), \cdot\rangle - y)\right))(\xi)$$

$$= T(\sigma_R\left(\exp(-\frac{1}{2}\langle\cdot L\cdot\rangle)\delta(\langle(\mathbb{1}_{[t_0,t)}, 0), \cdot\rangle - y)\right))(R\xi)$$

$$= \frac{1}{\sqrt{2\pi \det(Id + RLR)}} \exp(-\frac{1}{2}\langle R\xi, (Id + RLR)^{-1} R\xi\rangle)$$

$$\exp\left(\frac{1}{2\langle R(\mathbb{1}_{[t_0,t)}, 0), (Id + RLR)^{-1} R(\mathbb{1}_{[t_0,t)}, 0)\rangle}\right.$$

$$\left. \times (iy - \langle R\xi, (Id + RLR)^{-1} R(\mathbb{1}_{[t_0,t)}, 0)\rangle)^2\right).$$

Now with $R^2 = N^{-1}$ and since R is invertible with $R^{-1}R^{-1} = N$, we have
$$Id + RLR = RR^{-1}R^{-1}R + RLR = R(Id + K + L)R$$
and
$$(Id + RLR)^{-1} = R^{-1}(Id + K + L)^{-1}R^{-1}.$$
Thus
$$T\left(\sigma_R^{\dagger}\sigma_R\left(\exp(-\frac{1}{2}\langle\cdot L\cdot\rangle)\delta(\langle(\mathbb{1}_{[t_0,t)},0),\cdot\rangle - y)\right)\right)(\xi)$$
$$= \frac{1}{\sqrt{2\pi\det((N+L)N^{-1})}} \exp(-\frac{1}{2}\langle\xi,(N+L)^{-1}\xi\rangle)$$
$$\exp\left(\frac{1}{2\langle(\mathbb{1}_{[t_0,t)},0),(N+L)^{-1}(\mathbb{1}_{[t_0,t)},0)\rangle}(iy - \langle\xi,(N+L)^{-1}(\mathbb{1}_{[t_0,t)},0)\rangle)^2\right),$$
which equals the expression from Lemma 1. Hence we have that for a suitable quadratic potential
$$\sigma_R^{\dagger}\sigma_R\left(\exp(-\frac{1}{2}\langle\cdot L\cdot\rangle)\delta(\langle(\mathbb{1}_{[t_0,t)},0),\cdot\rangle - y)\right),$$
exists as a Hida distribution. Moreover for all quadratic potentials from the previous section 5, the T-transform obtained via scaling gives the generating functional as in the ansatz in equations (26) and (30) respectively. Since the T-transforms coincide, also the distributions are the same.

For the case of quadratic potentials we obtained the correct physics also by the scaling approach. Now we generalize this to more complicated potentials. Therefore we follow the way from[49] and[23] by the use of the Wick-formula for generalized function with Donsker's delta function and so-called finitely based Hida distributions, compare to[50].

First we have to list properties, which we demand from the potentials we investigate, compare also to[49].

Assumption 2. Let $0 < t < T < \infty$ and $\mathcal{O} \subset \mathbb{R}$ open such that $\mathbb{R} \setminus \mathcal{O}$ is of Lebesgue measure zero. We assume that the potential $V : \mathcal{D}_R \to \mathbb{C}$ is analytic, with
$$\mathcal{D}_R = \{x_0 + \langle(y_1,y_2), R\begin{pmatrix}1\\0\end{pmatrix}\rangle | y_1, y_2 \in \mathbb{R}\}$$
and there exist a constant $0 < A < \infty$, a locally bounded function $B : \mathcal{O} \to \mathbb{R}$ and some $\varepsilon < \frac{1}{8T}$ such that for all $x_0 \in \mathcal{O}$ and $y \in \mathbb{R}$ one has that
$$|\exp(-iV(x)| \leq A\exp(\varepsilon x^2),$$
and
$$|\exp(-iV(x_0 + \langle(y_1,y_2), R\begin{pmatrix}1\\0\end{pmatrix}\rangle))| \leq B(x_0)\exp(\varepsilon(y_1^2 + y_2^2)).$$

Furthermore we assume the following for the potential and its derivative:

Assumption 3. Let $0 < T < \infty$ and $V : \mathcal{D}_R \to \mathbb{C}$ such that Assumption 2 is fulfilled. We furthermore assume the existence of a locally bounded function $C : \mathcal{O} \times \mathcal{O} \to \mathbb{R}$ and some $0 < \varepsilon < \frac{1}{8T}$ such that for all $x_0, x_1 \in \mathcal{O}$ and $y_1, y_2 \in \mathbb{R}$ we have

$$| \exp(V(x_0 + \langle (y_1, y_2), R \begin{pmatrix} 1 \\ 0 \end{pmatrix} \rangle)) \exp(-iV(x_1 + \langle (y_1, y_2), R \begin{pmatrix} 1 \\ 0 \end{pmatrix} \rangle))|$$
$$\leq C(x_0, x_1) \exp(\varepsilon(y_1^2 + y_2^2))$$

and

$$|\frac{\partial}{\partial z} \exp(V(x_0 + \langle (y_1, y_2), R \begin{pmatrix} 1 \\ 0 \end{pmatrix} \rangle)) \exp(-iV(x_1 + \langle (y_1, y_2), R \begin{pmatrix} 1 \\ 0 \end{pmatrix} \rangle))|$$
$$\leq C(x_0, x_1) \exp(\varepsilon(y_1^2 + y_2^2)),$$

where $\frac{\partial}{\partial z}$ denotes the derivative of $z \to V(z)$ w.r.t. z.

Now we sketch how to give a meaning to a scaling in phase space for potentials V fulfilling Assumptions 2 and 3. For simplicity we consider the case $t_0 = x_0 = 0$.

First we consider the approximation by finitely based Hida distributions as in[23,49], compare also[50]. The main use of finitely based Hida distributions in this work is the fact, that they allow us to extend the generalized scaling operator.

Let $\eta_j \in L_2^2(\mathbb{R})$, $j = 1, \ldots, n$ a system of linear independent vectors and $G : \mathbb{R}^n \to \mathbb{C}$ such that $G \in L^p(\nu_M)$ for some $p > 1$, where ν_M denotes the measure on $\mathbb{R}^n$ with density

$$\exp(-\frac{1}{2} \sum_{k,l=1}^n x_k M_{k,j}^{-1} x_j),$$

w.r.t. the Lebesgue measure on $\mathbb{R}^n$, where $M_{k,j} := \langle \eta_k, \eta_l \rangle$. Then one can define compare[49]

$$\phi(\cdot) := G(\langle \eta_1, \cdot \rangle, \ldots, \langle \eta_n, \cdot \rangle) \in L^p(\mu).$$

Such elements are called finitely based Hida distributions, since they just depend on a finite number of basis elements $\langle \eta_k, \cdot \rangle$, $k \in \mathbb{N}$. The definition goes back to[36], see[23,49] for non-smooth η_k.

Lemma 2 ([49]). *If $G \in L^p(\nu_M)$ the following relation holds*

$$G(\langle \eta_1, \cdot \rangle, \ldots, \langle \eta_n, \cdot \rangle) = \int_{\mathbb{R}^n} G(x_1, \ldots, x_n) \prod_{j=1}^{n} \delta(\langle \eta_j, \cdot \rangle - x_j) \, dx_1 \ldots dx_n,$$

where the integral exists in $(S)'$ in the sense of Corollary 2 of Chapter 1.

The next lemma can be compared with the one in[49]

Lemma 3. *Let $\eta_j \subset L_2^2(\mathbb{R})$, $j = 1, \ldots, n$ be a system of linear independent vectors. Let $\nu_{M_{N-1}, \varepsilon}$ the measure having the density*

$$\exp\left(-\frac{1}{2} \begin{pmatrix} x_1 \\ \vdots \\ .x_n \end{pmatrix}^T \Re((M_{N-1})^{-1} - \varepsilon Id) \begin{pmatrix} x_1 \\ \vdots \\ .x_n \end{pmatrix} \right),$$

w.r.t. the Lebesgue measure on $\mathbb{R}^n$, where $(M_{N-1})_{k,l} = \langle \eta_k, \eta_l \rangle$. Let $G \in L^p(\nu_{M_{N-1}, \varepsilon})$. Then

$$\sigma_R \phi := \int_{\mathbb{R}^n} G(x_1, \ldots, x_n) \prod_{j=1}^{n} \sigma_R \delta(\langle \eta_j, \cdot \rangle - x_j) dx_1 \ldots dx_n,$$

is a well-defined Hida distribution as a Bochner integral in $(S)'$.

This can be proven analogously to the proof in[49]. Note that the density is analogue to the density in the complex scaling case.

The next assumption is based on[49].

Assumption 4. We consider the decomposition of the interval $[0, t)$ given by $t_k := t\frac{k}{n}$, $k = 1, \ldots, n$. We assume that the Riemann approximation

$$\phi_n := \exp\left(-i\frac{t}{n} \sum_{k=1}^{n-1} V(\langle R(\mathbb{1}_{[0, t_k)}), 0), \cdot \rangle) \right) \in L^2(\mu). \tag{41}$$

Then we have the following proposition.

Proposition 14. *The product of the generalized scaled Donsker's delta function with the Riemann approximation defined as in (41) can be defined as a Hida distribution, i.e.*

$$\Phi_n := \exp\left(-i\frac{t}{n} \sum_{k=1}^{n-1} V(\langle R(\mathbb{1}_{[0, t_k)}), 0), \cdot \rangle) \right) \sigma_R \delta(\langle (\mathbb{1}_{[0, t)}), 0) - y, \cdot \rangle) \in (S)', \tag{42}$$

for all $y \in \mathbb{R}$, $0 < t < \infty$ and $n \in \mathbb{N}$.

Proof. We define

$$G : \mathbb{R}^{n-1} \to \mathbb{C}$$

$$y = (y_1, \ldots, y_{n-1}) \mapsto \exp\left(-i\frac{t}{n}\sum_{k=1}^{n-1} V(y_k)\right). \tag{43}$$

Then we have

$$\phi_n = \int_{\mathbb{R}^{n-1}} G(y) \prod_{k=1}^{n-1} \delta((\langle R(\mathbb{1}_{[0,t_k)}), 0), \cdot \rangle - y_k) d^{n-1}y,$$

for all $n \in \mathbb{N}$ with $y = (y_1, \ldots, y_{n-1})$.

Since R is invertible and $(\mathbb{1}_{[0,t_k)}), 0)$, $k = 1, \ldots, n-1$ and $(\mathbb{1}_{[0,t]}, 0)$ form a linear independent system also their images under R form a linear independent system. Hence we have

$$\left(\prod_{k=1}^{n-1} \delta((\langle R(\mathbb{1}_{[0,t_k)}), 0), \cdot \rangle)\right) \delta((\langle R(\mathbb{1}_{[0,t)}), 0), \cdot \rangle),$$

is a Hida distribution for all $n \in \mathbb{N}$. Moreover we have

$$T(\exp\left(-i\frac{t}{n}\sum_{k=1}^{n-1} V(\langle R(\mathbb{1}_{[0,t_k)}), 0), \cdot \rangle)\right) \sigma_R \delta((\langle (\mathbb{1}_{[0,t)}), 0) - y, \cdot \rangle))(\xi)$$

$$= \int_{\mathbb{R}^{n-1}} G(y)T\left(\left(\prod_{k=1}^{n-1} \delta((\langle R(\mathbb{1}_{[0,t_k)}), 0), \cdot \rangle)\right) \delta((\langle R(\mathbb{1}_{[0,t)}), 0), \cdot \rangle)\right)(\xi)dy,$$

for all $\xi \in S_2(\mathbb{R})$. $\qquad\square$

Note that we can prove that the sequence converges in $(S)'$, since the T-transform cannot be estimated independently on n, see also[49] . The following assumption can be compared to[49] .

Assumption 5. Let $h_k := (\mathbb{1}_{[0,t_k)} - \frac{k}{n}\mathbb{1}_{[0,t)}, 0)$, $1 \le k \le n-1$. Then

$$\Psi_n := \exp\left(-i\frac{t}{n}\sum_{k=1}^{n-1} V(\frac{k}{n}y + \langle Rh_k, \cdot \rangle)\right) \in L^2(\mu)$$

and

$$\Phi_n = \Psi_n \diamond \sigma_R \delta((\langle (\mathbb{1}_{[0,t)}), 0), \cdot \rangle - y) \in (S)',$$

with Φ_n, $n \in \mathbb{N}$ as in Proposition 14.

Proposition 15. *Let ϕ_n and ψ_n, $n \in \mathbb{N}$, be defined as in Assumption 5. Then $\phi_n, \Psi_n \in L^2(\mu)$ and*

$$\lim_{n \to \infty} \phi_n = \exp\left(-i \int_0^t V(\langle R(\mathbb{1}_{[0,r)}, 0), \cdot \rangle) \, dr \right),$$

$$\lim_{n \to \infty} \psi_n = \exp\left(-i \int_0^t V(\langle R(\mathbb{1}_{[0,r)} - \frac{r}{t} \mathbb{1}_{[0,t)}, 0), \cdot \rangle - \frac{r}{t}y) \, dr \right).$$

This can be proven by Lebesgue dominated convergence for suitable potentials V.

Then we can as in[49] state the following crucial theorem.

Theorem 12. *Under Assumption 5 and for suitable potentials V the sequence Φ_n converges in $(S)'$. Then it is natural to identify the limit object with*

$$\Phi := \exp\left(-i \int_0^t V(\langle R(\mathbb{1}_{[0,r)}, 0), \cdot \rangle) \, dr \right) \sigma_R \delta(\langle (\mathbb{1}_{[0,t)}, 0), \cdot \rangle - y) := \lim_{n \to \infty} \Phi_n.$$

Moreover we have

$$S(\Phi)(\xi) = S(\psi)(\xi) S(\sigma_R \delta(\langle (\mathbb{1}_{[0,t)}, 0), \cdot \rangle - y))(\xi), \quad \xi \in S_2(\mathbb{R}).$$

This means we have given a meaning to the generalized scaled heat kernel as a Hida distribution. Note that here we just gave the ideas and the heuristics to achieve this goal. Moreover the limit here is strongly dependent on the sequence, since the projection operator, where the relation between ψ and ϕ is based on, is not closable on $L^2(\mu)$. The same holds for the generalized scaling operator. Nevertheless, the particular choice of the approximation converges to a well-defined Hida distribution. The object then is mathematical rigorously defined. The last step would now be to check if the so achieved integrands solve the Schrödinger equation, as it is done in[23,49] and[50].

7. Canonical Commutation Relations

Due to Heisenberg's uncertainty principle the operators of momentum and position in quantum mechanics are not commutative. However the Feynman integral is fully commutative in these two objects, which are functionals. Feynman and Hibbs[17] show, that $\mathbb{E}(\dot{x}(s + \varepsilon)x(x)I) - \mathbb{E}(\dot{x}(s - \varepsilon)x(x)I)$ converges to the commutator as $\varepsilon \to 0$. In a rigorous manner a similar derivation has been done in[50] for the Albeverio-Høegh-Krohn class in the configuration space setting and in[4] in the phase space setting for quadratic potentials. Here we use both techniques to prove a functional form of the canonical commutation relations for the class of Albeverio-Høegh-Krohn potentials in phase space expressed in space representation.

First we have the following lemma and proposition, compare e.g.[50].

Lemma 4. *Let* $\Phi \in (S)', \mathbf{k} \in S_2(\mathbb{R})$.

$$T(\langle \cdot, \mathbf{k} \rangle^n \Phi)(\boldsymbol{\xi}) = (-i)^n \frac{dx^n}{dx\lambda^n} T(\Phi)(\boldsymbol{\xi} + \lambda \mathbf{k}) \Big|_{\lambda=0}.$$

Proposition 16. *Let* $\mathbf{h}, \mathbf{k} \in L_2^2(\mathbb{R})$. *Then* $\langle \cdot, \mathbf{k} \rangle \cdot \langle \cdot, \mathbf{h} \rangle \cdot \Phi_n$ *are Hida distributions. Furthermore, both expressions converge in* $(S)'$, *as* $\varepsilon \to 0$, *and we have for the T-transform*

$$T(\Phi_n)(\boldsymbol{\xi}) \Bigg(\Bigg(\langle \mathbf{k}, N^{-1}\mathbf{h} \rangle - \Big(\langle \boldsymbol{\eta}, N^{-1}\mathbf{h} \rangle, M_{N^{-1}}^{-1} \langle \boldsymbol{\eta}, N^{-1}\mathbf{k} \rangle \Big) \Bigg)$$

$$- \Big(\langle \mathbf{f}, N^{-1}\mathbf{h} \rangle - \big((iy + \langle \boldsymbol{\eta}, N^{-1}\mathbf{f} \rangle), M_{N^{-1}}^{-1} \langle \boldsymbol{\eta}, N^{-1}\mathbf{h} \rangle \big) \Big)$$

$$\times \Big(\langle \mathbf{f}, N^{-1}\mathbf{k} \rangle - \big((iy + \langle \boldsymbol{\eta}, N^{-1}\mathbf{f} \rangle), M_{N^{-1}}^{-1} \langle \boldsymbol{\eta}, N^{-1}\mathbf{k} \rangle \big) \Big) \Bigg), \quad (44)$$

where

$$\mathbf{f} = \begin{pmatrix} \sum_{j=1}^n \alpha_j \mathbb{1}_{[0,t_j)} \\ 0 \end{pmatrix} + \boldsymbol{\xi}.$$

Proof: The proof is a simple application of Proposition 8.1.1 in[4]. ∎

For the commutator we have to consider $\mathbb{E}\Big((p(s+\varepsilon) - p(s-\varepsilon))x(s)\Phi_n \Big)$. Therefore Proposition 16 has to be extended to the case where $\mathbf{h} \in S_2'(\mathbb{R})$. In particular we have to define the formal expression

$$\Big\langle \cdot, \begin{pmatrix} 0 \\ \delta_{s+\varepsilon} - \delta_{s-\varepsilon} \end{pmatrix} \Big\rangle \Big\langle \cdot, \begin{pmatrix} \mathbb{1}_{[0,s)} \\ 0 \end{pmatrix} \Big\rangle I,$$

in a mathematically rigorous way.

Remark 8. Since the dual pairing between a Dirac delta function and an indicator function is defined as a limit object, we cannot expect that the approximation is independent of the choice of the approximating sequence. We choose the following approximation.

Definition 13. Let $\mathbf{h} \in L_d^2(\mathbb{R})$ and $\mathbf{k} \in S_d'(\mathbb{R})$ with compact support and let $(\psi_m)_{m \in \mathbb{N}}$ be a standard approximate identity. Since the convolution of a compactly supported smooth function with a compactly supported tempered distribution gives a Schwartz test function, i.e. $\psi_m * \mathbf{k} \in S_d(\mathbb{R})$, $m \in \mathbb{N}$, see e.g. [44, Chap. 9] we may define

$$\langle \cdot, \mathbf{k} \rangle \cdot \langle \cdot, \mathbf{h} \rangle \cdot \Phi_n := \lim_{m \to \infty} \langle \cdot, \psi_m * \mathbf{k} \rangle \cdot \langle \cdot, \mathbf{h} \rangle \cdot \Phi_n,$$

in the case the limit exists as a suitable distribution of white noise, i.e. the limit on the right-hand side is a U-functional.

Proposition 17. *Let $\psi_m, m \in \mathbb{N}$, be a standard approximate identity with compact support on the interval $[0, t]$. Then we have*

$$\left\langle \cdot, \begin{pmatrix} 0 \\ \psi_m * \delta_{s+\varepsilon} - \psi_m * \delta_{s-\varepsilon} \end{pmatrix} \right\rangle \left\langle \cdot, \begin{pmatrix} \mathbb{1}_{[0,s)} \\ 0 \end{pmatrix} \right\rangle \Phi_n \in (S)', \text{ with } s \in [0, t], \varepsilon > 0.$$

Moreover, the limit for $\varepsilon \to 0$ is a Hida distribution.

Proof: As in,[50] we just have to consider the terms involving the Dirac sequences. We have the following

$$\left\langle \begin{pmatrix} \mathbb{1}_{[0,s)} \\ 0 \end{pmatrix}, N^{-1} \begin{pmatrix} 0 \\ \psi_m * \delta_{s+\varepsilon} + \psi_m * \delta_{s-\varepsilon} \end{pmatrix} \right\rangle$$
$$= i \langle \mathbb{1}_{[0,s)}, \psi_m * \delta_{s+\varepsilon} - \psi_m * \delta_{x-\varepsilon} \rangle = i,$$

for m sufficiently large and $0 < \varepsilon < s$.

$$\left\langle \eta, N^{-1} \begin{pmatrix} 0 \\ \psi_m * \delta_{s+\varepsilon} - \psi_m * \delta_{x-\varepsilon} \end{pmatrix} \right\rangle = i \langle \mathbb{1}_{[0,t)}, \psi_m * \delta_{s+\varepsilon} - \psi_m * \delta_{x-\varepsilon} \rangle, \quad (45)$$

which vanishes for m sufficiently large and ε sufficiently small.

The term involving

$$\mathbf{f} = \begin{pmatrix} \sum_{j=1}^n \alpha_j \mathbb{1}_{[0,t_j)} \\ 0 \end{pmatrix} + \boldsymbol{\xi}$$

is more subtle. For this we have to assume that $t_j \neq s$ for all j. In this case we write

$$\lim_{m \to \infty} \left(\psi_m * \delta_{s+\varepsilon} - \psi_m * \delta_{x-\varepsilon}, \mathbb{1}_{[0,t_j)} \right)$$
$$= \mathbb{1}_{[0,t_j)}(s+\varepsilon) - \mathbb{1}_{[0,t_j)}(s-\varepsilon) = \mathbb{1}_{[s-\varepsilon,s+\varepsilon)}(t_j).$$

For $s \neq t_j$ this formula makes sense. However, in the point $s = t_j$ one can produce any value between 0 and 1. In total we obtain

$$
T\left(\left\langle \cdot, \begin{pmatrix} 0 \\ \psi_m * \delta_{s+\varepsilon} - \psi_m * \delta_{s-\varepsilon} \end{pmatrix} \right\rangle \left\langle \cdot, \begin{pmatrix} \mathbb{1}_{[0,s)} \\ 0 \end{pmatrix} \right\rangle \Phi_n \right)(\xi)
$$

$$
= iT(\Phi_n)(\xi) - T(\Phi_n)(\xi) \times \left(i \sum_{j=1}^{n} \alpha_j \mathbb{1}_{[s-\varepsilon, s+\varepsilon)}(t_j) + i\left(\xi(s+\varepsilon) - \xi(s-\varepsilon)\right) \right)
$$

$$
\times \left(i \sum_{j=1}^{n_{j,\max}} \alpha_j t_j + i(\xi(s) - \xi(0)) - \frac{s}{t}\left(iy + \langle \eta, N^{-1}\xi \rangle + i \sum_{j=1}^{n} \alpha_j t_j \right) \right),
$$

where $n_{j,\max} := \max\{j = 1, \ldots, n : t_j < s\}$. ∎

Lemma 5. *The series*

$$
\sum_{n=0}^{\infty} (-i)^n \int_{\Lambda_n} \int_{\mathbb{R}^{2n}} \left(p(s+\varepsilon)x(s) - x(s)p(s-\varepsilon) \right) \Phi_n \prod_{j=1}^{n} dx \, m_1(\alpha_j) \, dx \, m_2(\beta_j) \, dx t_j
$$

converges in $(S_2)^{-1}$. Moreover the limit $\varepsilon \to 0$ is a Kondratiev distribution and its expectation is given by the commutator.

Proof: The first term in (44) gives the required commutator in the sense of[17]. As in the proof of Lemma 122 in[50], the second term may be bounded by a function which is linearly dependent on ε and vanishes as $\varepsilon \to 0$. Hence the assertion is proved.

The convergence in the space of Kondratiev distributions is an analogue to[50], Lemma 122. ∎

Remark 9. It is possible to show the canonical commutation relations for different Feynman path integrals, also in phase space see e.g.[4].

References

1. S. Albeverio, G. Guatteri, and S. Mazzucchi. Phase Space Feynman Path Integrals. *J. Math. Phys.* 43(6):2847–2857, June 2002.
2. S. Albeverio, R. Høegh-Krohn, and S. Mazzucchi. *Mathematical Theory of Feynman Path Integrals: An Introduction*, volume 523 of *Lecture Notes in Mathematics*. Springer Verlag, Berlin, Heidelberg, New York, 2008.
3. W. Bock. Generalized Scaling Operators in White Noise Analysis and Applications to Hamiltonian Path Integrals with Quadratic Action. In *Stochastic and Infinite Dimensional Analysis*. Birkhäuser, Trends in Mathematics 2015.
4. W. Bock and M. Grothaus. A White Noise Approach to Phase Space Feynman Path Integrals. *Teor. Imovir. ta Matem. Statyst.*, (85):7–21, 2011.

5. W. Bock, M. Grothaus, and S. Jung. The Feynman Integrand for the Charged Particle in a Constant Magnetic Field as White Noise Distribution, *Comm. Stoch. Anal.*, Vol. 6, No. 4: 649-668, 2012.

6. R. H. Cameron. A Family of Integrals Serving to Connect the Wiener and Feynman integrals. *J. Math. and Phys.*, 39:126–140, 1960.

7. M. V. Fedoriuk and V. P. Maslov. Semi-Classical Approximation in Quantum Mechanics Springer Verlag, Berlin, Heidelberg, New York, 1981.

8. M. de Faria, J. Potthoff, and L. Streit. The Feynman integrand as a Hida distribution. *J. Math. Phys.*, 32:2123–2127, 1991.

9. Margarida de Faria, Maria João Oliveira, and Ludwig Streit. Feynman integrals for nonsmooth and rapidly growing potentials. *J. Math. Phys.*, 46(6):063505, 14, 2005.

10. C. DeWitt-Morette, A. Maheshwari, and B. Nelson. Path Integration in Phase Space. *General Relativity and Gravitation*, 8(8):581–593, 1977.

11. Paul A.M. Dirac. The Lagrangian in Quantum Mechanics. *Phys. Z.Sowjetunion*, 3:64–72, 1933.

12. H. Doss. Sur une Resolution Stochastique de l'Equation de Schrödinger à Coefficients Analytiques. *Communications in Mathematical Physics*, 73:247–264, October 1980.

13. I. Duru and H. Kleinert. Quantum Mechanics of H-Atom from Path Integrals. *Fortsch.d.Physik*, 30:401–435, 1982.

14. R. P. Feynman. Space-Time Approach to Non-Relativistic Quantum Mechanics. *Rev. Modern Physics*, 20:367–387, 1948.

15. D. Falco and D. C. Khandekar. Application of White Noise Calculus to the Computation of Feynman Integrals. *Stoch. Proc. and Appl.*, 29:257–266, 1988.

16. R.P. Feynman. An Operator Calculus having Applications in Quantum Electrodynamics. *Physical Review*, 84(1):108–124, 1951.

17. R.P. Feynman and A.R. Hibbs. *Quantum Mechanics and Path Integrals*. McGraw-Hill, London, New York, 1965.

18. R. Fleischmann, T.Geisel, R. Ketzmerich, and G. Petschel, Chaos und Fraktale Energiespektren in Antidot-Gittern. *Physikalische Blätter* 51, 177-181, 1995.

19. K. Gawedzki. *Construction of Quantum-Mechanical Dynamics by Means of Path Integrals in Path Space*. Rep. Math. Phys. Vol. 6 pp. 327342, 1974.

20. I. S. Gradshteyn and I. M. Ryzhik. *Table of Integrals, Series, and Products*. Fourth edition prepared by Ju. V. Geronimus and M. Ju. Ceïtlin. Translated from the Russian by Scripta Technica, Inc. Translation edited by Alan Jeffrey. Academic Press, New York, 1965.

21. M. Grothaus. White Noise Analysis and Feynman Integrals, Diploma thesis, University of Bielefeld.

22. M. Grothaus and L. Streit. Quadratic Actions, Semi-Classical Approximation, and Delta Sequences in Gaussian Analysis. *Rep. Math. Phys.*, 44(3):381–405, 1999.

23. M. Grothaus, L. Streit, and A. Vogel. The Complex Scaled Feynman–Kac Formula for Singular Initial Distributions. *Stochastics*, 84(2-3):347–366, April-June 2012.

24. M. Grothaus, F. Riemann, and H. P. Suryawan. A White Noise Approach to Feynman Integrand for Electrons in Random Media. *J.Math. Phys.*, 55, 2014
25. T. Hida, H.-H. Kuo, J. Potthoff, and L. Streit. *White Noise. An Infinite Dimensional Calculus.* Kluwer Academic Publisher, Dordrecht, Boston, London, 1993.
26. T. Hida and L. Streit. Generalized Brownian Functionals and the Feynman Integral. *Stoch. Proc. Appl.*, 16:55–69, 1983.
27. W. Hackenbroch and A. Thalmaier. *Stochastische Analysis.* Mathematische Leitfaden. [Mathematical Textbooks], B. G. Teubner, Stuttgart, 1994.
28. I. Karatzas and S. E. Shreve. *Brownian Motion and Stochastic Calculus, 2nd ed.* Springer, Berlin, Heidelberg, New York, 1998.
29. J. R. Klauder *A Modern Approach to Functional Integration.* Applied and Numerical Harmonic Analysis, Birkhäuser Boston, 2010.
30. D.C. Khandekar and S.V. Lawande. Feynman Path Integrals: Some Exact Results and Applications. *Physics Reports*, 137(2), 1986.
31. D.C. Khandekar and L. Streit. Constructing the Feynman Integrand. *Ann. Physik*, 1:46–55, 1992.
32. J. R. Klauder and I. Daubechies. Measures for Path Integrals. *Physical Review Letters*, 48(3):117–120, 1982.
33. J. R. Klauder and I. Daubechies. Quantum Mechanical Path Integrals with Wiener Measures for all Polynomial Hamiltonians. *Physical Review Letters*, 52(14):1161–1164, 1984.
34. John R. Klauder. The Feynman Path Integral: An Historical Slice, arXiv:quant-ph/0303034, 2003.
35. H. Kleinert. *Path Integrals in Quantum Mechanics, Statistics, Polymer Physics, and Financial Markets.* World Scientific, 2004.
36. I. Kubo and H.-H. Kuo. Finite-Dimensional Hida distributions. *J. Funct. Anal.*, 128(1):147, 1995.
37. T. Kuna, L. Streit, and W. Westerkamp. Feynman Integrals for a Class of Exponentially Growing Potentials. *J. Math. Phys.*, 39: 4476–4491, 1999.
38. Naoto Kumano-Go. Phase Space Feynman Path Integrals with Smooth Functional Derivatives by Time Slicing Approximation. *Bull. Sci. Math.*, 135(8):936–987, 2011.
39. H.-H. Kuo. *White Noise Distribution Theory.* CRC Press, Boca Raton, New York, London, Tokyo, 1996.
40. K. Itô , Generalized Uniform Complex Measures in the Hilbertian Metric Space with their Application to the Feynman Integral. In: *Proceedings of the 5th Berkley Symposium on Statistcs and Probability.* Vol. II, part 1, 145161, 1966.
41. A. Lascheck, P. Leukert, L. Streit, and W. Westerkamp. Quantum Mechanical Propagators in Terms of Hida Distributions. *Rep. Math. Phys.*, 33:221–232, 1993.
42. S. Mazzucchi. Functional Integral Solution for the Schrödinger Equation with Polynomial Potential: a White Noise Approach. *Infin. Dimens. Anal. Quantum. Probab. Relat. Top.* Vol. 14, Iss. 04, 2011.

43. N. Obata. *White Noise Calculus and Fock Spaces*, volume 1577 of LNM. Springer Verlag, Berlin, Heidelberg, New York, 1994.
44. M. Reed and B. Simon. *Methods of Modern Mathematical Physics*, volume I. Academic Press, New York, London, 1975.
45. O. G. Smolyanov and E. T. Shavgulidze. *Path Integrals*. Mosk. Gos. Univ., Moskow, 1990 [in Russian]
46. O. G. Smolyanov, A. G. Tokarev, and A. Truman, Hamiltonian Feynman Path Integrals via the Chernoff Formula, newblock *J. Math. Phys.* Vol. 43, Iss.10, pp. 51615171, 2002.
47. L.S. Schulman. *Techniques and Applications of Path Integration* newblock Wiley-Intersience, New York, 1981.
48. F. Scheck. *Theoretische Physik 1*. Springer-Verlag Berlin Heidelberg, Berlin, Heidelberg, achte auflage edition, 2007.
49. A. Vogel. A new Wick formula for products of white noise distributions and application to Feynman path integrands, 2010.
50. W. Westerkamp. *Recent Results in Infinite Dimensional Analysis and Applications to Feynman Integrals*. PhD thesis, University of Bielefeld, 1995.

Chapter 4

Local Times in White Noise Analysis

José Luís da Silva[1] and Martin Grothaus[2]

[1] *CIMA, University of Madeira,*
Campus da Penteada,
9020-105 Funchal, Portugal.
Email: luis@uma.pt

[2] *Department of Mathematics,*
University of Kaiserslautern,
P.O. Box 3049, 67653 Kaiserslautern, Germany
Email: grothaus@mathematik.uni-kl.de

In this chapter we give an overview of self intersection local times of Brownian motion paths and related processes in the framework of white noise analysis. More precisely, we show that self intersection local times are well defined Hida distributions as well as k-fold intersections. It turns out that the kernel functions of the chaos expansion are remarkably simple and exhibit clearly the dimension dependence. We use these kernel functions to study the regularity and convergence properties of self intersection local times.

1. Introduction

Much has already been written about intersections of Brownian motion (Bm) paths and more will be written. A comprehensive survey after its introduction by P. Lévy (more than seventy years ago) would be a major challenge and this chapter is not such a survey. Rather, it is an attempt to give the white noise analysis (WNA) point of view on the subject. More precisely, as WNA is based in the theory generalized functions and, local times "informally" are given by the integral of Donsker's delta function (cf. (2) below), then it seems natural to apply WNA to deal with local times. To this end, the very first step is to show that Donsker's delta function is a Hida distribution (cf. Ch. 1, Sec. 5), then the remaining integral is performed using the WNA machinery. Thus, except Section 2 below, where we recall the definition and mention some classic papers on the existence of self intersections local times, in this chapter we collect the main results on local times within WNA.

Recently the concepts of white noise analysis, or more generally Gaussian analysis, could be generalized to an infinite-dimensional non-Gaussian setting, see[28] and[27]. There in the characteristic function the exponential function is replaced by a parametrized family of Mittag-Leffler functions. The corresponding measures are called Mittag-Leffler measures. Using a concept of biorthogonal polynomials, called Appell system, with respect to the Mittag-Leffler measures a space of test functions and a distribution space could be constructed. These spaces are a generalization of the Kondratiev space to a non-Gaussian setting. In such distribution spaces in[28] Donsker's delta in a non-Gaussian setting is constructed. For a special choice of the inner product in the argument of the Mittag-Leffler function, one obtains Mittag-Leffler measures which enable to provide grey Brownian motion, see [69], as Brownian motion is constructed in white noise analysis. Hence the corresponding Mittag-Leffler measure is called grey noise measure and the corresponding Mittag-Leffler analysis is called grey noise analysis, see[27] for details. In this setting Donsker's delta of grey Brownian motion is obtained. In[22] the local time of grey Brownian motion is constructed. It corresponds to Donsker's delta of grey Brownian motion in the same way as in the case of classical Brownian motion. Hence the first steps of an analysis of local times in this non-Gaussian setting are provided. In Appendix A.3 we summarize the main results on that.

The chapter is organized as follows:

In Section 2 we collect the most important historical facts on the definition (see P. Lévy[45]) and existence properties (cf. A. Dvoretzky et al.[9-11]) of self intersection local times and its applications. In Section 3 we show how to treat self intersection and k-fold intersection of Bm in WNA as well as self intersections of independent Bm's. In particular, we compute the kernel functions (cf. Thm. 3 below) and their L^p properties, where p is dimension dependent. In addition, we also present the kernel functions of the regularized self intersection local times using a Gaussian regularization of the δ-function, cf. Thm. 4. Two important consequences are derived from these regularized kernel functions: At first, an improvement of the Varadhan estimate for the rate of convergence of the centered approximate self-intersection local time of planar Bm, see Thm. 6. The second, concerns the convergence in law of each chaos of the regularized local times to independent Bm's. Another important result in Section 3 is the convergence in mean square of the renormalized local times using the Clark-Haussmann-Ocone formula to represent it as a stochastic integral, see Thm. 8.

In Section 4 we present the main results of local times of fractional Brownian motion (fBm) in the framework of WNA. Two cases are considered, namely the self intersection of a d-dimensional fBm with Hurst parameter $H \in (0,1) \backslash \{\frac{1}{2}\}$ (cf. Subsection 4.1) and intersection local time of two independent d-dimensional multi-parameter fBm's, see Subsection 4.2. The WNA approach presented has the advantage that the underlying probability space does not depend on the Hurst coefficients. Thus, one may investigate the intersection local time of fBm (resp. any two independent fBm's), without any restriction on the Hurst coefficients. One of the main results of this section shows that the self intersection local times of fBm and their subtracted counterpart are Hida distributions. The kernel functions are explicitly computed. The renormalized procedure using the Gaussian approximation of the δ-function is investigated; the kernel functions and the strong convergence in the Hida distribution space is proved, see Thm. 16. In Subsection 4.2 we explore the intersection local times of two independent multi-parameter fBm's. Under suitable assumptions on the dimension d and Hurst parameters, it is shown that the truncated intersection local time is a Hida distribution and its kernel functions are computed, cf. Thm 18 and Cor. 1 below. The constraints on the existence of intersection local times in this case, namely for dimension $d > 2$, motivates the study of a regularized version. Hence, in Thm. 19 we prove that the regularized intersection local time is a Hida distribution and the kernel functions are explicitly computed.

Finally, in Appendix A.1 we summarize some useful results on regular generalized functions and the extensions of the Skorokhod and Itô integrals as well as the Clark-Haussmann-Ocone formula to these spaces. In Appendix A.2 we collect some well known facts about fractional calculus needed in Section 4.

2. Preliminaries on local times

In this section we review the concept of local times, self-intersections local times and k-fold intersections of Bm. We do not pretend to be mathematical rigorous in this section but instead present the problem and methods used to investigate these objects. In later sections we will present a rigorous treatment of local times in the framework of WNA. The fundamental question one addresses is

Question 1. *How to measure the amount of time spent by a Bm sample path at a given spatial point?*

In order to get an intuitive understanding of the problem, let us consider a one-dimensional Bm $B = (B_t)_{t \geq 0}$ defined in a probability space $(\Omega, \mathcal{F}, P)$ and $x \in \mathbb{R}$ fixed. The level set of B at x is the random variable defined, for any fixed $w \in \Omega$, by

$$\mathcal{L}_w(x) := \{t \geq 0 : B_t(w) = x\}.$$

Intuitively one may be tempted to define the local times of B at x as the Lebesgue measure of the level set $\mathcal{L}_w(x)$. But due to the measurability of $(t, w) \mapsto B_t(w)$ with respect to the product σ-algebra, an application of Fubini's theorem yields, cf. Chapter 2, Theorem 9.6 in[43] ,

$$\mathbb{E}\big(\lambda(\mathcal{L}.(x))\big) = \int_0^\infty P(B_t(\cdot) = x)\, dt = 0.$$

Here and below, λ denotes the Lebesgue measure on $\mathbb{R}_+ := [0, \infty)$ and, as usual, we denote $d\lambda(t) = dt$, etc. Therefore, w-almost surely (w-a.s. for short), we have $\lambda\big(\mathcal{L}_w(x)\big) = 0$. In addition, $\mathcal{L}_w(x)$ is unbounded, closed and a perfect set w-a.s. This shows that the Lebesgue measure of the level set $\mathcal{L}_w(x)$, does not provide significant information about the amount of time spent in the vicinity of a point x by the Bm path.

It was P. Lévy[46] who first suggested a non-trivial measure for this amount of time, namely the two-parameter random variable

$$L_t(x) := \lim_{\varepsilon \downarrow 0} \frac{1}{2\varepsilon} \lambda(\{s \in [0, t] : |B_s - x| < \varepsilon\}) = \lim_{\varepsilon \downarrow 0} \frac{1}{2\varepsilon} \int_0^t \mathbb{1}_{(x-\varepsilon, x+\varepsilon)}(B_s)\, ds$$

$$(1)$$

and showed that this limit exists (in L^2-sense and w-a.s.), is finite but not identically zero. P. Lévy called $L_t(x)$ the "mesure du voisinage" or "measure of the time spent by the Bm path in the vicinity of the point x". We refer to $L = \{L_t(x) : (t, x) \in [0, \infty) \times \mathbb{R}\}$ as *Brownian local times* or simply *local times*. It can be shown that there exists a jointly continuous version of L in time and space (t, x) by using the Kolmogorov criterion. In addition, for fixed x, $L_t(x)$ is nondecreasing in t and constant on each interval in the complement of the closed set $\mathcal{L}_w(x)$ and is uniquely determined by the property that for all $t \geq 0$ and $F \in \mathcal{B}(\mathbb{R})$,

$$\int_F L_t(x)\, dx = \int_0^t \mathbb{1}_F(B_s)\, ds,$$

where $\mathbb{1}_F$ denotes the indicator function of the set F.

Upon applying Fubini's theorem informally (i.e. not mathematically rigorous), we get

$$\int_F \int_0^t \delta(B_s - x)\, d\lambda(s)\, dx = \int_0^t \int_F \delta(B_s - x)\, dx\, ds$$

$$= \int_0^t \mathbb{1}_F(B_s)\, ds.$$

Hence, we conclude that the local times $L_t(x)$ is given (also informally) by

$$L_t(x) = \int_0^t \delta(B_s - x)\, ds. \tag{2}$$

This new concept provides a powerful tool for the study of Bm sample paths. In particular, it allows to generalize Itô's change-of-variable rule to convex but not necessarily differentiable functions, known as the Tanaka formula, cf.[7] . Namely,

$$|B_t - x| = |B_0 - x| + \int_0^t \mathrm{sgn}(B_s - x)\, dB_s + \frac{1}{2} L_t(x),$$

where $\mathrm{sgn}(0) = 0$ and $\mathrm{sgn}(x) = x/|x|$, $x \neq 0$.

Another way to define the local times is through a change of space and time integration which exhibits the local times as the true density of *occupation time*. More precisely, for any Borel set $F \in \mathcal{B}(\mathbb{R})$ we define the *occupation time* of F by the Bm path up to time t by

$$\mu_t(F) := \int_0^t \mathbb{1}_F(B_s)\, ds = \lambda(\{0 \leq s \leq t : B_s \in F\}), \quad t \geq 0.$$

Equation (1) suggest that the local times should serve as a *density with respect to the Lebesgue measure for occupation time*, namely

$$\mu_t(F) = \int_F L_t(x)\, dx, \quad t \geq 0,\ F \in \mathcal{B}(\mathbb{R}).$$

It was H. F. Trotter[75] who proved this partial extension of the result of P. Lévy about the existence of local times for one dimensional Bm. Moreover, if the local times $L_t(x)$ exist, then the following *occupation times formula* is valid, cf.[43] .

Theorem 1. *Let $f \in L^1_{\mathrm{loc}}(\mathbb{R})$ be a Borel measurable locally integrable function, then we have for each t, w-a.s.*

$$\int_0^t f(B_s)\, ds = \int_{\mathbb{R}} f(x) L_t(x)\, dx. \tag{3}$$

Remark 1.

(1) The occupation times formula is valid for more general processes than Bm, namely semimartingales. The Lebesgue measure on the left-hand side of Eq. (3) has to be replaced by the quadratic variation measure of the process, see[68].
(2) For a survey article about occupation densities for (non)random vector fields see[25] and references therein.

A simple criterium for the existence of local times for general processes X is due to S. M. Berman, see Lemma 3.1 in[3] (see also Theorem 21.9 in[25]) using harmonic analysis or Fourier analytic methods. Namely, if

$$\int_{\mathbb{R}} \left| \int_0^1 \int_0^1 \mathbb{E}\left[\exp\left(iu(X(t) - X(s))\right)\right] ds\, dt \right| du < \infty,$$

then, for almost all w, $L_t(x)$ exists and $L_t(\cdot) \in L^2(\mathbb{R})$. Other existence results of local times for general processes have been obtained using martingale (see[8,20,57]) and differentiation of measures (see[60]) methods in one and higher dimensions.

Remark 2. In the following sections we are going to use WNA to show the existence of local times as a well defined *generalized Brownian functional* (gBf) using the representation (2). Here we mention the two steps procedure.

(1) At first, we notice that the Donsker's delta function $\delta(B_s - x)$ in equation (2) is a gBf. This result is independent of local times itself and is an interesting example of a gBf with many applications, see for example[49] and references therein.
(2) Second, the integral in (2) is performed as a Bochner integration of a family of generalized functions using Corollary 21 of Ch. 1 proving that the local times $L_t(x)$ is a well defined gBf.

The same procedure may be used to define *self intersections local times* of Bm paths, namely through the informal expression

$$L := \int_0^t \int_0^t \delta(B_{t_2} - B_{t_1})\, dt_1\, dt_2 \tag{4}$$

which is intended to measure the amount of time spent by the Bm paths intersecting themselves. We can also speak about triple intersection local times of Bm paths informally given by the expression

$$\int_0^t \int_0^t \int_0^t \delta(B_{t_3} - B_{t_2})\delta(B_{t_2} - B_{t_1})\, dt_1\, dt_2\, dt_3.$$

In general, we have multiple self intersections local times, more precisely by a k-fold self intersections local times of Bm paths we understand a k-tuple $(t_1, t_2, \ldots, t_k)$ of times such that $B_{t_1} = B_{t_2} = \ldots = B_{t_k}$ which, informally, is given by the expression

$$L_k := \int_0^t \ldots \int_0^t \delta(B_{t_k} - B_{t_{k-1}}) \delta(B_{t_{k-1}} - B_{t_{k-2}}) \ldots \delta(B_{t_2} - B_{t_1}) \, dt_1 \ldots dt_k. \tag{5}$$

So far we have used the one dimensional Bm to introduce the concept of local times, self intersection local times and k-fold intersections local times of the Brownian sample paths. These concepts extend to any stochastic process $X = (X_t)_{t \geq 0}$ defined in a probability space $(\Omega, \mathcal{F}, P)$. Moreover, the above concepts are not restricted to one dimensional processes, in fact any process with (measurable) state space $(E, \mathcal{E})$. We have to assume that the map

$$\mathbb{R}_+ \times \Omega \ni (t, w) \mapsto X_t(w) \in E$$

is $\mathcal{B}(\mathbb{R}_+) \otimes \mathcal{F}/\mathcal{E}$ measurable. Below we choose $(E, \mathcal{E})$ as $(\mathbb{R}^d, \mathcal{B}(\mathbb{R}^d))$, $d \in \mathbb{N}$ and X is called d-dimensional stochastic process. More specifically, X will be a d-dimensional Bm $\boldsymbol{B}$ defined by

$$\boldsymbol{B} : [0, \infty) \times \Omega \longrightarrow \mathbb{R}^d, \ (t, w) \mapsto \boldsymbol{B}_t(w) := \left(B_t^1(w), \ldots, B_t^d(w) \right),$$

where $B^i = (B_t^i)_{t \geq 0}$, $i = 1, \ldots, d$, are independent one dimensional Bm. As one expects, self intersections local times become increasingly scarce as the dimension d increases. We close this section with an account on the existence of local times and k-fold self intersections local times of Bm paths.

- For planar Bm P. Lévy[45] showed that self intersection exists almost sure. In fact, planar Bm admits k-fold self intersections for any finite k, see.[10]
- A. Dvoretzky et al.[9] showed that a 3-dimensional Bm has self intersections almost surely but 4-dimensional Bm has no self intersections almost surely due to the transience of the process.
- In[11] is proved that 3-dimensional Bm has no k-fold self-intersection for $k \geq 3$.
- For $d = 5$ S. Kakutani[40] proved that almost all paths of a 5-dimensional Bm have no self intersections. This implies that almost all paths of Bm in $\mathbb{R}^d$, $d \geq 5$ have no double points.

3. Intersection local times of Bm

This section presents an overview of intersection local times of Bm paths in the framework of WNA. Although intersection local times have been studied for a long time, as shown in Section 2, we would like to refer in addition the works[23,26,36,37,47,48,54,59,65,66,76] and references therein. Between the various methods to treat rigorously the intersection local times of Bm paths we mention the approximation of the δ-function by a Gaussian sequence which produces increasingly singular objects as the dimension d increases. The different types of renormalizations needed to treat the divergences are dimension dependent, e.g., subtraction of the expectation[76], multiplicative renormalization[79], weak limit of measures to make a Gibbs factor well defined[78], a combination of additive functionals for a single Markov process and stochastic calculus[5] or the gap renormalization[70] and references therein. Here we choose the WNA approach to study intersection local times of Bm paths.

3.1. *Self intersection local time as generalized white noise functionals*

The first attempts to realize the self intersection local times of Bm paths in the framework of WNA appear in[41,42,71,72,74,77] and[36]. For a general Gaussian process, see[56]. Motivated by the work of M. Yor[80] on the stochastic integrals representation of local times, S. Watanabe[77] (see also[36] for a remark on a proper renormalization in case $d = 4$ or 5 in Thm. 2 below) established the following result for self intersection local times of d-dimensional Bm paths as a generalized Brownian functional in the sense of Hida.

Theorem 2 (cf.[77] Thm. 3.1 and 3.2). *Let $\varepsilon > 0$ be given. Then*

(1) for $d = 2$ or 3 the S-transform of the Brownian functional

$$\int_0^{1-\varepsilon} \int_{s+\varepsilon}^1 \left(\delta(\boldsymbol{B}_u - \boldsymbol{B}_s) - (2\pi(u-s))^{-d/2} \right) du\, ds \qquad (6)$$

converges as ε tends to zero. The limiting functional of (6) is a generalized Brownian functional in $(\mathcal{S})'$, cf. Ch. 1, Sec. 6.1.

(2) For d = 4 or 5 the S-transform of the Brownian functional

$$\int_0^{1-\varepsilon} \int_{s+\varepsilon}^1 \Big[\delta(\boldsymbol{B}_u - \boldsymbol{B}_s) - (2\pi(u-s))^{-d/2}$$

$$+\frac{d}{(2\pi)^{d/2}(u-s)^{d/2+1}} \sum_{i,j=1}^d \int_{s+\varepsilon}^u (B_r^j - B_s^j)\, dB_r^i \Big]\, du\, ds \qquad (7)$$

converges as ε tends to zero. Also, the limiting functional of (7) is a generalized Brownian functional in (S)'.

For simplicity one introduces the following notation which is used from now on. It corresponds to the tripe $S_d(\mathbb{R}) \subset L_d(\mathbb{R}) \subset S'_d(\mathbb{R})$ of Example 2(ii) in Ch. 1.

$$\boldsymbol{n} := (n_1, \ldots, n_d) \in \mathbb{N}_0^d, \qquad n := \sum_{i=1}^d n_i, \qquad \boldsymbol{n}! := \prod_{i=1}^d n_i!.$$

For any $\boldsymbol{f} = (f_1, \ldots, f_d) \in L_d^2(\mathbb{R})$ and $F_{\boldsymbol{n}} \in L_d^2(\mathbb{R})^{\otimes n}$

$$(\boldsymbol{f}, \boldsymbol{f}) = \sum_{i=1}^d \int_{\mathbb{R}} f_i^2(t)\, dt,$$

$$(F_{\boldsymbol{n}}, \boldsymbol{f}^{\otimes n}) = \int_{\mathbb{R}^n} F_{\boldsymbol{n}}(t) \bigotimes_{i=1}^d f_i^{\otimes n_i}(\boldsymbol{t})\, d\boldsymbol{t}$$

and similar for $\langle : \boldsymbol{w}^{\otimes n} :, F_{\boldsymbol{n}} \rangle$, where $: \boldsymbol{w}^{\otimes n} := \bigotimes_{i=1}^d : w^{\otimes n_i} :$. The elements of the Hilbert space (L^2) admits the chaos expansion

$$F(\boldsymbol{w}) = \sum_{\boldsymbol{n}} \langle : \boldsymbol{w}^{\otimes n} :, F_{\boldsymbol{n}} \rangle,$$

with kernels $F_{\boldsymbol{n}}$ from the Fock space $\mathrm{Exp}\big(L_d^2(\mathbb{R})_\mathbb{C}\big)$, see App. A.2 in Ch. 1. In addition, for any $k \in \mathbb{N}$, Δ_k denotes the simplex in $\mathbb{R}^k$ defined by

$$\Delta_k := \{\boldsymbol{x} \in \mathbb{R}^k : 0 < x_1 < x_2 < \ldots < x_k < t\} \qquad (8)$$

and for any element $\Phi \in (S)'$, $\Phi^{(k)}$ is the truncated Hida distribution, i.e., the projection onto the chaos of order $n \geq k$, defined by

$$\Phi^{(k)} := \sum_{n: n \geq k} \langle : .^{\otimes n} :, \Phi_{\boldsymbol{n}} \rangle.$$

We are now ready to announce the next important step in self intersections of Bm paths within WNA obtained by de Faria et. al.[14]. They consider (4) with $t = 1$ and their main result is the following:

Theorem 3 (cf.[14] Thm. 2). *Let* $d \in \mathbb{N}$, $N \in \mathbb{N}_0$ *be given such that* $2N > d - 2$. *Then the Bochner integral*

$$L^{(2N)} \equiv \int_{\Delta_2} \delta^{(2N)}(B(t_2) - B(t_1)) \, dt_1 \, dt_2$$

is an element of $(\mathcal{S})'$.

Moreover, they obtained the chaos expansion in terms of "Wick powers" of white noise (cf. Ch. 1, Section 3) an expansion which corresponds to that of multiple Wiener integrals when one considers the Wiener process as the fundamental random variable. It turns out that the kernel functions of the chaos expansion are remarkably simple and exhibit clearly the dimension dependent. We state this striking result in the following:

Theorem 4 (cf.[14] Thm. 3). *Let* $d \in \mathbb{N}$, $N \in \mathbb{N}_0$, $\boldsymbol{n} \in \mathbb{N}_0^d$ *be given such that* $2N > d - 2$ *and* $\varkappa := n + d/2 - 2$. *Then the kernel functions* $F_{\boldsymbol{n}}$ *of* $L^{(2N)}$ *are given by*

$$F_{2n}(u_1, \ldots, u_{2n}) = (-1)^n \left(\varkappa(\varkappa + 1)(2\pi)^{d/2} 2^n n! \right)^{-1} \Theta(u) \Theta(t - v)$$
$$\times \left(t^{-\varkappa} - v^{-\varkappa} - (t - u)^{-\varkappa} + (v - u)^{-\varkappa} \right),$$

with $v := \max_{1 \le k \le 2n} u_k$ *and* $v := \min_{1 \le k \le 2n} u_k$, *if* $n \ge N$, *except for* $2n = d = 2$, *where*

$$F_{2n}(u_1, u_2) = -\frac{1}{4\pi} \Theta(u) \Theta(t - v) \left(\ln(t) + \ln(v) + \ln(t - u) - \ln(v - u) \right).$$

All other $F_{\boldsymbol{n}}$ *vanish.* Θ *is the Heaviside function.*

These results, namely Thm. 3 and 4 are basis of a series of papers with further investigations on self intersections of Bm paths within WNA which we will give an account in the following. In particular, a systematic approach to renormalization as well as nonlinear properties. The L^p properties of the kernels F_{2n} of Thm. 4 are calculated as follows

Proposition 1 (cf.[14] Prop. 2). *The* F_{2n} *from Thm. 4 are in* $L^p(\mathbb{R}^{2n})$ *for all* p *with* $1 < p < 2 + (d - 3)/\varkappa$.

As mentioned in Section 2, page 1, Donsker's delta function as it stands is a well defined element in $(\mathcal{S})'$. But in the studies of self intersections of Brownian paths it is desirable to introduce regularizations with a view towards the construction of well defined renormalized generalized functionals of Brownian paths in higher dimension. In[14] de Faria et. al. discussed a simple regularization, namely for $\varepsilon > 0$

$$L_\varepsilon := \int_\varepsilon^t \int_0^{t_2} \delta_\varepsilon(\boldsymbol{B}_{t_2} - \boldsymbol{B}_{t_1})\, dt_1 dt_2,$$

where for $\varepsilon > 0$

$$\delta_\varepsilon(\boldsymbol{B}_{t_2} - \boldsymbol{B}_{t_1}) := (2\pi\varepsilon)^{-d/2} \exp\left(-\frac{1}{2\varepsilon}|\boldsymbol{B}_{t_2} - \boldsymbol{B}_{t_1}|^2\right), \tag{9}$$

in terms of its kernel functions. We quote this result explicitly as

Theorem 5 (cf.[14] Thm. 4). *For any $\varepsilon > 0$ and all dimensions d, $L_\varepsilon \in (\mathcal{S})'$, with kernel functions given by*

$$F_{\varepsilon,2n}(u_1,\ldots,u_{2n}) = (-1)^n \left(\varkappa(\varkappa+1)(2\pi)^{d/2}2^n n!\right)^{-1} \Theta(u)\Theta(t-v)$$

$$\times \left[(t+\varepsilon)^{-\varkappa} - (v+\varepsilon)^{-\varkappa} - (t-u+\varepsilon)^{-\varkappa} + (v-u+\varepsilon)^{-\varkappa}\right] \tag{10}$$

except for $2n = d = 2$, where

$$F_{\varepsilon,2n}(u_1, u_2) = -\frac{1}{4\pi}\Theta(u)\Theta(t-v)\big[\ln(t+\varepsilon) - \ln(v+\varepsilon) - \ln(t-u+\varepsilon)$$

$$+ \ln(v-u+\varepsilon)\big].$$

Furthermore,

$$s - \lim_{\varepsilon \searrow 0} L_\varepsilon^{(2N)} = L^{(2N)} \tag{11}$$

whenever $2N > d - 2$.

The limit in (11) is the strong limit in $(\mathcal{S})'$ as in Ch. 1, Corollary 1.

Remark 3. The white noise tools to successfully complete the above results, except Proposition 1, which is an integral estimation, are:

(1) At first it is shown that the Bochner integral

$$\delta(\boldsymbol{B}_t - \boldsymbol{B}_s) = \left(\frac{1}{2\pi}\right)^d \int_{\mathbb{R}^d} e^{i(\boldsymbol{\theta},(\boldsymbol{B}_t - \boldsymbol{B}_s))}\, d\boldsymbol{\theta}$$

is a generalized white noise functional if $t \neq s$ with S-transform given by (cf. Ch. 1, Ex. 3 for $d = 1$) with $z \in \mathbb{C}$ and $\boldsymbol{f} \in S_d(\mathbb{R})$

$$S\delta(\boldsymbol{B}_t - \boldsymbol{B}_s)(z\boldsymbol{f}) = \left(\frac{1}{\sqrt{2\pi|t-s|}}\right)^d \exp\left(-\frac{|z|^2}{2|t-s|}\left|\int_s^t \boldsymbol{f}(u)\, du\right|^2\right). \tag{12}$$

This is achieved showing that (12) obeys the conditions of Corollary 2 in Ch. 1.

(2) The kernel functions of the Donsker delta function can be read off their S-transform using equation (18) in Ch. 1.

A consequence of the simple form of the kernels $F_{\varepsilon,n}$ is an improvement of the Varadhan estimate for the rate of convergence of the centered approximate self-intersection local time of planar Bm, see[6]. More precisely, the Edwards model[24] for self-repelling d-dimensional Bm, with applications in polymer physics and quantum field theory, is informally given by a Gibbs factor

$$G = \frac{1}{Z} \exp\left(-g \int_0^t \int_0^{t_1} \delta(\boldsymbol{B}_{t_1} - \boldsymbol{B}_{t_2})\, dt_2 dt_1\right),$$

with $g > 0$ and Z a normalizing constant. For the planar Bm $d = 2$, Varadhan[76] has shown that centering

$$L_{\varepsilon,c} := L_\varepsilon - \mathbb{E}(L_\varepsilon) \qquad \text{and} \qquad L_c := \lim_{\varepsilon \searrow 0} L_{\varepsilon,c}$$

is sufficient to make the Gibbs factor $G = Z^{-1} \exp(-gL_c)$ well defined. The estimate for the rate of convergence

$$\|L_{\varepsilon,c} - L_c\|^2 \le C_{t,\alpha} \varepsilon^\alpha, \qquad C_{t,\alpha} < \infty,$$

for all $\alpha < 1/2$ is in Varadhan's words, "the most difficult step of all and requires considerable estimation". In[6] the authors improved the rate of convergence for all $\alpha < 1$ using the kernels (10) with a straightforward argument.

Theorem 6 (cf.[6] Thm. 2). *Given $t > 0$. Then, for any $\alpha < 1$, there is a constant $C_{t,\alpha} < \infty$ such that for all $\varepsilon > 0$*

$$\|L_{\varepsilon,c} - L_c\|^2 \le C_{t,\alpha} \varepsilon^\alpha,$$

the work of M. Yor[79] on the extension of the Varadhan renormalization of L for $d = 3$, C. Drumond et al.[12] investigated similar results for $d \ge 3$. Using the Gaussian regularization of the δ-function, for which the chaos expansion of the corresponding regularized L_ε was known, cf. Thm. 5 above, they obtained the convergence in law of each chaos to independent Bm's. In order to state these results, we need some preparation.

Each kernel function of L_ε is the sum of four terms, see (10), the most dominant one is

$$M_t(\varepsilon) := M_t(d, \boldsymbol{n}, \varepsilon) := \int_{[0,t]^n} (v - u + \varepsilon)^{-\varkappa} : \boldsymbol{w}^{\otimes \bar{\boldsymbol{n}}}(\boldsymbol{s}) : \, d\boldsymbol{s}, \quad t > 0$$

which is an element in (L^2) as well as a regular generalized function in $\mathcal{G}^{-1}$, see Definition 2 in Appendix A.1. An application of the generalized Clark-Haussmann-Ocone formula, cf. Thm. 20 in Appendix A.1, gives

$$M_t(\varepsilon) = \sum_{i=1}^d n_i \int_0^t m_i(\tau)\, dB_\tau^i =: \sum_{i=1}^d M_t^i(\varepsilon),$$

where

$$m_i(\tau) = \int_{[0,\tau]^{n-1}} (\tau - u + \varepsilon)^{-\kappa} : \boldsymbol{w}^{\otimes(\boldsymbol{n}-\boldsymbol{\delta}_i)}(\boldsymbol{s}) : d^{n-1}\boldsymbol{s}.$$

The other terms of the kernels (10) are less singular in the limit as ε goes to zero than $M_t(\varepsilon)$ and denoted by

$$N_t(\varepsilon) := N_t(d, \boldsymbol{n}, \varepsilon) := \int_{[0,t]^n} \big((t+\varepsilon)^{-\varkappa} - (v+\varepsilon)^{-\varkappa} - (t-u+\varepsilon)^{-\varkappa}\big) : \boldsymbol{w}^{\otimes n}(\boldsymbol{s}) : d^n\boldsymbol{s}.$$

All the processes $M_t^i(\varepsilon)$, $M_t(\varepsilon)$ and $N_t(\varepsilon)$ are continuous and the nth term of the chaos expansion of the regularized local times L_ε has the representation

$$K_t(\varepsilon) := K_t(d, \boldsymbol{n}, \varepsilon) := (-1)^n \left(\varkappa(\varkappa+1)(2\pi)^{d/2} 2^n \boldsymbol{n}!\right)^{-1} \big(M_t(\varepsilon) + N_t(\varepsilon)\big).$$

Remark 4. A key observation is that $M_t(\varepsilon)$ is more singular than $N_t(\varepsilon)$, as ε goes to zero, for $\varkappa$ positive and $M_t(\varepsilon)$ is a Brownian martingale.

Thus, the main result obtained for suitably subtracted and renormalized local times by Drumond et al.[12] is stated in the following:

Theorem 7 (cf.[12] Thm. 2.2). *For $d \geq 3$, the normalized $M^i(\varepsilon)$ converge in distribution to a Bm β^i independent of B^i, $i = 1, \ldots, d$, i.e.,*

$$(r(\varepsilon)M^i(\varepsilon),\ i = 1, \ldots, d) \xrightarrow{\text{law}} \left(\frac{n_i}{n} k_n \beta^i,\ i = 1, \ldots, d\right),\ \varepsilon \to 0$$

with

$$k_n^2 = n! \begin{cases} n(n-1), & \text{if } d = 3 \\ \frac{n!(d-4)!}{(n+d-5)!}, & \text{if } d > 3 \end{cases}$$

if

$$r(\varepsilon) = \begin{cases} |\ln(\varepsilon)|^{-1/2}, & \text{if } d = 3 \\ \varepsilon^{(d-3)/2}, & \text{if } d > 3. \end{cases} \tag{13}$$

Further studies on the square of the renormalized centered local times for $d \geq 3$ were investigated by de Faria et al.,[13,16] see also the Ph.D. by C. Drumond[21], Chapter 5. Let us define

$$L_{\varepsilon,\text{ren}} := r(\varepsilon) L_{\varepsilon,c} := r(\varepsilon)\big(L_\varepsilon - \mathbb{E}(L_\varepsilon)\big)$$

and state the results for $d = 3$ and $d = 4$ obtained using direct calculations.

Theorem 8 (cf.[21], Thm. 5.1.1 and Prop. 5.1.2).

(1) For $d = 3$ we have

$$\mathbb{E}(L_{\varepsilon,\text{ren}}^2) \longrightarrow \frac{t}{2\pi^2}, \ \varepsilon \to 0.$$

(2) For $d = 4$ we obtain

$$\mathbb{E}(L_{\varepsilon,\text{ren}}^2) \longrightarrow \frac{t}{2\pi^4}, \ \varepsilon \to 0.$$

For $d > 4$, we have to compute the following integrals

$$\lim_{\varepsilon \to 0} \mathbb{E}(L_{\varepsilon,\text{ren}}^2) = \frac{2}{(2\pi)^d} \lim_{\varepsilon \to 0} \int_0^{\frac{t}{\varepsilon}} \int_s^{\frac{t_2}{\varepsilon}} \int_0^{\frac{t}{\varepsilon}} \int_0^{\frac{s_2}{\varepsilon}} \left((1 + \tau + \tau\sigma - \delta^2)^{-\frac{d}{2}} \right.$$
$$\left. - ((1+\tau)(1+\sigma))^{-\frac{d}{2}} \right) ds_1 \, ds_2 \, dt_1 \, dt_2,$$

where the notation is

$$\sigma = s_2 - s_1, \quad \tau = t_2 - t_1, \quad \delta = \lambda([s_1, s_2] \cap [t_1, t_2])$$

and have assumed $s_1 < t_1$ compensating with the factor 2.

The integrals involve cancellation of divergences due to subtraction of the second term (square of the expectation) and they are complicated according to the type of overlap $[s_1, s_2] \cap [t_1, t_2]$ as partially or totally contained in $[s_1, s_2]$. Therefore, the result for $d > 4$ is obtained by a different route. More precisely, in[13] the Clark-Haussmann-Ocone formula is used in order to write $L_{\varepsilon,\text{ren}}$ as a stochastic integral with representation

$$L_{\varepsilon,\text{ren}} = \int_0^t \boldsymbol{\varphi}(\tau) \, d\boldsymbol{B}_\tau := \sum_{i=1}^d \int_0^t \varphi_i(\tau) \, dB_\tau^i,$$

where $\boldsymbol{\varphi} = (\varphi_1, \ldots, \varphi_d)$ is given (cf.[13]) by

$$\varphi(\tau) = -\frac{r(\varepsilon)}{(2\pi)^{d/2}} \int_\tau^t \left(\int_0^\tau (\varepsilon + t_2 - \tau)^{-1-d/2} e^{-\frac{|B_\tau - B_{t_1}|^2}{2(\varepsilon + t_2 - \tau)}} \langle \boldsymbol{w}, \mathbb{1}_{[t_1,\tau]} \rangle \, dt_1 \right) dt_2.$$

Then the Itô isometry is used. The closed form result is contained in the following:

Theorem 9 (cf.[13] Prop. 4.4). *Let $d > 4$ be given. If d is even, then*

$$\lim_{\varepsilon \to 0} \mathbb{E}\left(L^2_{\varepsilon,\mathrm{ren}}\right) = C(t,d)\left(\frac{2(d-4)}{(d-3)(d-2)} \sum_{k=1}^{\frac{d}{2}-2} \frac{(-1)^k k}{d-2k-2}\right.$$

$$\left. + \frac{2}{d-3} \sum_{k=1}^{\frac{d}{2}-3} \frac{(-1)^{k+1} k}{d-2k-4} + \frac{(-1)^{\frac{d}{2}}}{d-2} + \frac{(-1)^{\frac{d}{2}+1}(d-4)}{d-3} \ln(2)\right),$$

and for d odd we obtain

$$\lim_{\varepsilon \to 0} \mathbb{E}\left(L^2_{\varepsilon,\mathrm{ren}}\right) = C(t,d)\frac{d-4}{d-3}\left(2\sum_{k=0}^{\frac{d-7}{2}} \frac{(-1)^k}{d-2k-6} \frac{d-1}{(d-2)(d-4)} + \frac{(-1)^{\frac{d-1}{2}}\pi}{2}\right),$$

with $C(t,d) := 8t/((2\pi)^d(d-2))$.

Remark 5.

(1) In[16] de Faria et al. split the chaos of $L_{\varepsilon,c}$ into a dominant martingale part and a subdominant remainder as in $M_t(\varepsilon)$ and $N_t(\varepsilon)$. Then these two terms were expressed as stochastic integrals via the Clark-Haussmann-Ocone formula, cf. Thm. 3 in[16].

(2) On the other hand, representing $L_{\varepsilon,c} = M_t(\varepsilon) - N_t(\varepsilon)$, A. Rezgui and L. Streit (see e.g.[64] Thm. 2) showed the convergence in finite dimensional distribution for $d \geq 3$

$$r(\varepsilon)\big(M(\varepsilon) - N(\varepsilon)\big) \longrightarrow c\beta,$$

where c is a constant and β is Bm independent of $\boldsymbol{B}$, $r(\varepsilon)$ is defined in (13).

3.2. *Self intersection local time of independent Bm*

We continue the presentation in order to give an account on the self intersection local times of independent Bm's in $\mathbb{R}^d$, namely its chaos expansion and L^p properties. Informally, the expression

$$\tilde{L} := \int_0^t \int_0^t \delta(\boldsymbol{B}_{t_1} - \tilde{\boldsymbol{B}}_{t_2})\, dt_1\, dt_2 \tag{14}$$

supposes to sum up the contributions of each pair (t_1, t_2) for which the independent d-dimensional Bm's $\boldsymbol{B}$ and $\tilde{\boldsymbol{B}}$ arrive at the same point. We denote by $\tilde{L}_\varepsilon$, $\varepsilon > 0$, the regularization of $\tilde{L}$ given by

$$\tilde{L}_\varepsilon := \int_0^t \int_0^t \delta_\varepsilon(\boldsymbol{B}_{t_2} - \tilde{\boldsymbol{B}}_{t_1})\, dt_1\, dt_2,$$

where δ_ε is defined in (9). Using the notation

$$
\underline{u} := \begin{cases} \min_{1 \le i \le m} u_i & \text{if } m \ge 1, \\ 0 & \text{if } m = 0, \end{cases} \qquad \underline{v} := \begin{cases} \min_{1 \le i \le k} v_i & \text{if } k \ge 1, \\ 0 & \text{if } k = 0, \end{cases}
$$

$$
\overline{u} := \begin{cases} \max_{1 \le i \le m} u_i & \text{if } m \ge 1, \\ 0 & \text{if } m = 0, \end{cases} \qquad \overline{v} := \begin{cases} \max_{1 \le i \le k} v_i & \text{if } k \ge 1, \\ 0 & \text{if } k = 0, \end{cases}
$$

we obtain the following chaos expansion of $\tilde{L}_\varepsilon$, cf. Thm. 3.2 in[1].

Theorem 10. *For any positive integer d and $\varepsilon > 0$, $\tilde{L}_\varepsilon$ has chaos expansion*

$$
\tilde{L}_\varepsilon = \sum_m \sum_k \langle : \boldsymbol{w}_1^{\otimes m} : \otimes : \boldsymbol{w}_2^{\otimes k} :, G_{m,k,\varepsilon} \rangle, \tag{15}
$$

where the kernel functions $G_{m,k}$ of $\tilde{L}_\varepsilon$ are given by

$$
\begin{aligned}
G_{m,k,\varepsilon}(u_1, \ldots, u_m, v_1 \ldots, v_k) = {} & C_{m,k}(\kappa(\kappa+1))^{-1} \Theta(\underline{u})\Theta(\underline{v})\Theta(t-\overline{u})\Theta(t-\overline{v}) \\
& \times \left[(\overline{v}+\overline{u}+\varepsilon)^{-\kappa} - (\overline{v}+t+\varepsilon)^{-\kappa} - (\overline{u}+t+\varepsilon)^{-\kappa} + (2t+\varepsilon)^{-\kappa} \right],
\end{aligned}
$$

where

$$
C_{m,k} := (-1)^k \left(\frac{1}{2\pi} \right)^{d/2} \left(-\frac{1}{2} \right)^{(m+k)/2} \frac{1}{(\frac{m+k}{2})!} \binom{m+k}{m},
$$

$$
\kappa := \frac{d+m+k}{2} - 2,
$$

$$
m+k \ne 0
$$

and $\boldsymbol{m}+\boldsymbol{k}$ even, and zero otherwise except the case $d = 2$ and $m + k = 2$ where for $m = k = 1$ we have

$$
\begin{aligned}
G_{m,k,\varepsilon}(u_1, v_1) = \frac{1}{2\pi} \big[& \ln(\varepsilon + t + v_1) - \ln(\varepsilon + u_1 + v_1) - \ln(\varepsilon + 2t) \\
& + \ln(\varepsilon + t + u_1) \big] \Theta(u_1)\Theta(v_1)\Theta(t - u_1)\Theta(t - v_1),
\end{aligned}
$$

for $m = 2$ and $k = 0$ we obtain

$$
G_{m,0,\varepsilon}(u_1, u_2) = \frac{1}{4\pi} \big[\ln(\varepsilon+\overline{u}) - \ln(\varepsilon+t+\overline{u}) - \ln(\varepsilon+t) + \ln(\varepsilon+2t) \big] \Theta(\underline{u})\Theta(t-\overline{u}),
$$

and for $k = 2$ and $m = 0$ get

$$
G_{0,k,\varepsilon}(v_1, v_2) = \frac{1}{4\pi} \big[\ln(\varepsilon+\overline{v}) - \ln(\varepsilon+t+\overline{v}) - \ln(\varepsilon+t) + \ln(\varepsilon+2t) \big] \Theta(\underline{v})\Theta(t-\overline{v}).
$$

The proof consists in computing the S-transform of $\tilde{L}_\epsilon$ and then compare it with the general form of (15). In addition, it is shown that the Bochner integral, corresponding to the truncated intersection local time,

$$\tilde{L}^{(N)} := \int_0^t \int_0^t \delta^{(N)}(\boldsymbol{B}_{t_1} - \tilde{\boldsymbol{B}}_{t_2}) \, dt_1 \, dt_2$$

is a Hida distribution for $N \geq 0$ and $2N > d - 4$, cf.[1], Thm. 3.3. In particular for $d < 4$ the intersection local time $\tilde{L}$ given by (14) is a Hida distribution with S-transform given by

$$(S\tilde{L})(\boldsymbol{f}) = (2\pi)^{-d/2} \int_0^t \int_0^t \frac{1}{(t_1 + t_2)^{d/2}} \exp\left(-\frac{\left|\int_0^{t_1} \boldsymbol{f}(t) \, dt - \int_0^{t_2} \boldsymbol{f}(t) \, dt\right|^2}{2(t_1 + t_2)}\right),$$

for any $\boldsymbol{f} \in S_d(\mathbb{R})$, see Cor. 3.5 in[1]. For $d = 4$ or $d = 5$ it is sufficient to center $\tilde{L}$ in order to obtain a well defined object, or in other words, the local time $\tilde{L}$ becomes well defined if we subtract its divergent expectation. The kernels are computed and their L^p properties are studied, this is next two theorems, see[1] Thm 3.6 and Thm. 3.7.

Theorem 11. *For all natural numbers d and $N \geq 0$ such that $2N > d - 4$, $\tilde{L}^{(N)}$ has the chaos expansion*

$$\tilde{L}^{(N)} := \sum_m \sum_k \langle : \boldsymbol{w}_1^{\otimes m} : \otimes : \boldsymbol{w}_2^{\otimes k} :, G_{\boldsymbol{m},\boldsymbol{k}}\rangle,$$

where the kernels $G_{\boldsymbol{m},\boldsymbol{k}}$ of $\tilde{L}^{(N)}$ vanish for $m + k < 2N$ or $\boldsymbol{m} + \boldsymbol{k}$ not even, and for $m + k \neq 0$, $m + k \geq 2N$, $\boldsymbol{m} + \boldsymbol{k}$ even $G_{\boldsymbol{m},\boldsymbol{k}}$ are given by

$$G_{\boldsymbol{m},\boldsymbol{k}}(u_1, \ldots, u_m, v_1 \ldots, v_k)$$
$$= C_{\boldsymbol{m},\boldsymbol{k}}(\kappa(\kappa + 1))^{-1}\Theta(\underline{u})\Theta(\underline{v})\Theta(t - \overline{u})\Theta(t - \overline{v})$$
$$\times \left[(\overline{v} + \overline{u})^{-\kappa} - (\overline{v} + t)^{-\kappa} - (\overline{u} + t)^{-\kappa} + (2t)^{-\kappa}\right], \qquad (16)$$

with the exception of the case $d = 2$ and $m + k = 2$ where for $m = k = 1$ we have

$$G_{\boldsymbol{m},\boldsymbol{k}}(u_1, v_1) = \frac{1}{2\pi}\left[\ln(t + v_1) - \ln(u_1 + v_1) - \ln(2t)\right.$$
$$\left. + \ln(t + u_1)\right]\Theta(u_1)\Theta(v_1)\Theta(t - u_1)\Theta(t - v_1),$$

for $m = 2$ and $k = 0$ we obtain

$$G_{\boldsymbol{m},0}(u_1, u_2) = \frac{1}{4\pi}\left[\ln(\overline{u}) - \ln(t + \overline{u}) + \ln(2)\right]\Theta(\underline{u})\Theta(t - \overline{u}),$$

and for $k = 2$ and $m = 0$ get

$$G_{0,\boldsymbol{k}}(v_1, v_2) = \frac{1}{4\pi}\left[\ln(\overline{v}) - \ln(t + \overline{v}) + \ln(2)\right]\Theta(\underline{v})\Theta(t - \overline{v}).$$

Theorem 12. *Whenever $m \neq 0$ or $k \neq 0$, the kernels $G_{m,k}$ defined in (16) belong to $L^p(\mathbb{R}^{m+k})$ for all p such that $p < 2 - (d-4)/\kappa$. The case $d = 2$ and $m + k = 2$, $G_{m,k} \in L^p(\mathbb{R}^2)$ for each $p \geq 1$, independent of the particular form of $G_{m,k}$ in Thm. 11.*

We finish this subsection with a final remark on the particular case of the L^2 properties of $G_{m,k}$ and the strong convergence of $\tilde{L}_\varepsilon^{(N)}$ to $\tilde{L}^{(N)}$ as $\varepsilon \to 0$ in $(\mathcal{S})'$, see Cor. 3.8 and Thm. 3.9 in[1].

Remark 6.

(1) For each $d < 4$, the kernels $G_{m,k} \in L^2(\mathbb{R}^{m+k})$, whenever $m \neq 0$ or $k \neq 0$.
(2) For each d and $N \geq 0$ such that $2N > d - 4$, the truncated functional $L_\varepsilon^{(N)}$ converges strongly to the truncated local time $L^{(N)}$ and $(\mathcal{S})'$ when $\varepsilon \to 0$.

3.3. *Multiple intersection local times of Bm*

In this subsection we summarize the investigations on multiple intersection local times of Bm in the framework of WNA. These results are essentially due to S. Mendonça and L. Streit[52] and R. Jenane et al.[38], to which we address the interested reader more details and proofs. Let us point out that for $d = 2$, J. Rosen[66] carried out a systematic research on the renormalized local times of multiple intersections of planar Bm and A. Dvoretzky et al.[10] investigated the double points of the Bm paths.

Here, we are interested in the k-fold intersection local times of the d-dimensional Bm paths L_k, $d \geq 3$, cf. (5). More precisely:

• determine the number $r = r(k, d)$ of truncations in order to obtain a well defined generalized Brownian functional,
• compute the kernels of the chaos expansion of the regularized 3-fold intersections of Bm paths $L_{3,\varepsilon,c}$,
• give a multiplicative renormalization $\varrho(\varepsilon) = \varepsilon^{d-5/2}$ for $L_{3,\varepsilon,c}$ in order to obtain a non trivial limit,
• show that each nth chaos of $\varrho(\varepsilon) L_{3,\varepsilon,c}$ converges in law to a Bm.

To start with, let us introduce the following notation. The truncated exponential

$$\exp_r(s) := \sum_{n=r}^{\infty} \frac{x^n}{n!}$$

and define for any $\boldsymbol{f} \in S_d(\mathbb{R})$

$$\delta^{(r)}(\boldsymbol{B}_{t_2} - \boldsymbol{B}_{t_1})(\boldsymbol{f}) := S^{-1}\left((2\pi(t_2 - t_1))^{-d/2} \exp_r\left(-\frac{\left| \int_{t_1}^{t_2} \boldsymbol{f}(s)\, ds \right|^2}{2|t_2 - t_1|} \right) \right)$$

from it follows that

$$D(t_1, \ldots t_k) := \left(S\big(\delta^{(r)}(\boldsymbol{B}_{t_k} - \boldsymbol{B}_{t_{k-1}}) \ldots \delta^{(r)}(\boldsymbol{B}_{t_2} - \boldsymbol{B}_{t_1})\big) \right)(\boldsymbol{f})$$

$$= \prod_{j=1}^{k-1} \left((2\pi(t_{j+1} - t_j))^{-d/2} \exp_r\left(-\frac{\left| \int_{t_j}^{t_{j+1}} \boldsymbol{f}(s)\, ds \right|^2}{2|t_{j+1} - t_j|} \right) \right). \quad (17)$$

Using the easy bound $\exp_r(x) \leq |\exp_r(x)| \leq |x|^r \exp(|x|)$, there exist constants C_1, C_2 and a continuous norm $|\cdot|'$ on $S_d(\mathbb{R})$ such that

$$|D(t_1 \ldots, t_k)| \leq C_1 \exp(C_2 |f|') \prod_{j=1}^{k-1} (t_{j+1} - t_j)^{r-d/2}.$$

Finally we notice that

$$\int_{\Delta_k} \prod_{j=1}^{k-1} (t_{j+1} - t_j)^{r-d/2} dt_1 \ldots dt_k < \infty$$

whenever $r - d/1 + 1 > 0$. Thus, we have proved the following:

Proposition 2 (cf.,[52] Prop. 3.1). *Let $r > d/2 - 1$ be given, then the (truncated) k-fold intersection local times of Bm paths is a Hida distribution*

$$L_k := \int_{\Delta_k} \delta^{(r)}(\boldsymbol{B}_{t_k} - \boldsymbol{B}_{t_{k-1}}) \ldots \delta^{(r)}(\boldsymbol{B}_{t_2} - \boldsymbol{B}_{t_1})\, dt_1 \ldots dt_k \in (\mathcal{S})'.$$

The above proposition, in particular says that for $d = 2$ and $d = 3$ it is sufficient to center the δ-function by subtracting its expectation, choosing $r = 1$. Next we would like to compute the explicit kernels F_n of the chaos expansion

$$L_k(\boldsymbol{w}) = \sum_n \langle : \boldsymbol{w}^{\otimes n} :, F_n \rangle. \quad (18)$$

In order to shorten the notation, we will restrict to the case $k = 3$, i.e., the kernels of the 3-fold intersection local times of the d-dimensional Bm paths with $d \geq 3$. In addition, we use the regularization of L_3

$$L_{3,\varepsilon,c} := \int_{\Delta_3} \delta_{\varepsilon,c}(\boldsymbol{B}_{t_3} - \boldsymbol{B}_{t_2})\delta_{\varepsilon,c}(\boldsymbol{B}_{t_2} - \boldsymbol{B}_{t_1})\, dt_1\, dt_2\, dt_3,$$

where $\delta_{\varepsilon,c} := \delta_\varepsilon - \mathbb{E}(\delta_\varepsilon)$ and δ_ε is defined in (9). Hence, the above chaos expansion (18) becomes

$$L_{3,\varepsilon,c}(\boldsymbol{w}) = \sum_n \langle : \boldsymbol{w}^{\otimes n} :, F_{\varepsilon,n} \rangle.$$

Remark 7. For each $\varepsilon > 0$ the 3-fold intersection local times $L_{3,\varepsilon,c} \in (L^2)$.

It turns out that, the kernels $F_{\varepsilon,n}$ of $L_{3,\varepsilon,c}$ may be expressed in terms of certain functions of just four variables which we shall introduce in order to state the main results of this subsection.

Definition 1.

(1) Let $x = (x_1, \ldots, x_n) \in \mathbb{R}^n$ be given and σ any permutation such that $x_{\sigma(1)} \leq \ldots \leq x_{\sigma(n)}$.
(2) For $j \leq n$ we use the notation $(x)_j := x_{\sigma(j)}$. It is clear that $(x)_1 = \min_{1 \leq i \leq n} x_i$ and $(x)_n = \max_{1 \leq i \leq n} x_i$.
(3) Let $\varepsilon > 0$, $d \geq 3$ and $m, k > 0$ integers be given. For any $0 \leq x_1 \leq x_2 \leq x_3 \leq x_4$ we define the function g_{mk} by

$$g_{\varepsilon,mk}(x_1, x_2, x_3, x_4) := \int_{x_2}^{x_3} (s - x_1 + \varepsilon)^{1-m-d/2}(x_4 - s + \varepsilon)^{1-k-d/2}ds.$$

We are ready to announce the kernels $F_{\varepsilon,n}$ of $L_{3,\varepsilon,c}$, the proof consists in developing (17) (with $k = 3$) in Taylor series and a careful symmetrization of the variables. The details can be found in[52].

Theorem 13. *Let $\varepsilon > 0$ and $d \geq 3$ be given, then the kernels of $L_{3,\varepsilon,c}$ are given by*

$$F_{\varepsilon,2n}(u) = \sum_{2m+2k=2n} F_{\varepsilon,m,k}(u),$$

with

$$F_{\varepsilon,m,k}(u) = C(m,k)\mathbb{1}_{[0,t]}(u)\mathbb{1}_{E_{m,k}}(u)\big(g_{\varepsilon,mk}(x_1,x_2,x_3,x_4)$$
$$+g_{\varepsilon,mk}(0,x_2,x_3,t) - g_{\varepsilon,mk}(0,x_2,x_3,x_4) - g_{\varepsilon,mk}(x_1,x_2,x_3,t)\big),$$

where

$$u := (u^1, \ldots, u^d)$$

$$u^i := (u^i_1, \ldots, u^i_{2m_i}, u^i_{2m_i+1}, \ldots u^i_{2m_i+2k_i})$$

$$C(\boldsymbol{m}, \boldsymbol{k}) := \frac{(2\pi)^{-d}(2\boldsymbol{m})!(2\boldsymbol{k})!}{2^n \boldsymbol{m}! \boldsymbol{k}!(2n)!} \frac{1}{(1 - m - \frac{d}{2})(1 - k - \frac{d}{2})}$$

$$x_1 := \min_{i,j}(u^i_j) \geq 0$$

$$x_2 := \max_i(u^i)_{2m_i} \leq x_3 := \min_i(u^i)_{2m_i+1}$$

$$x_4 := \max_{i,j}(u^i_j) \leq t$$

$$E_{\boldsymbol{m},\boldsymbol{k}} := \{u : x_2 \leq x_3\} \subset [0,t]^{2(m+k)}.$$

The nth chaos of $L_{3,\varepsilon,c}$ may be written as a sum of two terms $M_t^{(3)}(\varepsilon)$ and $N_t^{(3)}(\varepsilon)$, namely

$$M_t^{(3)}(\varepsilon) := M_t^{(3)}(\varepsilon, \boldsymbol{n}, \boldsymbol{w}) := \sum_{2\boldsymbol{m}+2\boldsymbol{k}=2n} C(\boldsymbol{m}, \boldsymbol{k}) \langle : \boldsymbol{w}^{\otimes n} :, \mathbb{1}_{E_{\boldsymbol{m},\boldsymbol{k}}} g_{\varepsilon,mk}(x_1, x_2, x_3, x_4) \rangle,$$

$$N_t^{(3)}(\varepsilon) := N_t^{(3)}(\varepsilon, \boldsymbol{n}, \boldsymbol{w}) := \sum_{2\boldsymbol{m}+2\boldsymbol{k}=2n} C(\boldsymbol{m}, \boldsymbol{k}) \langle : \boldsymbol{w}^{\otimes n} :, \mathbb{1}_{E_{\boldsymbol{m},\boldsymbol{k}}} \big(g_{\varepsilon,mk}(0, x_2, x_3, t)$$
$$- g_{\varepsilon,mk}(0, x_2, x_3, x_4) - g_{\varepsilon,mk}(x_1, x_2, x_3, t) \big) \rangle$$

such that

$$K_t^{(3)}(\varepsilon) := K_t^{(3)}(\varepsilon, \boldsymbol{n}, \boldsymbol{w}) := M_t^{(3)}(\varepsilon) + N_t^{(3)}(\varepsilon).$$

Remark 8.

(1) The processes $M_t^{(3)}(\varepsilon)$ and $N_t^{(3)}(\varepsilon)$ are continuous and $M_t^{(3)}(\varepsilon)$ is more singular than $N_t^{(3)}(\varepsilon)$ as $\varepsilon \to 0$.

(2) Moreover, the process $M_t^{(3)}(\varepsilon)$ is a Brownian martingale. In fact, we can write it as an Itô integral, namely

$$M_t^{(3)}(\varepsilon) = \int_0^t \mathfrak{m}(\tau) \, d\boldsymbol{B}_\tau := \sum_{i=1}^d M_t^{(3),i}(\varepsilon) := \sum_{i=1}^d \int_0^t \mathfrak{m}_i(\tau) \, dB_t^i,$$

where $\mathfrak{m}$ is computed using the generalized Clark-Haussmann-Ocone formula, cf. Thm. 20 in Appendix A.1, and is equal to

$$\mathfrak{m}_i(\tau) := \sum_{2\boldsymbol{m}+2\boldsymbol{k}=2n} n_i C(\boldsymbol{m}, \boldsymbol{k}) \langle : \boldsymbol{w}^{\otimes(n-\delta_i)} :, \mathbb{1}_{[0,\tau]^n} \mathbb{1}_{E_{\boldsymbol{m},\boldsymbol{k}}} g_{\varepsilon,mk}(x_1, x_2, x_3, \tau) \rangle.$$

As the kernels $\mathfrak{m}$ do not depend on t (except on the boundaries of integration), then it is clear that $M_t^{(3),i}(\varepsilon)$ are Brownian martingale and as a result $M_t^{(3)}(\varepsilon)$ also, see for example Sec. 4 in[19] or Ch. 4 in[33].

(3) Finally, a crucial fact that the martingales $M^{(3),i}(\varepsilon)$ are orthogonal due to the orthogonality of the Wick polynomials, cf. Prop. 9 in Ch. 1.

The next theorem states the main results concerning the renormalization of the nth chaos of $L_{3,\varepsilon,c}$ and its convergence in law.

Theorem 14 (cf.[38], Thm. 2 and 3). *Let $\varepsilon > 0$, $d \geq 3$ and $\varrho(\varepsilon) = \varepsilon^{d-5/2}$ be given. Then the renormalized process*

(1) $\varrho(\varepsilon)M^{(3),i}(\varepsilon)$, $i = 1,\dots,d$ converges in law to a Bm $C^i\beta^i$, C^i constant which depends on other parameters,

(2) $\varrho(\varepsilon)K^{(3)}(\varepsilon)$ converges in law to a Bm $C\beta$, where C is a constant.

Taking into account Remark 8-2. and 3., to prove the above theorem, it is sufficient to study the convergence in probability of the quadratic variation process, namely

$$\lim_{\varepsilon \searrow 0} \mathbb{E}(\langle \varrho(\varepsilon)M^{(3),i}(\varepsilon), \varrho(\varepsilon)M^{(3),j}(\varepsilon)\rangle_t) = \text{const.}\delta_{ij}t$$

and then use Thm. VIII. 3.11 in[39] to conclude the result.

4. Self intersection local times of fBm

Intersection local times of fBm have been studied by many authors, see e.g. the works by Gradinaru et al.[32,34,35], Rosen[67], Grothaus et al.[31], and the references therein. In this section we give an account on the WNA approach to local times of fBm. We may consider self intersection of fBm paths as in[34] and[18] or intersection of independent fBm's as in[53] and[58].

We start by giving the definition of fBm in WNA using fractional operators, for the details see Appendix A.2. For an arbitrary Hurst parameter $H \in (0,1)\backslash\{\frac{1}{2}\}$ a version of a d-dimensional fBm $\boldsymbol{B}^H$ is given by

$$\boldsymbol{B}_t^H := (\langle w_1, M_H \mathbb{1}_{[0,t)}\rangle, \dots, \langle w_d, M_H \mathbb{1}_{[0,t)}\rangle), \quad \boldsymbol{w} \in S_d'(\mathbb{R}), \quad t > 0, \quad (19)$$

where $\mathbb{1}_A$ denotes the indicator function of the set A. The operator M_H is defined on $S(\mathbb{R})$ by

$$M_H\varphi := \begin{cases} K_H D_-^{-(H-\frac{1}{2})}\varphi, & H \in (0, \frac{1}{2}), \\ \varphi, & H = \frac{1}{2}, \\ K_H I_-^{H-\frac{1}{2}}\varphi, & H \in (\frac{1}{2}, 1), \end{cases} \quad (20)$$

and the normalizing constant $K_H := (2H \sin(\pi H)\Gamma(2H))^{1/2}$. Although the operator M_H is defined on $S(\mathbb{R})$, its domain is larger, in particular it can be applied to the indicator function $\mathbb{1}_{[0,t)}$, $t > 0$, see Lemma 1.1.3 in[51], and we have

$$M_H \mathbb{1}_{[0,t)} \in L^2(\mathbb{R}).$$

More details can be found in[4,51,63] and references therein. In a similar way we can define a d-dimensional multi-parameter fBm. Namely, for an arbitrary d-dimensional Hurst parameter $\boldsymbol{H} = (H_1, \ldots, H_d) \in (0,1)^d$, a version of a d-dimensional multi-parameter fBm $\boldsymbol{B}^{\boldsymbol{H}}$ is given by

$$\boldsymbol{B}_t^{\boldsymbol{H}} := (\langle w_1, M_{H_1} \mathbb{1}_{[0,t)} \rangle, \langle w_d, M_{H_d} \mathbb{1}_{[0,t)} \rangle), \quad \boldsymbol{w} \in S_d'(\mathbb{R}), \quad t > 0.$$

4.1. *Self intersection local time of fBm*

In this subsection we show the main results on self intersection local time of fBm $\boldsymbol{B}^H$ in WNA. Thus the object we are concern informally is given by

$$L_H := \int_0^t \int_0^t \delta(\boldsymbol{B}_{t_1}^H - \boldsymbol{B}_{t_2}^H) \, dt_1 dt_2.$$

The first result deals with Donsker's delta function of fBm $\boldsymbol{B}^H$.

Proposition 3 (cf. Prop. 5 in[18]). *For $s < t$ the Bochner integral*

$$\delta(\boldsymbol{B}_t^H - \boldsymbol{B}_s^H) := \left(\frac{1}{2\pi}\right)^d \int_{\mathbb{R}^d} e^{i\left(\boldsymbol{\theta}, \boldsymbol{B}_t^H - \boldsymbol{B}_s^H\right)} \, d\boldsymbol{\theta} \tag{21}$$

is a Hida distribution with S-transform given at each $\boldsymbol{\varphi} \in S_d(\mathbb{R})$ by

$$S\delta(\boldsymbol{B}_t^H - \boldsymbol{B}_s^H)(\boldsymbol{\varphi}) = \left(\frac{1}{\sqrt{2\pi}|t-s|^H}\right)^d \exp\left(\frac{-1}{2|t-s|^{2H}}\right.$$
$$\left. \times \left|\int_{\mathbb{R}} \varphi(x)(M_H \mathbb{1}_{[s,t)})(x) \, dx\right|^2\right).$$

The proof is a straightforward application of Cor. 2 in Ch. 1. to the S-transform of the integrand function in (21) with respect to the Lebesgue measure on $\mathbb{R}^d$.

The main result on self intersection local times L_H as well as on their subtracted counterparts $L_H^{(N)}$.

Theorem 15 (cf. Thm. 7 and Prop. 8 in[18]). *Let $H \in (0,1)$ and $d \geq 1$, $N \geq 0$ be given such that $2N(H-1) + dH < 1$.*

(1) Then, the Bochner integral

$$L_H^{(N)} := \int_{\Delta_2} \delta^{(N)}(\boldsymbol{B}_{t_1}^H - \boldsymbol{B}_{t_2}^H) \, dt_1 dt_2$$

is a Hida distribution.

(2) The kernels $F_{H,n}$ of $L_H^{(N)}$ are given by

$$F_{H,2n}(u_1,\ldots,u_{2n}) = \frac{1}{n!}\left(\frac{1}{2\pi}\right)^{d/2}\left(-\frac{1}{2}\right)^n \int_{\Delta_2} \frac{1}{|t_2 - t_1|^{2Hn+dH}}$$

$$\times \prod_{i=1}^{2n}(M_H \mathbb{1}_{[t_1,t_2]})(u_i)\, dt_1\, dt_2$$

for each $n \in \mathbb{N}^d$ such that $n > N$. All the other $F_{H,n}$ vanish.

Remark 9.

(1) The result of Thm. 15, (1) shows that for $d = 1$ all self intersection local times L_H are well-defined for all possible Hurst parameters $H \in (0,1)$.
(2) For $d \geq 2$, the intersection local times are well defined only for $H < 1/d$. In other words, for $H \geq 1/d$ and $d \geq 2$, the self intersection local times L_H becomes well defined once subtracted the divergent terms.

The above remark suggests a "renormalization" procedure which motivates the study of a regularization. As a computationally simple regularization we discuss

$$L_{H,\varepsilon} := \int_{\Delta_2} \delta_\varepsilon(\boldsymbol{B}_{t_1}^H - \boldsymbol{B}_{t_2}^H)\, dt_1\, dt_2, \qquad \varepsilon > 0.$$

Theorem 16 (cf. Thm. 9[18]). *Let $\varepsilon > 0$ be given. For all $H \in (0,1)$ and all dimensions $d \geq 1$ the functional $L_{H,\varepsilon}$ is a Hida distribution with kernel functions given by*

$$F_{H,\varepsilon,2n}(u_1,\ldots,u_{2n})$$

$$= \frac{1}{n!}\left(\frac{1}{2\pi}\right)^{d/2}\left(-\frac{1}{2}\right)^n \int_{\Delta_2} \frac{1}{(\varepsilon + |t_2 - t_1|^{2H})^{n+d/2}} \prod_{i=1}^{2n}(M_H \mathbb{1}_{[t_1,t_2]})(u_i)$$

for each $n \in \mathbb{N}^d$, and $F_{H,\varepsilon,n} = 0$ if at least one n_i is an odd number. Moreover, if $2N(H-1)+dH < 1$, then when ε goes to zero the (truncated) functional $L_{H,\varepsilon}^{(N)}$ converges strongly in $(\mathcal{S})'$ to the (truncated) intersection local time $L_H^{(N)}$.

The first part of the proof is an application of Cor. 2 in Ch. 1 and the convergence follows from Cor. 1 in Ch. 1.

4.2. *Self intersection local time of independent multi-parameter fBm*

We investigate the intersection local time of two independent d-dimensional multi-parameter fBm $\boldsymbol{B}^{\hat{H}}$ and $\boldsymbol{B}^{\tilde{H}}$. The WNA framework allows the definition of the intersection local time in terms of an integral over a Donsker's δ-function

$$L_{\hat{H},\tilde{H}} := \int_0^t \int_0^t \delta(\boldsymbol{B}_{t_1}^{\hat{H}} - \boldsymbol{B}_{t_2}^{\tilde{H}}) \, dt_1 dt_2.$$

At first we state that the Donsker's δ-function, the integrand of $L_{\hat{H},\tilde{H}}$, is a Hida distribution.

Theorem 17 (cf. Prop. 5 in[58]). *Let t and s be strictly positive real numbers, then the Bochner integral*

$$\delta(\boldsymbol{B}_t^{\hat{H}} - \boldsymbol{B}_s^{\tilde{H}}) = \left(\frac{1}{2\pi}\right)^d \int_{\mathbb{R}^d} e^{i(\lambda, B_t^{\hat{H}} - B_s^{\tilde{H}})} \, d\boldsymbol{\lambda}$$

is a Hida distribution with S-transform given by

$$S\delta(\boldsymbol{B}_t^{\hat{H}} - \boldsymbol{B}_s^{\tilde{H}})(\varphi) = \left(\frac{1}{2\pi}\right)^{d/2} \prod_{i=1}^d \frac{1}{\sqrt{t^{2\hat{H}_i} + s^{2\tilde{H}_i}}} \exp\left[-\frac{1}{2} \sum_{i=1}^d \frac{1}{t^{2\hat{H}_i} + s^{2\tilde{H}_i}}\right.$$

$$\left. \times \left(\int_{\mathbb{R}} \left(\hat{\varphi}_i(x)(M_{\hat{H}_i} \mathbb{1}_{[0,t)})(x) - \tilde{\varphi}_i(x)(M_{\tilde{H}_i} \mathbb{1}_{[0,s)})(x)\right) dx\right)^2\right],$$

for $\varphi = (\hat{\varphi}, \tilde{\varphi}) \in S_{2d}(\mathbb{R})$.

Again the proof of this result follows from an application of Cor. 2 in Ch. 1 to the S-transform of the integrand function

$$\Phi(\boldsymbol{w}) = e^{i(\lambda, B_t^{\hat{H}} - B_s^{\tilde{H}})}, \quad \boldsymbol{w} = (\hat{w}, \tilde{w}) \in S_{2d}'(\mathbb{R})$$

with respect to the Lebesgue measure on $\mathbb{R}^d$.

Before we state the main result for intersection local times $L_{\hat{H},\tilde{H}}$ we fix the following notation: for any d-dimensional Hurst parameter $\boldsymbol{H} = (H_1, \ldots, H_d) \in (0,1)^d$ we define

$$\bar{H} := \max_{1 < i < d} H_i.$$

Theorem 18 (cf. Thm. 8 in[58]). *Let $t > 0$ be given. For any pair of integer numbers $d \geq 1$, $N \geq 0$ and for any pair of d-dimensional Hurst parameters $\hat{H}$, $\tilde{H}$ such that*

$$\max\{\bar{\hat{H}}, \bar{\tilde{H}}\} \left(N + \frac{d}{2} - \frac{1}{2\min\{\bar{\hat{H}}, \bar{\tilde{H}}\}}\right) < N + \frac{1}{2},$$

the Bochner integral

$$L_{\hat{H},\tilde{H}}^{(N)} := \int_0^t \int_0^t \delta^{(N)}(B_t^{\hat{H}} - B_s^{\tilde{H}})\, ds dt$$

is a Hida distribution.

As a consequence one may derive the chaos expansion for the (truncated) intersection local times $L_{\hat{H},\tilde{H}}^{(N)}$.

Corollary 1 (cf. Prop. 9 in[58]). *Under the conditions of Thm. 18, $L_{\hat{H},\tilde{H}}^{(N)}$ has the chaos expansion*

$$L_{\hat{H},\tilde{H}}^{(N)}(\hat{w},\tilde{w}) = \sum_m \sum_k \langle :\hat{w}^{\otimes m}: \otimes :\tilde{w}^{\otimes k}:, F_{\hat{H},\tilde{H},m,k}\rangle$$

where the kernel functions $F_{\hat{H},\tilde{H},m,k}$ are given by

$$F_{\hat{H},\tilde{H},m,k} = \left(\frac{1}{\pi}\right)^{\frac{d}{2}} \frac{(-1)^{\frac{m+3k}{2}}}{\left(\frac{m+k}{2}\right)!} \left(\frac{1}{2}\right)^{\frac{m+k+d}{2}} \binom{m+k}{m}$$

$$\times \int_0^t \int_0^T \prod_{i=1}^d \left(\frac{1}{t^{2\hat{H}_i} + s^{2\tilde{H}_i}}\right)^{\frac{m_j+k_j+1}{2}}$$

$$\times \bigotimes_{i=1}^d \left((M_{\hat{H}_i}\mathbb{1}_{[0,t)})^{\otimes m_i} \otimes (M_{\tilde{H}_i}\mathbb{1}_{[0,s)})^{\otimes k_j}\right)\, ds\, dt$$

for each $m = (m_1,\ldots,m_d)$ and each $k = (k_1,\ldots,k_d)$ such that $m+k \geq 2N$ and all sums $m_j + k_j$, $j = 1,...,d$, are even numbers. All other kernel functions $F_{\hat{H},\tilde{H},m,k}$ are identically equal to zero.

Remark 10.

(1) The result of Thm. 18 shows that for $d = 1$ or $d = 2$ all intersection local times $L_{\hat{H},\tilde{H}}$ are well defined for all possible pairs of d-dimensional Hurst parameters $\hat{H}$, $\tilde{H}$ in $(0,1)^d$. For $d > 2$, intersection local times are well defined only for $1/\hat{H} + 1/\tilde{H} > d$.

(2) Under these conditions, Corollary 1, in addition, yields

$$\mathbb{E}(L_{\hat{H},\tilde{H}}) = F_{\hat{H},\tilde{H},0,0} = \left(\frac{1}{2\pi}\right)^{d/2} \int_0^t dt \int_0^t ds \prod_{i=1}^d \frac{1}{\sqrt{t^{2\hat{H}_i} + s^{2\tilde{H}_i}}}\, ds\, dt.$$

In other words, for $1/\hat{H} + 1/\tilde{H} \leq d$ with $d > 2$, the intersection local times only become well defined once subtracted the divergent terms. This motivates the study of a regularization. As a computationally simple regularization we discuss

$$L_{\hat{H},\tilde{H},\varepsilon} := \int_0^t \int_0^t \delta_\varepsilon(B_t^{\hat{H}} - B_s^{\tilde{H}}) \, ds \, dt, \quad \varepsilon > 0.$$

Theorem 19 (cf. Thm. 10 in[58]). *Let $\varepsilon > 0$ be given. For all $\hat{H}, \tilde{H} \in (0,1)^d$ and all dimensions $d \geq 1$ the intersection local time $L_{\hat{H},\tilde{H},\varepsilon}$ is a Hida distribution with kernel functions given by*

$$F_{\hat{H},\tilde{H},\varepsilon,m,k} = \left(\frac{1}{\pi}\right)^{\frac{d}{2}} \frac{(-1)^{\frac{m+3k}{2}}}{\left(\frac{m+k}{2}\right)!} \left(\frac{1}{2}\right)^{\frac{m+k+d}{2}} \binom{m+k}{m}$$

$$\times \int_0^T \int_0^T \prod_{i=1}^d \left(\frac{1}{\varepsilon + t^{2\hat{H}_i} + s^{2\tilde{H}_i}}\right)^{\frac{m_j + k_j + 1}{2}} ds \, dt$$

$$\times \bigotimes_{i=1}^d \left((M_{\hat{H}_i} 1\!\!1_{[0,t)})^{\otimes m_j} \otimes (M_{\tilde{H}_i} 1\!\!1_{[0,s)})^{\otimes k_j} \right)$$

for all $m = (m_1, ..., m_d)$, $k = (k_1, ..., k_d) \in \mathbb{N}_0$ such that all sums $m_i + k_j$, $j = 1, ..., d$, are even numbers, and $F_{\hat{H},\tilde{H},\varepsilon,m,k} = 0$ if at least one of the sums $m_i + k_i$ is an odd number. Moreover, if

$$\max\{\bar{\bar{\hat{H}}}, \bar{\bar{\tilde{H}}}\} \left(N + \frac{d}{2} - \frac{1}{2\min\{\bar{\bar{\hat{H}}}, \bar{\bar{\tilde{H}}}\}}\right) < N + \frac{1}{2},$$

then when ε goes to zero the (truncated) intersection local time $L_{\hat{H},\tilde{H},\varepsilon}^{(N)}$ converges strongly in $(\mathcal{S})'$ to the (truncated) local time $L_{\hat{H},\tilde{H}}^{(N)}$.

For the details of the proof of the above results we address the reader to the original work[58].

Remark 11 (Local times for multifractional Bm, cf.[2]). Most recently the local times for multi-fractional Bm (mBm) in d-dimensions have been studied in the framework of WNA. The expansion of local times of mBm in terms of Wick powers of white noises were computed and if a suitable number of kernels is subtracted, it exists in the sense of generalized white noise functionals. In addition, the convergence of the regularized truncated local times of mBm in the sense of Hida distributions is proved. For the details we refer to[2].

Acknowledgments

We would like to thank our mentor Prof. Ludwig Streit for the invitation to write a chapter for this book on the applications of WNA. The first author would like to thank the support of the project UID/MAT/04674/2013.

A.1. Regular generalized functions and the Clark-Haussmann-Ocone formula

In this appendix we recall certain concepts needed in this chapter, namely the spaces of regular generalized functions introduced by J. Potthoff and M. Timpel[62] and later generalized in[30]. A characterization of such spaces is given in,[29][30]. The striking property of these spaces lies on the fact that their kernels are elements of the symmetric tensor powers of $L^2_d(\mathbb{R})$ or an abstract Hilbert space $\mathcal{H}_\mathbb{C}$. The concepts of Skorokhod and Itô integrals as well the Clark-Haussmann-Ocone formula were extended to these spaces, cf.[15],[55].

Let $F \in (L^2)$ be a square integrable function with chaos expansion (cf. Ch. 1, Eq. (9))

$$F(\boldsymbol{w}) = \sum_n \langle : \boldsymbol{w}^{\otimes n} :, F_n \rangle, \quad F_n \in \left(L^2_d(\mathbb{R}) \right)^{\hat{\otimes} n},$$

such that for any $q \in \mathbb{N}_0$ the series converges

$$\|F\|^2_{q,1} := \sum_n (n!)^2 2^{nq} |F_n|^2 < \infty.$$

Define the set $\mathcal{G}^1_q$ by

$$\mathcal{G}^1_q := \left\{ F \in (L^2) \mid F = \sum_n \langle : \cdot^{\otimes n} :, F_n \rangle, \ \|F\|_{q,1} < \infty \right\}.$$

Definition 2 (Regular test/generalized functions, cf.[30],[62]).

(1) We define the space of regular test functions by

$$\mathcal{G}^1 := \operatorname*{proj\,lim}_{q \in \mathbb{N}_0} \mathcal{G}^1_q.$$

(2) Denoting the dual space of $\mathcal{G}^1_q$ w.r.t. (L^2) by $\mathcal{G}^{-1}_{-q}$, then the space of regular generalized functions is given by

$$\mathcal{G}^{-1} := \operatorname*{ind\,lim}_{q \in \mathbb{N}_0} \mathcal{G}^{-1}_{-q}.$$

As sets, the spaces $\mathcal{G}^1$ and $\mathcal{G}^{-1}$ coincides, respectively, with

$$\mathcal{G}^1 = \bigcap_{q \in \mathbb{N}_0} \mathcal{G}^1_q$$

$$\mathcal{G}^{-1} = \bigcup_{q \in \mathbb{N}_0} \mathcal{G}^{-1}_{-q}.$$

For any regular test function $\psi \in \mathcal{G}^1$ we may define the gradient of ψ, denoted by $\nabla \psi = (\partial^i \psi)^d_{i=1}$ which is a continuous linear operator $\nabla : \mathcal{G}^1_q \longrightarrow \mathbb{R}^d \otimes L^2(\mathbb{R}) \otimes \mathcal{G}^1_q$. This operator can be extended to the space of regular generalized functions $\mathcal{G}^{-1}$ such that for any $\Phi \in \mathcal{G}^{-1}$, $\nabla \Phi = (\partial^i \Phi)^d_{i=1}$ is a bounded linear operator $\nabla : \mathcal{G}^{-1}_{-q} \longrightarrow \mathbb{R}^d \otimes L^2(\mathbb{R}) \otimes \mathcal{G}^{-1}_{-p}$ for $p > q$. For any $\Phi \in L^2(\mathbb{R}) \otimes \mathcal{G}^{-1}_{-q}$, $q \in \mathbb{N}_0$ characterized by the sequence $\Phi_n(\cdot, \cdot) \in L^2(\mathbb{R}) \otimes L^2(\mathbb{R}^n)$, $n \in \mathbb{N}_0$, we define for any $1 \leq i \leq d$ the functional characterized by the sequence

$$\Psi^i_0 := 0,$$
$$\Psi^i_n := \tilde{\Phi}_{n-\delta_i} \in L^2(\mathbb{R}^n), \ \boldsymbol{n} = (n_1, \dots, n_d), \ n \in \mathbb{N},$$

where $\tilde{\Phi}_{n-\delta_i}$ denotes the symmetrization of $\Phi_{n-\delta_i}$ in the variables $t, s^i_1, \dots, s^i_{n_i-1}$. The sequence (Ψ_n), $n \in \mathbb{N}_0$, defines an element in $\mathcal{G}^{-1}_{-q}$ which we denote by $I_i(\Phi)$. We have the following duality relation: for any regular test function $\psi \in \mathcal{G}^1$ with kernels (ψ_n), $n \in \mathbb{N}_0$ and each $1 \leq i \leq d$,

$$\langle\!\langle I_i(\Phi), \psi \rangle\!\rangle = \langle\!\langle \Phi, \partial^i_\cdot \psi \rangle\!\rangle.$$

The functional $I_i(\Phi)$ is the unique element in $\mathcal{G}^{-1}$ for which the above equality holds for every regular test function $\psi \in \mathcal{G}^1$. The details of these concepts and proofs may be founded in Subs. 2.2 in[15].

Definition 3 (Generalized Skorokhod integral). Let $\Phi \in \mathbb{R}^d \otimes L^2(\mathbb{R}) \otimes \mathcal{G}^{-1}_{-q}$ be given for some $q \in \mathbb{N}_0$. We call generalized Skorokhod integral of Φ the regular generalized function in $\mathcal{G}^{-1}$, denoted by $I(\Phi)$ and defined by

$$I(\Phi) := \sum_{i=1}^{d} I_i(\Phi_i),$$

where, for each $i = 1, \dots, d$, $I_i(\Phi_i)$ is the unique regular generalized function in $\mathcal{G}^{-1}$ such that

$$\langle\!\langle I_i(\Phi_i), \psi \rangle\!\rangle = \langle\!\langle \Phi_i, \partial^i_\cdot \psi \rangle\!\rangle$$

holds for each regular test function $\psi \in \mathcal{G}^1$.

Now we come to the generalized Clark-Haussmann-Ocone formula, cf. Thm. 3.1 in[15] and Thm. 4.2 in[55].

Theorem 20 (Generalized Clark-Haussmann-Ocone formula). *Let*
Φ *be a regular generalized function,* $\Phi \in \mathcal{G}^{-1}$, *then it can be written as a*
generalized Itô integral

$$\Phi = \mathbb{E}(\Phi) + \sum_{i=1}^{d} I_i \big(\mathrm{Exp}(\mathbb{1}_{[0,\cdot]}) \partial_\cdot^i \Phi \big),$$

where $\mathrm{Exp}(\mathbb{1}_{[0,\cdot]})$ *is the second quantization operator, cf. Subs. 8.3 in*
Ch. 1.

A.2. Elements of fractional calculus

In this appendix we collect certain notions and notations from fractional
calculus needed in Section 4, a detailed exposition on the subject may be
seen in[17,44,50,61,73] and references therein.

For any $\alpha > 0$ we denote by g_α the function defined by

$$g_\alpha(t) := \frac{t^{\alpha-1}}{\Gamma(\alpha)}, \quad t > 0,$$

where Γ denotes the gamma function. The *left-side (right-side) Riemann-*
Liouville fractional integral of order $0 < \alpha < 1$, *is defined for any* $f \in$
$L^p(\mathbb{R})$, $1 < p < 1/\alpha$ by

$$(I_+^\alpha f)(x) := \int_{-\infty}^{x} f(t) g_\alpha(x - t)\, dt, \quad x \in \mathbb{R},$$

and

$$(I_-^\alpha f)(x) := \int_{x}^{\infty} f(t) g_\alpha(t - x)\, dt, \quad x \in \mathbb{R}.$$

In addition, the Hardy–Littlewood theorem holds: $I_\pm^\alpha : L^p(\mathbb{R}) \longrightarrow L^q(\mathbb{R})$ is
continuous for any $q = p/(1 - \alpha p)$, cf. Thm. 5.3 in[73]. This is sufficient to
guarantee that if $f \in I_\pm(L^p(\mathbb{R}))$, then there exists a unique $h \in L^p(\mathbb{R})$ such
that $(I_\pm f)(x) = h(x)$. For $0 < \alpha < 1$, h coincides with the left-(right-) side
Riemann-Liouville fractional derivative of f of order α, denote by $D_\pm^\alpha f$,
and given, for a.a. $x \in \mathbb{R}$, by

$$(D_+^\alpha f)(x) := \frac{1}{\Gamma(1-\alpha)} \frac{d}{dx} \int_{-\infty}^{x} f(t)(x-t)^{-\alpha}\, dt,$$

$$(D_-^\alpha f)(x) := \frac{-1}{\Gamma(1-\alpha)} \frac{d}{dx} \int_{-\infty}^{x} f(t)(x-t)^{-\alpha}\, dt.$$

A.3. Local time of grey Brownian motion

In this appendix, we summarize the main results on local times of generalized grey Brownian motion (ggBm) within the grey noise analysis. They extend some results of this chapter for a non-Gaussian setting. More details and proofs can be founded in[28] and[27]. The starting point of grey noise analysis is the nuclear triple from white noise analysis, cf. Ex. 2 in Ch. 1. For any $\beta \in (0,1]$, the grey noise reference measure μ_β is defined as the unique probability measure on $S'(\mathbb{R})$ such that for all $\varphi \in S(\mathbb{R})$

$$\int_{S'(\mathbb{R})} e^{i\langle w,\varphi\rangle} d\mu_\beta(w) = E_\beta\left(-\frac{1}{2}\langle\varphi,\varphi\rangle\right),$$

where E_β is the Mittag-Leffler function given by its power series by $E_\beta(z) = \sum_{n=0}^{\infty} z^n/\Gamma(\beta n + 1)$. Here Γ denotes the usual Gamma function. Furthermore, a second complex valued parameter $0 < \alpha < 2$ allows a generalization as

$$E_{\beta,\alpha}(z) = \sum_{n=0}^{\infty} \frac{z^n}{\Gamma(\beta n + \alpha)}$$

which is called generalized Mittag-Leffler function. Hence, we obtain the corresponding L^p spaces, denoted by $L^p(\mu_\beta) := L^p(S'(\mathbb{R}), \mu_\beta; \mathbb{C})$ for $p \geq 1$. It is easy to show that the moments of μ_β, for any $n \in \mathbb{N}$ and $\varphi \in S(\mathbb{R})$, are given by

$$\int_{S'(\mathbb{R})} \langle w, \varphi\rangle^{2n+1} d\mu_\beta(w) = 0,$$

$$\int_{S'(\mathbb{R})} \langle w, \varphi\rangle^{2n} d\mu_\beta(w) = \frac{(2n)!}{2^n \Gamma(\beta n + 1)} \langle\varphi,\varphi\rangle^n.$$

The bilinear pairing $\langle\cdot,\varphi\rangle$, $\varphi \in S(\mathbb{R})$ may be defined to include $\varphi = f \in L^2(\mathbb{R})$ as a $L^2(\mu_\beta)$-limit.

The grey noise measure μ_β belongs to the class for which the Appell system exist, see[28]. Therefore, we may construct the test function space $(S(\mathbb{R}))^1$ and a distribution space $(S(\mathbb{R}))_{\mu_\beta}^{-1}$ such that we obtain a nuclear triple

$$(S(\mathbb{R}))^1 \subset L^2(\mu_\beta) \subset (S(\mathbb{R}))_{\mu_\beta}^{-1}.$$

The distribution space $(S(\mathbb{R}))_{\mu_\beta}^{-1}$ is characterized using an integral transform, called S_{μ_β}-transform, and spaces of holomorphic functions in the same spirit as Sec. 6.2 in Ch. 1. The S_{μ_β}-transform of an element $\Phi \in (S(\mathbb{R}))_{\mu_\beta}^{-1}$ is defined, for any φ in a neighborhood $\mathcal{U} \subset S_{\mathbb{C}}(\mathbb{R})$ of zero, by

$$S_{\mu_\beta}\Phi(\varphi) := \frac{1}{E_\beta(\frac{1}{2}\langle\varphi,\varphi\rangle)} \langle\!\langle \Phi, e^{\langle\cdot,\varphi\rangle}\rangle\!\rangle_{\mu_\beta},$$

where $\langle\!\langle \cdot, \cdot \rangle\!\rangle_{\mu_\beta}$ is the dual pairing between $(S(\mathbb{R}))^1_{\mu_\beta}$ and $(S(\mathbb{R}))^{-1}_{\mu_\beta}$. Later we use other related integral transform, called T_{μ_β}-transform, defined for any $\varphi \in \mathcal{U}$, by

$$T_{\mu_\beta}\Phi(\varphi) := E_\beta\left(-\frac{1}{2}\langle\varphi,\varphi\rangle\right) S_{\mu_\beta}\Phi(i\varphi), \qquad \Phi \in (S(\mathbb{R}))^{-1}_{\mu_\beta}.$$

For any $0 < \alpha < 2$ and $0 < \beta \leq 1$ we define the process

$$S'(\mathbb{R}) \ni w \mapsto B^{\alpha,\beta}_t(w) := \langle w, M_{\alpha/2}\mathbb{1}_{[0,t)}\rangle, \qquad t > 0,$$

as an element in $L^2(\mu_\beta)$ and call this process ggBm. The operator $M_{\alpha/2}$ is defined in (20). For $0 < \alpha = \beta \leq 1$, $B^\alpha_t := B^{\alpha,\alpha}_t$, $t \geq 0$, is called grey Brownian motion and choosing $\beta = 1$, $B^{\alpha,1}_t$, $t \geq 0$, is a fBm with Hurst parameter $H = \alpha/2$. For $\alpha = \beta = 1$, $B^{1,1}_t$, $t \geq 0$, is a Bm. As a consequence of the estimate, cf. Prop. 3.8 in[27],

$$\mathbb{E}_{\mu_\beta}\left(|B^{\alpha,\beta}_t - B^{\alpha,\beta}_s|^{2p}\right) \leq K|t-s|^{\alpha p}, \qquad t,s \geq 0,$$

and Kolmogorov's continuity theorem, the ggBm process has a continuous version.

The S_{μ_β}-transform of a ggBm $B^{\alpha,\beta}_t = \langle \cdot, M_{\alpha/2}\mathbb{1}_{[0,t)}\rangle \in L^2(\mu_\beta)$ is well-defined and, for any $\varphi \in \mathcal{U}$, is given by

$$\left(S_{\mu_\beta}B^{\alpha,\beta}_t\right)(\varphi) = \langle\varphi, M_{\alpha/2}\mathbb{1}_{[0,t)}\rangle \frac{E_{\beta,\beta}\left(\frac{1}{2}\langle\varphi,\varphi\rangle\right)}{\beta E_\beta\left(\frac{1}{2}\langle\varphi,\varphi\rangle\right)}.$$

Now we would like to define the Donsker delta function in grey noise analysis, namely using the integral representation of the Dirac delta distribution

$$\delta_a(B^{\alpha,\beta}_t) = \frac{1}{2\pi}\int_{\mathbb{R}} e^{ix\langle\cdot, M_{\alpha/2}\mathbb{1}_{[0,t)}\rangle}\, dx, \qquad a \in \mathbb{R},$$

and give sense to the expression. To this end, we introduce the concept of integrable mappings (in a weak sense) with values in $(S(\mathbb{R}))^{-1}_{\mu_\beta}$ which is a consequence of the characterization theorem of $(S(\mathbb{R}))^{-1}_{\mu_\beta}$, see Thm. 4.9 in[28].

Theorem 21. *Let $(T, \mathcal{B}, \nu)$ be a measure space and $\Phi_t \in (S(\mathbb{R}))^{-1}_{\mu_\beta}$ for all $t \in T$. Let $\mathcal{U} \subset S_{\mathbb{C}}(\mathbb{R})$ be an appropriate neighborhood of zero and $C < \infty$ such that:*

(1) $S_{\mu_\beta}\Phi_\cdot(\varphi)\colon T \longrightarrow \mathbb{C}$ is measurable for all $\varphi \in \mathcal{U}$.
(2) $\int_T |S_{\mu_\beta}\Phi_t(\varphi)|\, d\nu(t) \leq C$ for all $\varphi \in \mathcal{U}$.

Then there exists $\Psi \in (S(\mathbb{R}))^{-1}_{\mu_\beta}$ such that for all $\varphi \in \mathcal{U}$

$$S_{\mu_\beta}\Psi(\varphi) = \int_T S_{\mu_\beta}\Phi_t(\varphi)\,d\nu(t).$$

We denote Ψ by $\int_T \Phi_t\,d\nu(t)$ and call it the weak integral of Φ.

We are ready to state the main result on local times of ggBm within grey noise analysis, defined for any $t > 0$ and $a \in \mathbb{R}$ by

$$L_{\alpha,\beta}(a,t) := \int_0^t \delta_a(B_s^{\alpha,\beta})\,ds.$$

Theorem 22 (cf. Thm. 5.3 in[27]). *The local time $L_{\alpha,\beta}(a,t)$ of ggBm with $0 < \alpha < 2$, $0 < \beta < 1$, $0 < t < \infty$ and $a \in \mathbb{R}$ exists as a weak integral in $(S(\mathbb{R}))^{-1}_{\mu_\beta}$ with T_{μ_β}-transform*

$$\left(T_{\mu_\beta}L_{\alpha,\beta}(a,t)\right)(\varphi) = \int_0^t \left(T_{\mu_\beta}\delta_a(B_s^{\alpha,\beta})\right)(\varphi)\,ds$$

for all $\varphi \in \mathcal{U}$.

As a corollary we compute explicitly the expectation of $L_{\alpha,\beta}(0,t)$.

Corollary 2 (cf. Rem. 5.4 in[27]). *The expectation of $L_{\alpha,\beta}(0,t)$ can be calculated explicitly:*

$$\mathbb{E}(L_{\alpha,\beta}(0,t)) = \int_0^t \mathbb{E}(\delta_0(B_s^{\alpha,\beta}))\,ds = \int_0^t \frac{1}{\sqrt{2\pi s^\alpha}}\frac{1}{\Gamma(1-\beta/2)}\,ds$$

$$= \frac{1}{\Gamma(1-\beta/2)\sqrt{2\pi}}\frac{2}{2-\alpha}t^{1-\alpha/2}.$$

This result extends the existence result for ggBm local time in[22].

References

1. S. Albeverio, M. J. Oliveira, and L. Streit. Intersection local times of independent Brownian motions as generalized white noise functionals. *Acta Appl. Math.*, 69(3):221–241, 2001.
2. W. Bock, J. L. Da Silva, and H. P. Suryawan. Local times for multifractional Brownian motion in higher dimensions: A white noise approach. Submitted to IDA-QP, August 2016.
3. S. M. Berman. Local times and sample function properties of stationary Gaussian processes. *Trans. Amer. Math. Soc.*, 137:277–299, 1969.
4. F. Biagini, Y. Hu, B. Øksendal, and T. Zhang. *Stochastic Calculus for Fractional Brownian Motion and Applications*. Probability and Its Applications. Springer, 2007.

5. R. F. Bass and D. Khoshnevisan. Intersection local times and Tanaka formulas. *Ann. Inst. H. Poincaré Probab. Statist.*, 29(3):419–451, 1993.
6. W. Bock, M. J. Oliveira, J. L. da Silva, and L. Streit. Polymer measure: Varadhan's renormalization revisited. 27(3):1550009 (5 pages), June 2015.
7. K. L. Chung and R. J. Williams. *Introduction to Stochastic Integration*. Probability and its Applications. Birkhäuser, 2 edition, 1990.
8. Y. A. Davydov. On local times for random processes. 21(1):171–178, 1976.
9. A. Dvoretzky, P. Erdös, and S. Kakutani. Double points of paths of Brownian motion in *n*-space. *Acta Sci. Math. Szeged*, 12(Leopoldo Fejer et Frederico Riesz LXX annos natis dedicatus, Pars B):75–81, 1950.
10. A. Dvoretzky, P. Erdös, and S. Kakutani. Multiple points of paths of Brownian motion in the plane. *Bull. Res. Council Israel*, 3:364–371, 1954.
11. A. Dvoretzky, P. Erdős, S. Kakutani, and S. J. Taylor. Triple points of Brownian paths in 3-space. *Proc. Cambridge Philos. Soc.*, 53:856–862, 1957.
12. M. de Faria, C. Drumond, and L. Streit. The renormalization of self-intersection local times. I. The chaos expansion. *Infin. Dimens. Anal. Quantum Probab. Relat. Top.*, 3(2):223–236, 2000.
13. M. de Faria, C. Drumond, and L. Streit. The square of self intersection local time of Brownian motion. In F. Gesztesy, H. Holden, J. Jost, S. Paycha, M. Röckner, and S. Scarlatti, editors, *Stochastic processes, physics and geometry: new interplays, I (Leipzig, 1999)*, volume 28 of *CMS Conf. Proc.*, pages 115–122. American Mathematical Soc., Amer. Math. Soc., Providence, RI, 2000.
14. M. de Faria, T. Hida, L. Streit, and H. Watanabe. Intersection local times as generalized white noise functionals. *Acta Appl. Math.*, 46:351–362, 1997.
15. M. de Faria, M. J. Oliveira, and L. Streit. A generalized Clark-Ocone formula. *Random Oper. Stoch. Equ.*, 8(2):163–174, 2000.
16. M. de Faria, A. Rezgui, and L. Streit. Martingale approximation for self-intersection local time of Brownian motion. In S. Albeverio, A. B. de Monvel, and H. Ouerdiane, editors, *Proceedings of the International Conference on Stochastic Analysis and Applications*, pages 95–106. Springer Netherlands, 2004.
17. K. Diethelm. *The analysis of fractional differential equations: An application-oriented exposition using differential operators of Caputo type*, volume 2004. 2010.
18. C. Drumond, M. J. Oliveira, and J. L. Silva. Intersection local times of fractional Brownian motions as generalized white noise functionals. In C. C.. Bernido and V.Č. Bernido, editors, *Stochastic and Quantum Dynamics of Biomolecular Systems*, volume 1021 of *AIP Conference Proceedings*, pages 34–45, Melville, NY: American Institute of Physics (AIP), 2008. AIP Conference Proceedings 1021.
19. T. Deck, J. Potthoff, and G. Våge. A review of white noise analysis from a probabilistic standpoint. *Acta Appl. Math.*, 48(1):91–112, 1997.
20. Y. A. Davydov and A. L. Rozin. On sojourn times for functions and random processes. 23(3):633–637, 1979.
21. C. Drumond. *Tempos locais das auto-intersecções do movimento Browniano.*

PhD thesis, University of Madeira, 2000.

22. J. L. Da Silva and M. Erraoui. Generalized grey brownian motion local time: existence and weak approximation. *Stochastics*, 87(2):347–361, October 2015.

23. E. B. Dynkin. Regularized self-intersection local times of planar Brownian motions. 1:58–74, 1988.

24. S. F. Edwards. The statistical mechanics of polymers with excluded volume. *Proc. Phys. Soc.*, 85:613–624, 1965.

25. D. Geman and J. Horowitz. Occupation densities. 8(1):1–67, 1980.

26. D. Geman, J. Horowitz, and J. Rosen. A local time analysis of intersections of Brownian paths in the plane. 12(1):86–107, 1984.

27. M. Grothaus and F. Jahnert. Mittag-Leffler Analysis II: Application to the fractional heat equation. 270(7):2732–2768, April 2016.

28. M. Grothaus, F. Jahnert, F. Riemann, and J. L. Silva. Mittag-Leffler Analysis I: Construction and characterization. 268(7):1876–1903, April 2015.

29. M. Grothaus, Y. G. Kondratiev, and L. Streit. Complex Gaussian analysis and the Bargmann-Segal space. *Methods Funct. Anal. Topology*, 3(2):46–64, 1997.

30. M. Grothaus, Y. G. Kondratiev, and L. Streit. Regular generalized functions in Gaussian analysis. 2(1):1–25, March 1999.

31. M. Grothaus, M.J. Oliveira, J.L. da Silva, and L. Streit. Self-avoiding fractional brownian motionthe edwards model. *Journal of Statistical Physics*, 145(6):1513–1523, 2011.

32. M. Gradinaru, F. Russo, and P. Vallois. Generalized covariations, local time and Stratonovich Itô's formula for fractional Brownian motion with Hurst index $H \geq \frac{1}{4}$. 31(4):1772–1820, 2003.

33. T. Hida. *Brownian Motion*, volume 11 of *Applications of Mathematics*. New York-Berlin, 1980.

34. Y. Hu and D. Nualart. Renormalized self-intersection local time for fractional Brownian motion. 33(3):948–983, 2005.

35. Y. Hu and D. Nualart. Regularity of renormalized self-intersection local time for fractional Brownian motion. *Commun. Inf. Syst.*, 7(1):21–30, 2007.

36. S. W. He, W. Q. Yang, R. Q. Yao, and J. G. Wang. Local times of self-intersection for multidimensional Brownian motion. *Nagoya Math. J.*, 138:51–64, 1995.

37. P. Imkeller, V. Perez-Abreu, and J. Vives. Chaos expansions of double intersection local time of Brownian motion in $\mathbb{R}^d$ and renormalization. *Stochastic Process. Appl.*, 56(1):1–34, 1995.

38. R. Jenane, R. Hachaichi, and L. Streit. Renormalisation du temps local des points triples du mouvement Brownien. 9(04):547–566, 2006.

39. J. Jacod and A.Ñ. Shiryaev. *Limit theorems for stochastic processes*. Springer-Verlag, Berlin, second edition, 2003.

40. S. Kakutani. On Brownian motions in n-space. *Proc. Imp. Acad. Tokyo*, 20(9):648–652, 1944.

41. Y. G. Kondratiev, P. Leukert, and L. Streit. Wick calculus in Gaussian analysis. *Acta Appl. Math.*, 44:269–294, 1996.

42. H.-H. Kuo and N.-R. Shieh. A generalized Itô's formula for multidimensional

Brownian motions and its applications. *Chinese J. Math.*, 15(3):163–174, 1987.

43. I. Karatzas and S.É. Shreve. *Brownian motion and stochastic calculus.* Graduate Texts in Mathematics. New York Berlin, 2 edition, 1991.

44. A. A. Kilbas, H. M. Srivastava, and J. J. Trujillo. *Theory and applications of fractional differential equations*, volume 204 of *North-Holland Mathematics Studies*. Elsevier Science B.V., Amsterdam, 2006.

45. M.P̃. Lévy. Le mouvement Brownien plan. *Amer. J. Math.*, 62(1):487–550, 1940.

46. P. Lévy. *Processus Stochastiques et Mouvement Brownien. Suivi d'une note de M. Loève.* Gauthier-Villars, Paris, 1948.

47. J.-F. Le Gall. Sur le temps local d'intersection du mouvement Brownien plan et la méthode de renormalisation de Varadhan. 1123:314–331, 1985.

48. J.-F. Le Gall. Sur la saucisse de Wiener et les points multiples du mouvement Brownien. 14(4):1219–1244, 1986.

49. A Lascheck, Peter Leukert, Ludwig Streit, and W. Westerkamp. More about Donsker's delta function. *Soochow J. Math.*, 20(math-ph/0303011):401–418, March 1994.

50. F. Mainardi. Fractional calculus: some basic problems in continuum and statistical mechanics. In *Fractals and fractional calculus in continuum mechanics (Udine, 1996)*, volume 378 of *CISM Courses and Lectures*, pages 291–348. Springer, Vienna, 1997.

51. Y. S. Mishura. *Stochastic Calculus for Fractional Brownian Motion and Related Processes.* Lecture Notes in Mathematics. Springer-Verlag Berlin Heidelberg, Berlin, Heidelberg, 2008.

52. S. Mendonça and L. Streit. Multiple intersection local times in terms of white noise. *Infin. Dimens. Anal. Quantum Probab. Relat. Top.*, 4(4):533–543, 2001.

53. D. Nualart and S. Ortiz-Latorre. Intersection local time for two independent fractional Brownian motions. *J. Theoret. Probab.*, 20(4):759–767, 2007.

54. D. Nualart and J. Vives. Smoothness of local time and related Wiener functional. In V. Houdre, C. Perez-Abreu, editor, *Chaos Expansions, Multiple Wiener-Itô Integrals and their Applications*, volume 1 of *Probability and Stochastics Series*, page 317. CRC Press, 1994.

55. B. Oksendal, K. Aase, N. Privault, and J. Ubøe. White noise generalizations of the Clark-Haussmann-Ocone theorem with application to mathematical finance. *Finance Stoch.*, 4(4):465–496, July 2000.

56. H. Ouerdiane and A. Rezgui. Distributions gaussiennes et temps locaux d'intersection. *Rev. Mat. Complut.*, 13(2):351–366, 2000.

57. S. Orey. Gaussian sample functions and the Hausdorff dimension of level crossings. *Z. Wahrsch. verw. Gebiete*, 15(3):249–256, 1970.

58. M. J. Oliveira, J. L. Silva, and L. Streit. Intersection local times of independent fractional Brownian motions as generalized white noise functionals. *Acta Appl. Math.*, 113(1):17–39, 2011.

59. V. Pérez-Abreu. Chaos expansions: A review. *Resenhas do Instituto de Matemática e Estatística da Universidade de São Paulo*, 1(2/3):335–359, 1994.

60. L.D. Pitt. Local times for Gaussian vector fields. *Indiana Univ. Math. J.*, 27(2):309–330, 1978.

61. I. Podlubny. *Fractional differential equations. An introduction to fractional derivatives, fractional differential equations, to methods of their solution and some of their applications*, volume 198 of *Mathematics in Science and Engineering*. Academic Press Inc., San Diego, CA, 1999.

62. J. Potthoff and M. Timpel. On a dual pair of spaces of smooth and generalized random variables. *Potencial Anal.*, 4(6):637–654, 1995.

63. V. Pipiras and M.S. Taqqu. Integration questions related to fractional Brownian motion. *Probab. Theory Related Fields*, 118(2):251–291, 2000.

64. A. Rezgui. The renormalization of self intersection local times of Brownian motion. *Math. Sci. Q. J.*, 2(1):19–32, 2008.

65. J. Rosen. A local time approach to the self-intersections of Brownian paths in spaces. 88:327–338, 1983.

66. J. Rosen. A renormalized local time for multiple intersections of planar Brownian motions. 1204:515–531, 1986.

67. J. Rosen. The intersection local time of fractional Brownian motion in the plane. *J. Multivar. Anal.*, 23(1):37–46, 1987.

68. D. Revuz and M. Yor. *Continuous martingales and Brownian motion*, volume 293 of *Grundlehren der Mathematischen Wissenschaften [Fundamental Principles of Mathematical Sciences]*. Berlin, 3rd edition, 1999.

69. W.R. Schneider. Grey noise. In S. Albeverio, J.-E. Fenstad, H. Holden, and T. Lindstrøm, editors, *Ideas and methods in mathematical analysis, stochastics, and applications (Oslo, 1988)*, pages 261–282. Cambridge Univ. Press, Cambridge, 1992.

70. Lothar Schäfer. *Excluded Volume effects in Polymer Solutions: as Explained by the Renormalization Group*. Springer, 2012.

71. N.-R. Shieh. A W.N.C. viewpoint on intersection local times. In K. Itô, editor, *Gaussian random fields (Nagoya, 1990)*, volume 1, pages 346–353. World Sci. Publ., River Edge, NJ, 1991.

72. N.-R. Shieh. White noise analysis and Tanaka formula for intersections of planar Brownian motion. *Nagoya Math. J.*, 122:1–17, June 1991.

73. S. G. Samko, A. A. Kilbas, and O. I. Marichev. *Fractional integrals and derivatives*. Gordon and Breach Science Publishers, Yverdon, 1993. Theory and applications, Edited and with a foreword by S. M. Nikol'skiĭ, Translated from the 1987 Russian original, Revised by the authors.

74. L. Streit. Introduction to white noise analysis. In A.L. Cardoso, M. de Faria, J. Potthoff, R. Sénéor, and L. Streit, editors, *Stochastic Analysis and Applications*, pages 415–440, Drodrecht, 1994.

75. H.F. Trotter. A property of Brownian motion paths. *Illinois J. Math.*, 2(3):425–433, 1958.

76. S. R.S. Varadhan. Appendix to "Euclidean quantum field theory" by K. Symanzik. In R. Jost, editor, *Local Quantum Theory*, New York, 1969.

77. H. Watanabe. The local time of self-intersections of Brownian motions as generalized Brownian functionals. *Lett. Math. Phys.*, 23(1):1–9, 1991.

78. M. J. Westwater. On Edwards' model for long polymer chains. *Comm. Math. Phys.*, 72(2):131–174, 1980.

79. M. Yor. Renormalisation et convergence en loi pour les temps locaux d'intersection du mouvement Brownien dans $\mathbb{R}^3$. In *Séminaire de probabilités, XIX, 1983/84*, volume 1123 of *Lecture Notes in Math.*, pages 350–365. Springer, Berlin, 1985.

80. M. Yor. Sur la représentation comme intégrales stochastiques des temps d'occupation du mouvement Brownien dans $\mathbb{R}^d$. In J. Azéma and M. Yor, editors, *Séminaire de Probabilités, XX, 1984/85*, volume 1204 of *Lecture Notes Math.*, pages 543–552. Springer, 1986.

Chapter 5

White Noise Analysis and the Chern-Simons Path Integral

Atle Hahn

Grupo de Física Matemática da Universidade de Lisboa
Av. Prof. Gama Pinto, 2
PT-1649-003 Lisboa, Portugal
Email: atle.hahn@gmail.com

1. Introduction

In the 1980s and the beginning of the 1990s Jones, Witten, Drinfeld, Turaev, Reshetikhin, Kontsevich, and others revolutionized and created a whole new area which is now called "Quantum Topology". Quantum Topology is both a deep and a beautiful theory, beautiful in the sense that it naturally connects a large number of branches of Mathematics and Physics[a] like Algebra (Lie algebras, affine Lie algebras, quantum groups, ..., and the corresponding representation theories), low-dimensional Topology & Knot Theory, Riemannian Geometry & Global Analysis, Infinite Dimensional Analysis, and Quantum Field Theory (Gauge Field Theory, Conformal Field Theory, Quantum Gravity, String Theory).

The first major step towards Quantum Topology was the discovery of the Jones polynomial and its generalizations (in particular, the HOMFLY and the Kauffman polynomials) in 1984 and 1985. In 1988, Witten demonstrated in a celebrated paper[28] that the heuristic Feynman path integral associated to a certain 3-dimensional gauge field theory can be used to give a very elegant and intrinsically 3-dimensional "definition" of the Jones polynomial and the other knot polynomials mentioned above. The

---

[a]This is reflected also in the present chapter: Quantum Gauge Field Theory plays the main role in Sec. 2.2 and Sec. 2.3, Riemannian Geometry/Global Analysis in Sec. 2.5, Infinite Dimensional Analysis in Sec. 3, and Algebra and low-dimensional Topology in Appendix A

aforementioned gauge theory is the so-called (pure) "Chern-Simons model" which is specified by a triple (M, G, k) where M is an oriented connected 3-dimensional manifold (usually compact), G is a semi-simple Lie group (often compact and simply-connected), and $k \in \mathbb{N}$ is a fixed parameter ("the level"). In the very important special case $G = SU(N)$, $N \geq 2$, the action function S_{CS} of the Chern-Simons model is given explicitly by

$$S_{CS}(A) = k \int_M \text{Tr}(A \wedge dA + \tfrac{2}{3} A \wedge A \wedge A), \qquad A \in \mathcal{A} \qquad (1)$$

where $\mathcal{A}$ is the space of all $su(N)$-valued 1-forms A on M, $\wedge$ the wedge product wedge product associated to the multiplication of $\text{Mat}(N, \mathbb{C})$, and $\text{Tr} : \text{Mat}(N, \mathbb{C}) \to \mathbb{C}$ the suitably normalized trace.

The Chern-Simons "path space measure" is the informal complex measure $\exp(iS_{CS}(A))DA$ where DA is the (ill-defined) Lebesgue measure on the infinite dimensional space $\mathcal{A}$. The Chern-Simons path integral (functional) $\int \cdots \exp(iS_{CS}(A))DA$ not only gives a unifying framework for the aforementioned knot polynomials[b] but also leads naturally to a generalization of these invariants to all closed 3-manifolds, the "Jones-Witten invariants".

Using several heuristic arguments, some of them from Conformal Field Theory, Witten was able to evaluate the Jones-Witten invariants explicitly. Two years later Reshetikhin and Turaev finally found a rigorous (and equivalent) version of the Jones-Witten invariants, the so-called "Reshetikhin-Turaev invariants",[22,23]. The approach in[22,23] is algebraic and very different from the path integral approach. In particular, it is based on the representation theory of quantum groups and "surgery operations" on the relevant 3-manifolds. Turaev[26] later found an equivalent approach called the "shadow world" approach, which is also based on quantum group representations but uses certain finite "state sums" instead of the surgery operations.

Many open questions in the field of Quantum Topology are closely related to the following problem which is generally considered to be one of the major open problems in the field (cf.[18] und[24]):

(P1) Find a rigorous realization[c] of the (original or gauge fixed) Chern-Simons path integral expressions for all simply-connected compact Lie

---

[b]The polynomial invariants mentioned above are given by WLO(L) as defined in Eq. (6) below. In the special case $M = S^3$, $G = SU(2)$, and where each representation ρ_i appearing in Eq. (6) is the fundamental representation of $SU(2)$ WLO(L) is given by an explicit expression involving the Jones polynomial of L.
[c]Here the word "realization" is meant to imply that the values of the rigorously defined

groups G and all (compact oriented 3-dimensional) base manifolds M.

Since (P1) seems to be a very hard problem it makes sense to restrict oneself first to the following weakened version of (P1).

(P1)' Find a rigorous realization of the (suitably gauge fixed) Chern-Simons path integral expressions for all compact and simply-connected Lie groups G and some fixed base manifold M like, e.g., $M = S^3$ or $M = S^2 \times S^1$.

Note that the quadratic part of the Chern-Simons action function S_{CS} is degenerate. Moreover, there is a cubic term, which is clearly not semi-bounded. Because of this we cannot hope to be able to transform the (original or gauge-fixed) Chern-Simons path space measure into a bounded positive (heuristic) measure by applying a so-called "Wick rotation"[d] This means that the well-established theory of positive bounded measures on infinite dimensional topological vector spaces is not as useful as in standard Constructive QFT. Instead of working with bounded measures one can try to work with suitable distributions, for example Hida distributions.

And indeed, as we will show in the present chapter, White Noise Analysis can be used to make progress towards the solution of Problem (P1)'. More precisely, we consider the special manifold $M = S^2 \times S^1$ and we will combine the "torus gauge fixing" approach to Chern-Simons theory on $M = S^2 \times S^1$ which was developed in[2–4,8–10,12] with several ideas from[1,7,19] (cf. also[8,10,11] for earlier work in this direction). By doing so we will obtain a rigorous realization of the corresponding Jones-Witten invariants (= the "Wilson loop observables" WLO(L) of Chern-Simons theory, cf. Footnote b). We expect that at least for a simple class of (ribbon) links L the aforementioned rigorous realization reproduces the Reshetikhin-Turaev invariants, cf. Conjecture 2 in Sec. 3 and Remark 2 in Sec. 2.4 below.

The present chapter is organized as follows:

In Secs. 2.1–2.3 we recall the relevant heuristic formulas for the Wilson loop observables in Chern-Simons theory, first the original formula Eq. (6) and later the modified formula which was obtained in[12] by applying torus gauge fixing in the special case where $M = \Sigma \times S^1$, cf. Eq. (8). In Secs.

Chern-Simons path integral expressions coincide with the corresponding Reshetikhin-Turaev invariant.

[d]This is different from the situation in many other bosonic QFTs where the path space measure can be transformed into a bounded positive (heuristic) measure by means of a Wick rotation.

2.4–2.6 we then rewrite the heuristic formula Eq. (8) in a suitable way (for the special case $\Sigma = S^2$) and after doing so arrive at another heuristic formula, namely Eq. (46), on which the rest of this chapter is based.

In Sec. 3 we then explain how one can make rigorous sense of the RHS of the aforementioned Eq. (46). In Sec. 4 we conclude the main part of this chapter with a list of open questions.

In Appendix A we briefly recall the definition of Turaev's shadow invariant in the special case relevant for us, i.e. $M = S^2 \times S^1$. In Appendix B we fill in some technical details which were omitted in Sec. 3.

2. The heuristic Chern-Simons path integral in the torus gauge

We fix a simple simply-connected compact Lie group G and a maximal torus T of G. By $\mathfrak{g}$ and $\mathfrak{t}$ we will denote the Lie algebras of G and T and by $\langle \cdot, \cdot \rangle_{\mathfrak{g}}$ or simply by $\langle \cdot, \cdot \rangle$ the unique Ad-invariant scalar product on $\mathfrak{g}$ satisfying the normalization condition $\langle \check{\alpha}, \check{\alpha} \rangle = 2$ for every short coroot $\check{\alpha}$ w.r.t. $(\mathfrak{g}, \mathfrak{t})$.

Moreover, we will fix a compact oriented 3-manifold M of the form $M = \Sigma \times S^1$ where Σ is a compact oriented surface. Finally, we fix an (ordered and oriented) "link" L in M, i.e. a finite tuple $L = (l_1, \ldots, l_m)$, $m \in \mathbb{N}$, of pairwise non-intersecting knots l_i and we equip each l_i with a "color", i.e. a finite-dimensional complex representation ρ_i of G. Recall that a "knot" in M is an embedding $l : S^1 \to M$. Using the surjection $[0,1] \ni t \mapsto e^{2\pi i t} \in S^1 \cong \{z \in \mathbb{C} \mid |z| = 1\}$ we can consider each knot as a loop $l : [0,1] \to M$, $l(0) = l(1)$, in the obvious way.

2.1. Basic spaces

As in[12] we will use the following notation

$$\mathcal{B} = C^\infty(\Sigma, \mathfrak{t}) \tag{2a}$$

$$\mathcal{A} = \Omega^1(M, \mathfrak{g}) \tag{2b}$$

$$\mathcal{A}_\Sigma = \Omega^1(\Sigma, \mathfrak{g}) \tag{2c}$$

$$\mathcal{A}_{\Sigma, \mathfrak{t}} = \Omega^1(\Sigma, \mathfrak{t}), \quad \mathcal{A}_{\Sigma, \mathfrak{k}} = \Omega^1(\Sigma, \mathfrak{k}) \tag{2d}$$

$$\mathcal{A}^\perp = \{A \in \mathcal{A} \mid A(\partial/\partial t) = 0\} \tag{2e}$$

$$\check{\mathcal{A}}^\perp = \{A^\perp \in \mathcal{A}^\perp \mid \int A^\perp(t)dt \in \mathcal{A}_{\Sigma, \mathfrak{k}}\} \tag{2f}$$

$$\mathcal{A}_c^\perp = \{A^\perp \in \mathcal{A}^\perp \mid A^\perp \text{ is constant and } \mathcal{A}_{\Sigma, \mathfrak{t}}\text{-valued}\} \tag{2g}$$

Here $\Omega^p(N, V)$ denotes the space of V-valued p-forms on a smooth manifold N, $\mathfrak{k}$ is the orthogonal complement of $\mathfrak{t}$ in $\mathfrak{g}$ w.r.t. $\langle \cdot, \cdot \rangle$, dt is the normalized translation-invariant volume form on S^1, and $\partial/\partial t$ is the vector field on $M = \Sigma \times S^1$ obtained by "lifting" in the obvious way the normalized translation-invariant vector field $\partial/\partial t$ on S^1. Moreover, in Eqs. (2f) and (2g) we used the "obvious" identification (cf. Sec. 2.3.1 in[12])

$$\mathcal{A}^\perp \cong C^\infty(S^1, \mathcal{A}_\Sigma) \tag{3}$$

where $C^\infty(S^1, \mathcal{A}_\Sigma)$ is the space of maps $f : S^1 \to \mathcal{A}_\Sigma$ which are "smooth" in the sense that $\Sigma \times S^1 \ni (\sigma, t) \mapsto (f(t))(X_\sigma) \in \mathfrak{g}$ is smooth for every smooth vector field X on Σ. It follows from the definitions above that

$$\mathcal{A}^\perp = \check{\mathcal{A}}^\perp \oplus \mathcal{A}_c^\perp. \tag{4}$$

2.2. *The heuristic Wilson loop observables*

The Chern-Simons action function $S_{CS} : \mathcal{A} \to \mathbb{R}$ associated to M, G, and the "level" $k \in \mathbb{Z}\backslash\{0\}$ is given by[e]

$$S_{CS}(A) = -k\pi \int_M \langle A \wedge dA \rangle + \tfrac{1}{3}\langle A \wedge [A \wedge A]\rangle, \quad A \in \mathcal{A} \tag{5}$$

where $[\cdot \wedge \cdot]$ denotes the wedge product associated to the Lie bracket $[\cdot, \cdot] : \mathfrak{g} \times \mathfrak{g} \to \mathfrak{g}$ and where $\langle \cdot \wedge \cdot \rangle$ denotes the wedge product associated to the scalar product $\langle \cdot, \cdot \rangle : \mathfrak{g} \times \mathfrak{g} \to \mathbb{R}$.

Recall that the heuristic Wilson loop observable $\text{WLO}(L)$ of a link $L = (l_1, l_2, \ldots, l_m)$ in M with "colors" $(\rho_1, \rho_2, \ldots, \rho_m)$ is given by the informal "path integral" expression

$$\text{WLO}(L) := \int_\mathcal{A} \prod_i \text{Tr}_{\rho_i}(\text{Hol}_{l_i}(A)) \exp(iS_{CS}(A))DA \tag{6}$$

where $\text{Hol}_l(A) \in G$ is the holonomy of $A \in \mathcal{A}$ around the loop $l \in \{l_1, \ldots, l_m\}$ and where DA is the (ill-defined) "Lebesgue measure" on the infinite-dimensional space $\mathcal{A}$. We will use the following explicit formula for $\text{Hol}_l(A)$

$$\text{Hol}_l(A) = \lim_{n\to\infty} \prod_{j=1}^n \exp\left(\tfrac{1}{n}A(l'(t))\right)_{|t=j/n} \tag{7}$$

where $\exp : \mathfrak{g} \to G$ is the exponential map of G.

---

[e]Eq. (5) generalizes Eq. (1) in Sec. 1. In Eq. (1) the factor $-\pi$ is hidden in the trace functional $\text{Tr} : \text{Mat}(N, \mathbb{C}) \to \mathbb{C}$. Moreover, in Eq. (1) we have a factor $2/3$ instead of $1/3$ because the wedge product $\wedge$ in Eq. (1) differs from each of the two wedge products in Eq. (5).

2.3. *The basic heuristic formula from* [12]

The starting point for the main part of[12] was a second heuristic formula for
WLO(L) which one obtains from Eq. (6) above after applying a suitable
gauge fixing.

Let $\pi_\Sigma : \Sigma \times S^1 \to \Sigma$ be the canonical projection. For each loop l_i
appearing in the link L we set $l^i_\Sigma := \pi_\Sigma \circ l_i$. Moreover, we fix $\sigma_0 \in \Sigma$ such
that

$$\sigma_0 \notin \text{arc}(L_\Sigma) := \bigcup_i \text{arc}(l^i_\Sigma).$$

By applying "abstract torus gauge fixing" (cf. Sec. 2.2.4 in[12]) and a suitable
change of variable one can derive at a heuristic level (cf. Eq. (2.53) in[12])

$$\begin{aligned}
\text{WLO}(L) \sim \sum_{y \in I} \int_{\mathcal{A}^\perp_c \times \mathcal{B}} \Big\{ & 1_{\mathcal{B}_{reg}}(B)\text{Det}_{FP}(B) \\
& \times \Big[\int_{\check{\mathcal{A}}^\perp} \prod_i \text{Tr}_{\rho_i}\left(\text{Hol}_{l_i}(\check{A}^\perp + A^\perp_c, B)\right) \exp(iS_{CS}(\check{A}^\perp, B)) D\check{A}^\perp \Big] \\
& \times \exp\big(-2\pi ik\langle y, B(\sigma_0)\rangle\big) \Big\} \exp(iS_{CS}(A^\perp_c, B))(DA^\perp_c \otimes DB) \quad (8)
\end{aligned}$$

where "$\sim$" denotes equality up to a multiplicative "constant"[f] C, where
$I := \ker(\exp_{|\mathfrak{t}}) \subset \mathfrak{t}$, and where DB and $DA^\perp_c$ are the informal "Lebesgue
measures" on the infinite-dimensional spaces $\mathcal{B}$ and $\mathcal{A}^\perp_c$. Moreover, we have
set

$$\mathcal{B}_{reg} := \{B \in \mathcal{B} \mid \forall \sigma \in \Sigma : B(\sigma) \in \mathfrak{t}_{reg}\} = C^\infty(\Sigma, \mathfrak{t}_{reg}) \quad (9)$$

with $\mathfrak{t}_{reg} := \exp^{-1}(T_{reg})$ where T_{reg} is the set of "regular" elements of
T, i.e. T_{reg} is the set of all $t \in T$ which are not contained in a different
maximal torus T'.

Moreover, we have set for each $B \in \mathcal{B}$, $A^\perp \in \mathcal{A}^\perp$

$$S_{CS}(A^\perp, B) := S_{CS}(A^\perp + Bdt) \quad (10)$$

$$\text{Hol}_l(A^\perp, B) := \text{Hol}_l(A^\perp + Bdt) \quad (11)$$

Here dt is the real-valued 1-form on $M = \Sigma \times S^1$ obtained by pulling back
the 1-form dt on S^1 by means of the canonical projection $\pi_{S^1} : \Sigma \times S^1 \to S^1$.
Finally, $\text{Det}_{FP}(B)$ is the informal expression given by

$$\text{Det}_{FP}(B) := \det\big(1_{\mathfrak{k}} - \exp(\text{ad}(B))_{|\mathfrak{k}}\big) \quad (12)$$

---

[f] "Constant" in the sense that C does not depend on L. By contrast, C may depend on
G, Σ, and k.

where $1_{\mathfrak{k}} - \exp(\mathrm{ad}(B))_{|\mathfrak{k}}$ is the linear operator on $C^{\infty}(\Sigma, \mathfrak{k})$ given by

$$(1_{\mathfrak{k}} - \exp(\mathrm{ad}(B))_{|\mathfrak{k}} \cdot f)(\sigma) = (\mathrm{id}_{\mathfrak{k}} - \exp(\mathrm{ad}(B(\sigma)))_{|\mathfrak{k}}) \cdot f(\sigma)$$

for all $f \in C^{\infty}(\Sigma, \mathfrak{k})$ and $\sigma \in \Sigma$ where on the RHS $\mathrm{id}_{\mathfrak{k}}$ is the identity on $\mathfrak{k}$.

For the rest of this chapter we will now fix an auxiliary Riemannian metric $\mathbf{g} = \mathbf{g}_{\Sigma}$ on Σ. After doing so we obtain scalar products $\ll \cdot, \cdot \gg_{\mathcal{A}_{\Sigma}}$ and $\ll \cdot, \cdot \gg_{\mathcal{A}^{\perp}}$ on $\mathcal{A}_{\Sigma}$ and $\mathcal{A}^{\perp} \cong C^{\infty}(S^1, \mathcal{A}_{\Sigma})$ in a natural way. Moreover, we obtain a well-defined Hodge star operator $\star : \mathcal{A}_{\Sigma} \to \mathcal{A}_{\Sigma}$ which induces an operator $\star : C^{\infty}(S^1, \mathcal{A}_{\Sigma}) \to C^{\infty}(S^1, \mathcal{A}_{\Sigma})$ in the obvious way, i.e. by $(\star A^{\perp})(t) = \star(A^{\perp}(t))$ for all $A^{\perp} \in \mathcal{A}^{\perp}$ and $t \in S^1$. We have the following explicit formula (cf. Eq. (2.48) in[12])

$$S_{CS}(A^{\perp}, B) = \pi k \ll A^{\perp}, \star(\tfrac{\partial}{\partial t} + \mathrm{ad}(B))A^{\perp} \gg_{\mathcal{A}^{\perp}} + 2\pi k \ll \star A^{\perp}, dB \gg_{\mathcal{A}^{\perp}} \tag{13}$$

for all $B \in \mathcal{B}$ and $A^{\perp} \in \mathcal{A}^{\perp}$, which implies

$$S_{CS}(\check{A}^{\perp}, B) = \pi k \ll \check{A}^{\perp}, \star(\tfrac{\partial}{\partial t} + \mathrm{ad}(B))\check{A}^{\perp} \gg_{\mathcal{A}^{\perp}} \tag{14}$$

$$S_{CS}(A_c^{\perp}, B) = 2\pi k \ll \star A_c^{\perp}, dB \gg_{\mathcal{A}^{\perp}} \tag{15}$$

for $B \in \mathcal{B}$, $\check{A}^{\perp} \in \check{\mathcal{A}}^{\perp}$, and $A_c^{\perp} \in \mathcal{A}_c^{\perp}$.

The following informal definitions will be useful in Sec. 2.4 below: For each $B \in \mathcal{B}$ we set

$$\check{Z}(B) := \int \exp(iS_{CS}(\check{A}^{\perp}, B)) D\check{A}^{\perp}, \tag{16}$$

$$d\mu_B^{\perp} := \tfrac{1}{\check{Z}(B)} \exp(iS_{CS}(\check{A}^{\perp}, B)) D\check{A}^{\perp} \tag{17}$$

Moreover, we will denote by $d_{\mathbf{g}}$ the distance function on Σ and by $d\mu_{\mathbf{g}}$ the volume measure on Σ which are associated to the Riemannian metric $\mathbf{g}$.

2.4. *Ribbon version of Eq.* (8)

It is well-known in the mathematics and physics literature on quantum 3-manifold invariants that rather than working with links one actually has to work with framed links or, equivalently, with ribbon links (see below) if one wants to get meaningful results.

From the knot theory point of view the framed link picture and the ribbon link picture are equivalent. However, the ribbon picture seems to be better suited for the study of the Chern-Simons path integral in the torus gauge.

For every (closed) ribbon R in $\Sigma \times S^1$, i.e. every smooth embedding $R : S^1 \times [0,1] \to \Sigma \times S^1$, we define

$$\text{Hol}_R(A) := \lim_{n \to \infty} \prod_{j=1}^{n} \exp\left(\tfrac{1}{n} \int_0^1 A(R'_u(t)) du\right)_{|t=j/n} \in G$$

where R_u, for $u \in [0,1]$, is the loop in $\Sigma \times S^1$ given by $R_u(t) = R(t,u)$ for all $t \in S^1$.

A ribbon link in $\Sigma \times S^1$ is a finite tuple of non-intersecting closed ribbons in $\Sigma \times S^1$. We will replace the link $L = (l_1, l_2, \ldots, l_m)$ by a ribbon link $L_{ribb} = (R_1, R_2, \ldots, R_m)$ where each R_i, $i \leq m$, is chosen such that $l_i(t) = R_i(t, 1/2)$ for all $t \in S^1$.

Moreover, we "scale" each R_i, i.e. for each $s \in (0,1)$ we introduce the ribbon $R_i^{(s)}$ by $R_i^{(s)}(t,u) := R_i(t, s \cdot (u - 1/2) + 1/2)$ for all $t \in S^1$ and $u \in [0,1]$.

Then we replace $\text{Hol}_{l_i}(A)$ appearing in Eq. (8) (with $A = \check{A}^\perp + A_c^\perp + Bdt$) by $\text{Hol}_{R_i^{(s)}}(A)$. Moreover, we include a $s \to 0$ limit.

Convention 1. We will usually write simply L instead of L_{ribb} when no confusion can arise.

Remark 1. The inclusion of the limit $s \to 0$ above is the formal implementation of the intuitive idea that our ribbons should have "infinitesimal width". If one does not include this limit $s \to 0$ it may still be possible to derive a result like Eq. (70) below for ribbon links L fulfilling Assumption 1 but most probably not for general ribbon links.

After these preparations we arrive at the following ribbon analogue of Eq. (8) above

$$\text{WLO}(L) \sim \lim_{s \to 0} \sum_{y \in I} \int_{\mathcal{A}_c^\perp \times \mathcal{B}} \left\{ 1_{\mathcal{B}_{reg}}(B) \text{Det}_{FP}(B) \check{Z}(B) \right.$$

$$\times \left[\int_{\check{\mathcal{A}}^\perp} \prod_i \text{Tr}_{\rho_i} \left(\text{Hol}_{R_i^{(s)}}(\check{A}^\perp, A_c^\perp, B) \right) d\mu_B^\perp(\check{A}^\perp) \right]$$

$$\left. \times \exp\left(-2\pi i k \langle y, B(\sigma_0) \rangle\right) \right\} \exp(i S_{CS}(A_c^\perp, B))(DA_c^\perp \otimes DB) \quad (18)$$

where we have set (cf. Sec. 2.3 above)

$$\text{Hol}_{R_i^{(s)}}(\check{A}^\perp, A_c^\perp, B) := \text{Hol}_{R_i^{(s)}}(\check{A}^\perp + A_c^\perp + Bdt).$$

In the following we set

$$R_\Sigma^i := \pi_\Sigma \circ R_i.$$

From now on we will restrict ourselves to ribbon links $L = (R_1, R_2, \ldots, R_m)$ fulfilling the following assumption.

Assumption 1. The maps R^i_Σ, $i \leq m$, neither intersect themselves nor each other. More precisely: Each R^i_Σ, $i \leq m$, is an injection $S^1 \times [0,1] \to \Sigma$ and we have $\mathrm{Image}(R^i_\Sigma) \cap \mathrm{Image}(R^j_\Sigma) = \emptyset$ if $i \neq j$.

Remark 2. It is interesting to consider also the weakened version of Assumption 1 where instead of demanding that for each $i \leq m$ the map $R^i_\Sigma :$ $S^1 \times [0,1] \to \Sigma$ is an injection we only demand that $R^i_\Sigma(t_1, u_1) \neq R^i_\Sigma(t_2, u_2)$ for all $t_1, t_2 \in S^1$ and all $u_1, u_2 \in [0,1]$ fulfilling $u_1 \neq u_2$ and that for fixed $u \in [0,1]$ the Image of $S^1 \ni t \mapsto R^i_\Sigma(t, u) \in \Sigma$ lies on an embedded circle in Σ.

Observe that this includes a certain class of torus (ribbon) knots. I expect that the obvious generalization of Conjecture 1 below will also hold for the aforementioned weakened version of Assumption 1 (if combined with a suitably modified version of Assumption 2 below). Moreover, I expect that Conjecture 2 below can be generalized to this more general situation and that by doing so one can obtain a continuum analogue of Theorem 5.7 in[14].

2.5. *Definition of* $\mathrm{Det}_{rig}(B)$

We will now explain how, using a suitable "heat kernel regularization", one can make rigorous sense of the expression

$$\mathrm{Det}(B) := \mathrm{Det}_{FP}(B)\check{Z}(B) \tag{19}$$

and how one can evaluate the rigorous version $\mathrm{Det}_{rig}(B)$ of $\mathrm{Det}(B)$ explicitly. Here $B \in \mathcal{B}$ is fixed and $\mathrm{Det}_{FB}(B)$ and $\check{Z}(B)$ are given as in Eq. (12) and Eq. (16) above.

Remark 3. The approach which we use here is a simplified version of the approach in Sec. 6 in[2]. The main difference is that we use the exponentials $e^{-\epsilon \Delta_i}$ of the original (="plain") Hodge Laplacians Δ_i while in Sec. 6 in[2] "covariant Hodge Laplacians" are used. The use of the covariant Hodge Laplacians produces an additional term containing the dual Coxeter number $c_\mathfrak{g}$ of $\mathfrak{g}$. The overall effect in the simple situation in[2] where only "vertical links" (see the paragraph after Remark 4 below) are used is a "shift" $k \to k + c_\mathfrak{g}$, in agreement with the shift predicted in Witten's original paper[28]. In the case of general links it is doubtful that the use of covariant Hodge Laplacian can produce a shift $k \to k + c_\mathfrak{g}$ in all places where this would be necessary. On the other hand, the fact that by working with the "plain" Hodge Laplacians we do not get a shift $k \to k + c_{\mathfrak{g}}$ should not be a cause for concern. It seems to be generally accepted nowadays that the occurrence and magnitude of the shift in k will depend on the regularization procedure and renormalization prescription which is applied (cf. Remark 3.2 in[12]).

Informally, we have

$$\check{Z}(B) \sim \left| \det\left(\tfrac{\partial}{\partial t} + \operatorname{ad}(B) \right) \right|^{-1/2} = \det\left(1_{\mathfrak{k}} - \exp(\operatorname{ad}(B))_{|\mathfrak{k}} \right)^{-1/2} \tag{20}$$

where $\tfrac{\partial}{\partial t} + \operatorname{ad}(B)$ is as in Eq. (14) above and where $1_{\mathfrak{k}} - \exp(\operatorname{ad}(B))_{|\mathfrak{k}}$ is the linear operator on $\mathcal{A}_{\Sigma,\mathfrak{k}} = \Omega^1(\Sigma, \mathfrak{k})$ given by

$$\forall \alpha \in \mathcal{A}_{\Sigma,\mathfrak{k}} : \forall \sigma \in \Sigma : \forall X_\sigma \in T_\sigma \Sigma :$$
$$(1_{\mathfrak{k}} - \exp(\operatorname{ad}(B))_{|\mathfrak{k}} \cdot \alpha)(X_\sigma) = (\operatorname{id}_{\mathfrak{k}} - \exp(\operatorname{ad}(\mathrm{B}(\sigma))_{|\mathfrak{k}}) \cdot \alpha(X_\sigma) \tag{21}$$

with $\operatorname{id}_{\mathfrak{k}}$ and $\exp(\operatorname{ad}(B(\sigma)))$ as in Sec. 2.3 above.

Now observe that for $b \in \mathfrak{t}$ we have (cf. Eq. A.2 in Appendix A below)

$$\det(\operatorname{id}_{\mathfrak{k}} - \exp(\operatorname{ad}(b))_{|\mathfrak{k}}) = \prod_{\alpha \in \mathcal{R}+} 4 \sin^2(\pi\alpha(\mathrm{b})) \tag{22}$$

where $\mathcal{R}_+$ is the set of positive real roots of $(\mathfrak{g}, \mathfrak{t})$.

In view of Eqs. (12), (20), and (22) we now rewrite the informal determinant $\operatorname{Det}(B)$ in Eq. (19) as

$$\operatorname{Det}(B) = \prod_{\alpha \in \mathcal{R}_+} \det(O_\alpha^{(0)}(B))^2 \det(O_\alpha^{(1)}(B))^{-1}$$

where for each fixed $\alpha \in \mathcal{R}_+$ the operators $O_\alpha^{(i)}(B) : \Omega^i(\Sigma, \mathbb{R}) \to \Omega^i(\Sigma, \mathbb{R})$, $i = 0, 1$, are the multiplication operators obtained by multiplication with the function $\Sigma \ni \sigma \mapsto 2 \sin(\pi\alpha(B(\sigma))) \in \mathbb{R}$.

Let us now equip the two spaces $\Omega^i(\Sigma, \mathbb{R})$, $i = 0, 1$, with the scalar product which is induced by the Riemannian metric $\mathbf{g}$ on Σ fixed in Sec. 2.3 above. By $\overline{\Omega^i(\Sigma, \mathbb{R})}$ we will denote the completion of the pre-Hilbert space $\Omega^i(\Sigma, \mathbb{R})$, $i = 0, 1$.

Let us now define

$$\operatorname{Det}_{rig}(B) := \prod_{\alpha \in \mathcal{R}_+} \operatorname{Det}_{rig,\alpha}(B) \tag{23a}$$

with

$$\operatorname{Det}_{rig,\alpha}(B) := \lim_{\epsilon \to 0} \left[\det_\epsilon(O_\alpha^{(0)}(B))^2 \det_\epsilon(O_\alpha^{(1)}(B))^{-1} \right] \tag{23b}$$

where for $i = 0, 1$ we have set[g]

$$\det_\epsilon(O_\alpha^{(i)}(B)) := \exp\left(\text{Tr}\left(e^{-\epsilon\Delta_i}\log(O_\alpha^{(i)}(B))\right)\right). \tag{23c}$$

Here Δ_i is the Hodge Laplacian on $\overline{\Omega^i(\Sigma, \mathbb{R})}$ w.r.t. the Riemannian metric $\mathbf{g}$ on Σ and $\log : \mathbb{R}\backslash\{0\} \to \mathbb{C}$ is the restriction to $\mathbb{R}\backslash\{0\}$ of the principal branch of the complex logarithm. We remark that in the special case $B \in \mathcal{B}_{reg}$, which we will assume in the following, each of the bounded operators $O_\alpha^{(i)}(B)$, $i = 0, 1$, is a symmetric operator whose spectrum is bounded away from zero (cf. Eq. (A.1) in Appendix A) so $\log(O_\alpha^{(i)}(B))$ is a well-defined bounded operator. Moreover, since $e^{-\epsilon\Delta_i}$ is trace-class the product $e^{-\epsilon\Delta_i}\log(O_\alpha^{(i)}(B))$ is also trace-class and the expression $\text{Tr}(e^{-\epsilon\Delta_i}\log(O_\alpha^{(i)}(B)))$ is therefore well-defined. Explicitly, we have

$$\text{Tr}(e^{-\epsilon\Delta_i}\log(O_\alpha^{(i)}(B))) = \int_\Sigma \text{Tr}(K_\epsilon^{(i)}(\sigma, \sigma))\log(2\sin(\pi\alpha(B(\sigma))))d\mu_\mathbf{g}(\sigma) \tag{24}$$

where $K_\epsilon^{(0)} : \Sigma \times \Sigma \to \mathbb{R} \cong \text{End}(\mathbb{R})$ is the integral kernel of $e^{-\epsilon\Delta_0}$ and $K_\epsilon^{(1)} : \Sigma \times \Sigma \to \bigcup_{\sigma_1, \sigma_2 \in \Sigma} \text{Hom}(T_{\sigma_1}\Sigma, T_{\sigma_2}\Sigma)$ is the integral kernel of $e^{-\epsilon\Delta_1}$. According to a famous result in[20] the negative powers of ϵ that appear in the asymptotic expansion of $K_\epsilon^{(i)}$, $i = 0, 1$ as $\epsilon \to 0$ cancel each other (= the "fantastic cancellations") and we obtain

$$[2\text{Tr}(K_\epsilon^{(0)}(\sigma, \sigma)) - \text{Tr}(K_\epsilon^{(1)}(\sigma, \sigma))] \to \frac{1}{4\pi}R_\mathbf{g}(\sigma) \quad \text{uniformly in } \sigma \text{ as } \epsilon \to 0 \tag{25}$$

where $R_\mathbf{g}$ is the scalar curvature (= twice the Gaussian curvature) of $(\Sigma, \mathbf{g})$.

From Eqs. (23c), (24), and (25) it follows that the $\epsilon \to 0$ limit in Eq. (23b) really exists and that we have

$$\text{Det}_{rig,\alpha}(B) = \exp\left(\int_\Sigma \log(2\sin(\pi\alpha(B(\sigma))))\frac{1}{4\pi}R_\mathbf{g}(\sigma)d\mu_\mathbf{g}(\sigma)\right). \tag{26}$$

In the special case where $B \equiv b$ (with $b \in \mathbf{t}_{reg}$) we can apply the classical Gauss-Bonnet Theorem

$$4\pi\chi(\Sigma) = \int_\Sigma R_\mathbf{g}d\mu_\mathbf{g} \tag{27}$$

where $\chi(\Sigma)$ is the Euler characteristic of Σ and obtain

$$\text{Det}_{rig,\alpha}(B) = (2\sin(\pi\alpha(b)))^{\chi(\Sigma)} \tag{28}$$

---

[g]This ansatz is, of course, motivated by the rigorous formula $\det(A) = \exp(\text{Tr}(\ln(A)))$ which holds for every strictly positive (self-adjoint) operator A on a finite-dimensional Hilbert-space.

and therefore

$$\mathrm{Det}_{rig}(B) = (\det(\mathrm{id}_{\mathfrak{k}} - \exp(\mathrm{ad}(b))_{|\mathfrak{k}}))^{\chi(\Sigma)/2} \tag{29}$$

So in particular, the value of $\mathrm{Det}_{rig}(B)$ is independent of the auxiliary Riemannian metric **g** in this special case.

Remark 4. i) In the special case where B is constant the calculation of $\mathrm{Det}_{rig}(B)$ just described can be simplified considerably. In particular, in this case Eq. (25) is not necessary for evaluating $\mathrm{Det}_{rig}(B)$ and proving Eq. (29). Indeed, for constant B we only have to show that $\lim_{\epsilon \to 0}(2\mathrm{Tr}(e^{-\epsilon\triangle_0}) - \mathrm{Tr}(e^{-\epsilon\triangle_1})) = \chi(\Sigma)$. But this follows because $\lim_{\epsilon \to 0}(2\mathrm{Tr}(e^{-\epsilon\triangle_0}) - \mathrm{Tr}(e^{-\epsilon\triangle_1})) \overset{(*)}{=} 2\dim(\ker(\triangle_0)) - \dim(\ker(\triangle_1)) \overset{(**)}{=} 2\dim(H^0(\Sigma, \mathbb{R})) - \dim(H^1(\Sigma, \mathbb{R})) = \chi(\Sigma)$. Here step $(*)$ follows from another famous argument in[20] and step $(**)$ follows because according to the Hodge theorem we have $\ker(\triangle_i) \cong H^i(\Sigma, \mathbb{R})$.

ii) We also mention that in the special case where B is constant there is an alternative way of defining and computing $\mathrm{Det}(B)$ using the Ray-Singer Torsion (which makes use of a suitable ζ-function regularization), cf. Sec. 3 in[2].

The special case $B \equiv b$ mentioned above was the only case which was relevant in[2] where only links consisting of "vertical" loops were studied (at a heuristic level). Here a "vertical" loop in $M = \Sigma \times S^1$ is a loop $l : S^1 \to \Sigma \times S^1$ which is "parallel" to S^1 (or in other words: $\mathrm{arc}(l_\Sigma)$ is just a point in Σ).

By contrast we will work with more general (ribbon) links which means that step functions B of the type

$$B = \sum_{i=1}^{r} b_i 1_{Y_i}, \qquad r \in \mathbb{N} \tag{30}$$

will appear later during the explicit evaluation of $\mathrm{WLO}_{rig}(L)$ defined in Sec. 3.4 below, namely, after the $\epsilon \to 0$ and $s \to 0$-limits on the RHS of Eq. (69) below have been carried out. The regions $(Y_i)_{i \leq r}$ here are the r connected components of

$$\Sigma \backslash \mathrm{arc}(L_\Sigma) = \Sigma \backslash \bigcup_{i=1}^{m} \mathrm{arc}(l_\Sigma^i). \tag{31}$$

In view of Eq. (29) above one would expect that for $B : \Sigma \to \mathfrak{t}$ of the form (30) one has[h]

$$\mathrm{Det}_{rig}(B) = \prod_{i=1}^{r} (\det{}^{1/2} (\mathrm{id}_{\mathfrak{t}} - \exp(\mathrm{ad}(b_i))_{|\mathfrak{t}}))^{\chi(Y_i)} \tag{32}$$

where $\det^{1/2}(\mathrm{id}_{\mathfrak{t}} - \exp(\mathrm{ad}(\cdot))_{|\mathfrak{t}}) : \mathfrak{t} \to \mathbb{R}$ is given by

$$\det{}^{1/2}(\mathrm{id}_{\mathfrak{t}} - \exp(\mathrm{ad}(b))_{|\mathfrak{t}}) = \prod_{\alpha \in \mathcal{R}+} 2 \sin(\pi \alpha(b)). \tag{33}$$

And in fact, Eq. (32) is exactly the formula which is necessary for Conjecture 2 below to be true. The obvious question now is whether Eq. (32) follows from Eq. (23a) and Eq. (26) above[i].

In order to answer this question recall the following, more general version[j] of the classical Gauss-Bonnet Theorem mentioned above: Let $Y \subset \Sigma$ be such that the boundary ∂Y is (either empty or) a smooth 1-dimensional submanifold of Σ. We equip ∂Y with the Riemannian metric induced by $\mathbf{g} = \mathbf{g}_\Sigma$ and denote by ds the corresponding "line element" on ∂Y. Then we have

$$4\pi\chi(Y) = \int_Y R_{\mathbf{g}} d\mu_{\mathbf{g}} + 2 \int_{\partial Y} k_{\mathbf{g}} ds \tag{34}$$

where $k_{\mathbf{g}}(p)$ for $p \in \partial Y$ is the geodesic curvature of ∂Y in the point p.

Let us now go back to the question whether Eq. (32) follows from Eq. (26). The short answer is: not for an arbitrary choice of $\mathbf{g}$ but for a natural subclass of the possible choices, cf. Assumption 2 and Remark 5 below. Observe that in Eq. (26) there is no term involving geodesic curvature. It is conceivable that such a term appears during the explicit evaluation of the RHS of Eq. (65) in Step 3 below as a result of "self linking". However, even in this case the "self linking" expressions which we obtain will depend on the precise regularization procedure which is used. We plan to study this issue in more detail in the near future (cf. Question 3 in Sec. 4 below).

---

[h]Observe that in contrast to $\chi(\Sigma)$, which is always an even number, $\chi(Y_i)$ can be odd, so in general we do *not* have $(\det^{1/2}(\mathrm{id}_{\mathfrak{t}} - \exp(\mathrm{ad}(b_i))_{|\mathfrak{t}}))^{\chi(Y_i)} = (\det(\mathrm{id}_{\mathfrak{t}} - \exp(\mathrm{ad}(b_i))_{|\mathfrak{t}}))^{\chi(Y_i)/2}$

[i]Observe that the RHS of Eq. (26) makes sense not only if $B : \Sigma \to \mathfrak{t}$ is smooth but also if B is measurable, bounded, and bounded away from $\mathfrak{t}_{sing} := \mathfrak{t} \backslash \mathfrak{t}_{reg}$, cf. Eq. (A.1) in Appendix A.

[j]In view of Remark 6 below we also remark that Eq. (34) can be generalized further to the situation where the boundary ∂Y is only a piecewise smooth (rather than a smooth) submanifold of Σ. In the generalized formula there will be an extra term on the RHS involving a sum over the finite number of points p of ∂Y where ∂Y is not smooth (and containing the corresponding "angle" of ∂Y at p).

In this chapter we will bypass this issue by restricting ourselves to the situation where the auxiliary Riemannian metric **g** chosen above fulfills a suitable condition. One sufficient condition would be to assume that **g** is chosen such that the geodesic curvature of each of the sets arc(l_{Σ}^i) vanishes. In this case Eq. (32) follows indeed from Eq. (26). However, since at first look this condition may seem somewhat unnatural we will use the following assumption (which according to Remark 5 below is definitely natural) in the following:

Assumption 2. From now on we will assume that the auxiliary Riemannian metric **g** on Σ was chosen such that on each Image($\pi_\Sigma \circ R_i$), $i \le m$, it coincides with the Riemannian metric "induced" by $\pi_\Sigma \circ R_i : S^1 \times (0,1) \to \Sigma$. Here we have equipped $S^1 \times (0,1)$ with the product of the standard (normalized) Riemannian metrics on S^1 and on $(0,1)$.

We expect that Assumption 2 will lead to the correct values for the rigorous implementation $\mathrm{WLO}_{rig}(L)$ of $\mathrm{WLO}(L)$, which we will give below, cf. Conjectures 1 and 2 in Sec. 3.4 below. Moreover, the use of Assumption 2 eliminates the regularization dependence of the "self linking terms" we referred to above.

Remark 5. Recall that the original heuristic path integral expression in Eq. (6) above is topologically invariant. In particular, it does not involve a Riemannian metric. However, for technical reasons, we later introduced an auxiliary Riemannian metric **g** breaking topological invariance.

Clearly, in a situation where one introduces an auxiliary object $\mathcal{O}$ in order to make sense of a heuristic expression one of the following two principles (or a combination of them) ought to be fulfilled in order to be able to claim to have a natural treatment:

(1) The auxiliary object $\mathcal{O}$ can be chosen arbitrarily and the final result does not depend on it.
(2) There is a distinguished/canonical choice of $\mathcal{O}$ and that is the choice which we use.

Our Assumption 2 is a combination of these two principles. The restriction of **g** on $S := \bigcup_{i=1}^m \mathrm{Image}(\pi_\Sigma \circ R_i)$ is given canonically. On the other hand, the restriction of **g** on $S^c := \Sigma \backslash S$ can essentially be chosen arbitrarily (as long as $\mathbf{g}_{|S}$ and $\mathbf{g}_{|S^c}$ "fit together" smoothly, i.e. induce a smooth Riemannian metric on all of Σ).

Remark 6. If one wants to study the situation of general (ribbon) links L, i.e. links which need not fulfill Assumption 1 above, then Assumption 2 must be modified. This is because when $U := \text{Image}(\pi_\Sigma \circ R_i) \cap \text{Image}(\pi_\Sigma \circ R_j) \neq \emptyset$ for $i \neq j$ then in general each of the two maps $\pi_\Sigma \circ R_i$ and $\pi_\Sigma \circ R_j$ will induce a different Riemannian metric on U. One way to deal with this complication is to use, instead of Assumption 2, the aforementioned weaker condition[k] that $\mathbf{g}$ is chosen such that the geodesic curvature of each of the sets $\text{arc}(l_\Sigma^i)$ vanishes. Moreover, the generalization of Eq. (34) mentioned in Footnote j above will then be relevant.

2.6. *The final heuristic formula (in the special case $\Sigma = S^2$)*

Observe that

$$\mathcal{A}_c^\perp \cong \mathcal{A}_{\Sigma,\mathfrak{t}}.$$

For simplicity we will assume in the following that $\Sigma \cong S^2$ and therefore $H^1(\Sigma) = \{0\}$. In this case the Hodge decomposition of $\mathcal{A}_{\Sigma,\mathfrak{t}}$ (w.r.t. the metric $\mathbf{g}$ fixed above) is given by

$$\mathcal{A}_{\Sigma,\mathfrak{t}} = \mathcal{A}_{ex} \oplus \mathcal{A}_{ex}^* \tag{35}$$

where

$$\mathcal{A}_{ex} := \{df \mid f \in C^\infty(\Sigma,\mathfrak{t})\} \tag{36}$$
$$\mathcal{A}_{ex}^* := \{\star df \mid f \in C^\infty(\Sigma,\mathfrak{t})\} \tag{37}$$

$\star$ being the relevant Hodge star operator. According to Eq. (35) we can replace the $\int \cdots D\mathcal{A}_c^\perp$ integration in Eq. (18) by the integration $\int \int \cdots D\mathcal{A}_{ex} D\mathcal{A}_{ex}^*$ where $D\mathcal{A}_{ex}$, $D\mathcal{A}_{ex}^*$ denote the heuristic "Lebesgue measures" on $\mathcal{A}_{ex}$ and $\mathcal{A}_{ex}^*$.

Taking this into account and replacing the heuristic expression $\text{Det}(B)$ in Eq. (19) by $\text{Det}_{rig}(B)$ as given by Eqs. (23) we see that we can rewrite Eq. (18) as

$$\text{WLO}(L) \sim \lim_{s \to 0} \sum_{y \in I} \int_{\mathcal{A}_{ex}^* \times C^\infty(\Sigma,\mathfrak{t})} \left\{ \int_{\mathcal{A}_{ex}} I^{(s)}(L)(\mathcal{A}_{ex} + \mathcal{A}_{ex}^*, B) \right\} D\mathcal{A}_{ex} \right\}$$
$$\times \exp(-2\pi i k \langle y, B(\sigma_0) \rangle) 1_{\mathcal{B}_{reg}}(B) \text{Det}_{rig}(B)$$
$$\times \exp(2\pi i k \ll \star \mathcal{A}_{ex}^*, dB \gg_{\mathcal{A}^\perp})(D\mathcal{A}_{ex}^* \otimes DB) \tag{38}$$

---

[k]This condition is arguably quite natural as well since it arises from Assumption 2 (which is natural according to Remark 5 above) by applying a suitable limit procedure.

where we have set

$$I^{(s)}(L)(A_c^\perp, B) := \int_{\check{A}^\perp} \prod_i \mathrm{Tr}_{\rho_i}(\mathrm{Hol}_{R_i^{(s)}}(\check{A}^\perp, A_c^\perp, B)) d\mu_B^\perp(\check{A}^\perp) \qquad (39)$$

for $A_c^\perp \in \mathcal{A}_c^\perp \cong \mathcal{A}_{\Sigma,\mathfrak{t}}$ and $B \in \mathcal{B}$ and where we have used that

$$\ll \star A_c^\perp, d\check{B} \gg_{\mathcal{A}^\perp} = \ll \star A_c^\perp, d\check{B} \gg_{\mathcal{A}_\Sigma} = - \ll \star dA_c^\perp, \check{B} \gg_{L_{\mathfrak{t}}^2(\Sigma, d\mu_{\mathbf{g}})} \qquad (40)$$

(which implies that $\ll \star A_{ex}, d\check{B} \gg_{\mathcal{A}^\perp} = 0$ if $A_{ex} \in \mathcal{A}_{ex}$).

Using heuristic methods one can show[1]

$$I^{(s)}(L)(A_{ex} + A_{ex}^*, B) = I^{(s)}(L)(A_{ex}^*, B) \qquad (41)$$

so, informally,

$$\int_{\mathcal{A}_{ex}} I^{(s)}(L)(A_{ex} + A_{ex}^*, B) \, DA_{ex} \sim I^{(s)}(L)(A_{ex}^*, B). \qquad (42)$$

Moreover, if we introduce the decomposition $\mathcal{B} = \check{\mathcal{B}} \oplus \mathcal{B}_c$ where

$$\check{\mathcal{B}} := \{B \in \mathcal{B} \mid \int_\Sigma B \, d\mu_{\mathbf{g}} = 0\} \qquad (43)$$

$$\mathcal{B}_c := \{B \in \mathcal{B} \mid B \text{ is constant}\} \cong \mathfrak{t} \qquad (44)$$

we can also replace $\int \cdots DB$ by $\int \int \cdots D\check{B}db$ where $D\check{B}$ is the heuristic "Lebesgue measure" on $\check{\mathcal{B}}$ and db is the (rigorous) normalized Lebesgue measure on $\mathcal{B}_c \cong \mathfrak{t}$. Taking this into account we obtain from Eqs. (38) and (42)

$$\mathrm{WLO}(L) \sim \lim_{s \to 0} \sum_{y \in I} \left[\int_{\mathcal{B}_c} db \left[\int_{\mathcal{A}_{ex}^* \times \check{\mathcal{B}}} I^{(s)}(L)(A_{ex}^*, \check{B} + b) \right. \right.$$
$$\times \exp(-2\pi i k \langle y, \check{B}(\sigma_0) + b \rangle) 1_{\mathcal{B}_{reg}}(\check{B} + b) \mathrm{Det}_{rig}(\check{B} + b)$$
$$\left. \left. \times \exp(-2\pi i k \ll \star A_{ex}^*, d\check{B} \gg_{\mathcal{A}^\perp})(DA_{ex}^* \otimes D\check{B}) \right] \right]. \qquad (45)$$

Observe that the operator $\star d : \check{\mathcal{B}} \to \mathcal{A}_{ex}^*$ is a linear isomorphism. We can therefore make the change of variable $\check{B}_1 := (\star d)^{-1} A_{ex}^*$ and $\check{B}_2 := \check{B}$ and rewrite Eq. (45) as

$$\mathrm{WLO}(L) \sim \lim_{s \to 0} \sum_{y \in I} \left[\int_{\mathcal{B}_c} \left[\int_{\check{\mathcal{B}} \times \check{\mathcal{B}}} J_{b,y}^{(s)}(L)(\check{B}_1, \check{B}_2) \, d\nu(\check{B}_1, \check{B}_2) \right] db \right] \qquad (46)$$

---

[1]This is easy if L fulfills Assumption 1, cf. also Remark 7 below for a rigorous argument. For general (ribbon) links L this point is not yet clear, cf. Sec. 4 below.

where we have set

$$J_{b,y}^{(s)}(L)(\check{B}_1, \check{B}_2) := I^{(s)}(L)(\star d\check{B}_1, \check{B}_2 + b) \exp(-2\pi i k \langle y, \check{B}_2(\sigma_0) + b \rangle)$$

$$\times 1_{\mathcal{B}_{reg}}(\check{B}_2 + b) \mathrm{Det}_{rig}(\check{B}_2 + b) \quad (47)$$

and (cf. Eq. (40) above)

$$d\nu(\check{B}_1, \check{B}_2) := \tfrac{1}{Z} \exp(-2\pi i k \ll \star d \star d\check{B}_1, \check{B}_2 \gg_{L_t^2(\Sigma, d\mu_{\mathbf{g}})})(D\check{B}_1 \otimes D\check{B}_2) \quad (48)$$

with[m]

$$Z := \int \exp(-2\pi i k \ll \star d \star d\check{B}_1, \check{B}_2 \gg_{L_t^2(\Sigma, d\mu_{\mathbf{g}})})(D\check{B}_1 \otimes D\check{B}_2)$$

3. Rigorous realization of the RHS of Eq. (46)

We will now explain how one can make rigorous sense of the path integral expression appearing on the RHS of Eq. (46) within the framework of White Noise Analysis. In order to do so we will proceed in four steps:

Step 1: We make rigorous sense of the integral functional $\int \cdots d\mu_B^\perp$ appearing in Eq. (39).

Step 2: We make rigorous sense of the integral functional $\int \cdots d\nu$ appearing in Eq. (46).

Step 3: We make rigorous sense of the integral expression appearing in Eq. (39) above, i.e. of

$$I^{(s)}(L)(A_c^\perp, B) = \int_{\check{\mathcal{A}}^\perp} \prod_i \mathrm{Tr}_{\rho_i}(\mathrm{Hol}_{R_i^{(s)}}(\cdot, A_c^\perp, B))) d\mu_B^\perp \quad (49)$$

for $A_c^\perp \in \mathcal{A}_c^\perp$ and $B \in \mathcal{B}$.

Step 4: We make rigorous sense of the total expression on the RHS of Eq. (46).

3.1. *Step 1*

We will now give a rigorous implementation of the integral functional $\int \cdots d\mu_B^\perp$ appearing in Eq. (39) above as a generalized distribution $\Phi_B^\perp$ on a suitable extension $\overline{\check{A}}^\perp$ of $\check{A}^\perp$.

---

[m]Clearly, we could drop the factor $1/Z$ in Eq. (48) since the symbol $\sim$ in Eq. (46) denotes equality up to a multiplicative constant. In Sec. 3.2 we introduce a rigorous version of the integral functional $\int \cdots d\nu$ and there it is convenient that $d\nu$ is normalized.

(1) First we will choose a suitable nuclear triple $(\mathcal{N}, \mathcal{H}_{\mathcal{N}}, \mathcal{N}')$ and set $\overline{\breve{\mathcal{A}}^{\perp}} :=$ $\mathcal{N}'$.

Before we do this recall that

$$\mathcal{A}^{\perp} \cong C^{\infty}(S^1, \mathcal{A}_{\Sigma})$$

$$\breve{\mathcal{A}}^{\perp} = \{A^{\perp} \in \mathcal{A}^{\perp} \mid \int A^{\perp}(t)dt \in \mathcal{A}_{\Sigma,\mathfrak{t}}\}$$

We set $\mathcal{H}_{\Sigma} := L^2\text{-}\Gamma(\mathrm{Hom}(T\Sigma, \mathfrak{g}), d\mu_{\mathbf{g}})$, i.e. $\mathcal{H}_{\Sigma}$ is the Hilbert space of L^2-sections (w.r.t. the measure $d\mu_{\mathbf{g}}$) of the bundle $\mathrm{Hom}(T\Sigma, \mathfrak{g}) \cong T^*\Sigma \otimes \mathfrak{g}$ equipped with the fiber metric induced by $\mathbf{g}$ and $\langle \cdot, \cdot \rangle$. Moreover, we set[n]

$$\mathcal{H}^{\perp} := L^2_{\mathcal{H}_{\Sigma}}(S^1, dt)$$

$$\breve{\mathcal{H}}^{\perp} := \{H^{\perp} \in \mathcal{H}^{\perp} \mid \int H^{\perp}(t)dt \in \mathcal{H}_{\Sigma,\mathfrak{t}}\}$$

where $\mathcal{H}_{\Sigma,\mathfrak{t}}$ is the Hilbert space defined in a completely analogous way as $\mathcal{H}_{\Sigma}$ but with $\mathfrak{t}$ playing the role of $\mathfrak{g}$.

The nuclear triple $(\mathcal{N}, \mathcal{H}_{\mathcal{N}}, \mathcal{N}')$ we choose is given by

$$\mathcal{N} := \breve{\mathcal{A}}^{\perp} \tag{50}$$

$$\mathcal{H}_{\mathcal{N}} := \breve{\mathcal{H}}^{\perp} \tag{51}$$

where we have equipped $\breve{\mathcal{A}}^{\perp}$ with a suitable family of semi-norms. (More precisely, the family of semi-norms must be chosen such that $\mathcal{N} = \breve{\mathcal{A}}^{\perp}$ is nuclear and the inclusion map $\mathcal{N} \to \mathcal{H}_{\mathcal{N}}$ is continuous.)

As explained in Ch. 1, the nuclear triple $(\mathcal{N}, \mathcal{H}_{\mathcal{N}}, \mathcal{N}')$ gives rise to another nuclear triple $((\mathcal{N})^0, (L^2), (\mathcal{N})^{-0})$ (via the Ito-Segal-Wiener isomorphism).

(2) Next we evaluate the Fourier transform $\mathcal{F}\mu_B^{\perp}$ of the informal measure $\mu_B^{\perp}$. From Eq. (14) and Eq. (17) we obtain immediately

$$\forall j \in \mathcal{N} : \mathcal{F}\mu_B^{\perp}(j) = \int \exp(i \ll \cdot, j \gg_{\mathcal{H}_{\mathcal{N}}})d\mu_B^{\perp} = \exp(-\tfrac{1}{2} \ll j, C_B j \gg_{\mathcal{H}_{\mathcal{N}}})$$

where C_B is given informally by

$$C_B = \left(-2\pi k \star (\partial/\partial t + \mathrm{ad}(B))\right)^{-1} = \tfrac{1}{2\pi k} \star (\partial/\partial t + \mathrm{ad}(B))^{-1}. \tag{52}$$

For each fixed $B \in \mathcal{B}_{reg}$ we will now make sense of $(\partial/\partial t + \mathrm{ad}(B))^{-1}$ as a densely defined linear operator on $\breve{\mathcal{H}}^{\perp}$. In order to do so we first

---

[n]In other words: $\mathcal{H}^{\perp}$ is the space of $\mathcal{H}_{\Sigma}$-valued (measurable) functions on S^1 which are square-integrable w.r.t. dt.

introduce the space $\check{C}^\infty(S^1, \mathfrak{g}) := \{f \in C^\infty(S^1, \mathfrak{g}) | \int f(t)dt \in \mathfrak{k}\}$. It is not difficult to see that for $b \in \mathfrak{t}_{reg}$ the operator $\partial/\partial t + \text{ad}(b)$: $\check{C}^\infty(S^1, \mathfrak{g}) \to \check{C}^\infty(S^1, \mathfrak{g})$ is invertible.[o]

Now let $(\partial/\partial t + \text{ad}(B))^{-1} : \check{\mathcal{A}} \to \check{\mathcal{A}} \subset \check{\mathcal{H}}^\perp$ be the linear operator given by

$$\left(\left((\partial/\partial t + \text{ad}(B))^{-1} \cdot \check{A}^\perp\right)(t)\right)(X_\sigma) = (\partial/\partial t + \text{ad}(B(\sigma)))^{-1} \cdot \left(\check{A}^\perp(t)(X_\sigma)\right)$$
(54)

for all $\check{A}^\perp \in \check{\mathcal{A}}^\perp$, $t \in S^1$, $\sigma \in \Sigma$, and $X_\sigma \in T_\sigma \Sigma$. Observe that $(\partial/\partial t + \text{ad}(B))^{-1} \cdot \check{A}^\perp$ is indeed a well-defined element of $\check{\mathcal{A}}^\perp$ because, by the assumption on B we have $B \in \mathcal{B}_{reg}$, i.e. $B(\sigma) \in \mathfrak{t}_{reg}$ for all $\sigma \in \Sigma$.

It is easy to check that $(\partial/\partial t + \text{ad}(B))^{-1}$ is bounded and anti-symmetric. Since $\star$ is bounded and anti-symmetric as well we have now found a rigorous realization of the operator C_B in Eq. (52) as a bounded and symmetric operator on $\check{\mathcal{H}}^\perp$.

For technical reasons we need to define[p] C_B also for $B \notin \mathcal{B}_{reg}$. In view of the indicator function $1_{\mathcal{B}_{reg}}$ appearing in Eq. (69) it seems[q] that we are entitled to define C_B in an arbitrary way if $B \notin \mathcal{B}_{reg}$. For simplicity we will take C_B to be trivial (i.e. $C_B = 0$) if $B \notin \mathcal{B}_{reg}$.

---

[o]Its inverse $(\partial/\partial t + \text{ad}(b))^{-1}$ is given explicitly by (cf. Eq. (5.8) in[12])

$$\left(((\partial/\partial t + \text{ad}(b))^{-1}f)(t) = T(b) \cdot \int_0^1 e^{s\text{ad}(b)} f(t + i_{S^1}(s))ds\right.$$
(53)

for all $f \in \check{C}^\infty(S^1, \mathfrak{g})$ and $t \in S^1$ where $i_{S^1} : [0, 1] \ni s \mapsto e^{2\pi is} \in U(1) \cong S^1$ and where $T(b) \in \text{End}(\mathfrak{g})$ is given by $T(b)(X) = (e^{\text{ad}(b)} - \pi_\mathfrak{k})^{-1}(X)$ if $X \in \mathfrak{k}$ and $T(b)(X) = X$ if $X \in \mathfrak{t}$. Here $\pi_\mathfrak{k} : \mathfrak{g} \to \mathfrak{k}$ is the orthogonal projection. We remark that in the special case where f takes only values in $\mathfrak{t}$ Eq. (53) reduces to $((\partial/\partial t + \text{ad}(b))^{-1}f)(t) = ((\partial/\partial t)^{-1}f)(t) = \int_0^1 f(t + i_{S^1}(s))ds$.

[p]Observe that in Sec. 3.4 below we replace the indicator function $1_{\mathcal{B}_{reg}}$ by regularized versions $1_{\mathcal{B}_{reg}}^{(n)}$ and the condition $1_{\mathcal{B}_{reg}}^{(n)}(B) \neq 0$ (where $n \in \mathbb{N}$ is fixed) does no longer guarantee that $B \in \mathcal{B}_{reg}$.

[q]In fact, in spite of this indicator function $1_{\mathcal{B}_{reg}}$ some of the functions B which appear during the explicit evaluation of the RHS of Eq. (69) will not be elements of $\mathcal{B}_{reg}$ (for reasons explained in the previous footnote). It would be more satisfactory to modify our approach in a suitable way, for example by using the regularization procedure described in Appendix B already now (and not only in Step 4), and working with $\Psi_{B^{(n)}}^\perp$ instead of $\Psi_B^\perp$ with $B^{(n)}$ given by Eq. (B.2) below. In the situation of ribbon links L fulfilling Assumption 1 above we can bypass this problem by simply defining C_B by $C_B := 0$ if $B \in \mathcal{B}_{reg}$. With this choice Eq. (66) in Proposition 2 will hold for all $B \in \mathcal{B}$ and we can expect Conjectures 1 and 2 to be true.

(3) For fixed $B \in \mathcal{B}$ let $U_B : \mathcal{N} \to \mathbb{C}$ be the well-defined continuous function given by

$$U_B(j) = \exp(-\tfrac{1}{2} \ll j, C_B j \gg_{\mathcal{H}_{\mathcal{N}}}) \tag{55}$$

for every $j \in \mathcal{N}$. It is straightforward to show that the function $U_B : \mathcal{N} \to \mathbb{C}$ is a "U-functional" in the sense of Ch. 1, Definition 7. In view of Theorem 2 in Ch. 1 (rewritten in terms of the T-transform) the integral functional $\Phi_B^{\perp} := \int \cdots d\mu_B^{\perp}$ can be defined rigorously as the unique element $\Phi_B^{\perp}$ of $(\mathcal{N})^{-0}$ such that

$$\Phi_B^{\perp}(\exp(i(\cdot, j)_{\mathcal{N}})) = U_B(j) \tag{56}$$

holds for all $j \in \mathcal{N}$. Here $(\cdot, \cdot)_{\mathcal{N}} : \mathcal{N}' \times \mathcal{N} \to \mathbb{R}$ is the canonical pairing.

Convention 2. Let $\pi : \mathcal{A}^{\perp} = \check{\mathcal{A}}^{\perp} \oplus \mathcal{A}_c^{\perp} \to \check{\mathcal{A}}^{\perp}$ be the canonical projection. The map $\pi' : (\check{\mathcal{A}}^{\perp})' \to (\mathcal{A}^{\perp})'$ which is dual to π is an injection. Using π' we will identify $(\check{\mathcal{A}}^{\perp})'$ with a subspace of $(\mathcal{A}^{\perp})'$.

Moreover, we will identify each element $A^{\perp}$ of $(\mathcal{A}^{\perp})'$ with the continuous map $f_{A^{\perp}} : \mathcal{A}_{\mathbb{R}}^{\perp} \to \mathfrak{g}$ given by $f_{A^{\perp}}(\psi^{\perp}) = \sum_a T_a(A^{\perp}, T_a \psi^{\perp})$ for all $\psi^{\perp} \in \mathcal{A}_{\mathbb{R}}^{\perp}$ where $\mathcal{A}_{\mathbb{R}}^{\perp} := C^{\infty}(S^1, \mathcal{A}_{\Sigma, \mathbb{R}})$ and where $(T_a)_a$ is any fixed $\langle \cdot, \cdot \rangle$-orthonormal basis of $\mathfrak{g}$.

3.2. Step 2

In order to make rigorous sense of the heuristic integral functional $\int_{\check{\mathcal{B}} \times \check{\mathcal{B}}} \cdots d\nu$ as a generalized distribution on a suitable extension $\overline{\check{\mathcal{B}} \times \check{\mathcal{B}}}$ of the space $\check{\mathcal{B}} \times \check{\mathcal{B}}$ we will proceed in a similar way as in Step 1 above.

(1) First we choose a suitable nuclear triple $(\mathcal{E}, \mathcal{H}_{\mathcal{E}}, \mathcal{E}')$ and set

$$\overline{\check{\mathcal{B}} \times \check{\mathcal{B}}} := \mathcal{E}'.$$

More precisely, we choose

$$\mathcal{E} := \check{\mathcal{B}} \times \check{\mathcal{B}} \tag{57}$$

$$\mathcal{H}_{\mathcal{E}} := \check{L}_t^2(\Sigma, d\mu_{\mathbf{g}}) \oplus \check{L}_t^2(\Sigma, d\mu_{\mathbf{g}}) \tag{58}$$

where we have equipped $\mathcal{E}$ with a suitable family of semi-norms and where we have set $\check{L}_t^2(\Sigma, d\mu_{\mathbf{g}}) := \{f \in L_t^2(\Sigma, d\mu_{\mathbf{g}}) \mid \int f d\mu_{\mathbf{g}} = 0\}$. As explained in Ch. 1, the nuclear triple $(\mathcal{E}, \mathcal{H}_{\mathcal{E}}, \mathcal{E}')$ gives rise to another nuclear triple $((\mathcal{E})^0, (L^2), (\mathcal{E})^{-0})$ (via the Ito-Segal-Wiener isomorphism).

(2) Next we evaluate the Fourier transform $\mathcal{F}\nu$ of the heuristic "measure" ν at an informal level. Clearly, ν is of "Gauss type" with the well-defined covariance operator

$$C := -\frac{1}{2\pi k}\begin{pmatrix} 0 & (\triangle_{|\check{\mathcal{B}}})^{-1} \\ (\triangle_{|\check{\mathcal{B}}})^{-1} & 0 \end{pmatrix}, \qquad \text{where } \triangle := \star d \star d$$

(Observe that the kernel of $\triangle : \mathcal{B} \to \mathcal{B}$ equals $\mathcal{B}_c$, so $\triangle_{|\check{\mathcal{B}}}$ is injective.) Taking this into account we obtain

$$\forall j \in \mathcal{E}: \quad \mathcal{F}\nu(j) = \int \exp(i \ll \cdot, j \gg_{\mathcal{H}_\mathcal{E}})d\nu = \exp(-\tfrac{1}{2} \ll j, Cj \gg_{\mathcal{H}_\mathcal{E}})$$

We remark that C is a (densely defined) bounded and symmetric linear operator on $\mathcal{H}_\mathcal{E}$.

(3) Let $U : \mathcal{E} \to \mathbb{C}$ be given by

$$U(j) = \exp(-\tfrac{1}{2} \ll j, Cj \gg_{\mathcal{H}_\mathcal{E}}) \tag{59}$$

for every $j \in \mathcal{E}$. Clearly, $U : \mathcal{E} \to \mathbb{C}$ is a "U-functional" in the sense of Ch. 1, Definition 7, so in view of Theorem 2 in Ch. 1 (rewritten in terms of the T-transform rather than the S-Transform) the integral functional $\Psi := \int_{\check{\mathcal{B}} \times \check{\mathcal{B}}} \cdots d\nu$ can be defined rigorously as the unique element Ψ of $(\mathcal{E})^{-0}$ such that

$$\Psi(\exp(i(\cdot, j)_\mathcal{E})) = U(j) \tag{60}$$

holds for all $j \in \mathcal{E}$. Here $(\cdot, \cdot)_\mathcal{E} : \mathcal{E}' \times \mathcal{E} \to \mathbb{R}$ is the canonical pairing.

Convention 3. Let $\pi : \mathcal{B} = \check{\mathcal{B}} \oplus \mathcal{B}_c \to \check{\mathcal{B}}$ be the canonical projection. The map $\pi' : \check{\mathcal{B}}' \to \mathcal{B}'$ which is dual to π is an injection. Using π' we will identify $\check{\mathcal{B}}'$ with a subspace of $\mathcal{B}'$.

Moreover, we will identify each element B of $\mathcal{B}'$ with the continuous map $f_B : C^\infty(\Sigma, \mathbb{R}) \to \mathfrak{t}$ given by $f_B(\psi) = \sum_a T_a(B, T_a\psi)_\mathcal{B}$ for all $\psi \in C^\infty(\Sigma, \mathbb{R})$ where $(\cdot, \cdot)$ is the canonical pairing $\mathcal{B} \times \mathcal{B}' \to \mathbb{R}$ and $(T_a)_a$ is a fixed orthonormal basis of $\mathfrak{t}$.

3.3. *Step 3*

Let us now make rigorous sense of the heuristic integral

$$I^{(s)}(L)(A_c^\perp, B) = \int_{\check{\mathcal{A}}^\perp} \prod_i \mathrm{Tr}_{\rho_i}(\mathrm{Hol}_{R_i^{(s)}}(\check{A}^\perp, A_c^\perp, B))d\mu_{\check{B}}^\perp(\check{A}^\perp) \tag{61}$$

for $A_c^\perp \in \mathcal{A}_c^\perp$ and $B \in \mathcal{B}$. We already have a rigorous version $\Phi_B^\perp$ of the heuristic integral functional $\int \cdots d\mu_B^\perp$. However, clearly we cannot just consider

$$\Phi_B^\perp\Big(\prod_i \mathrm{Tr}_{\rho_i}\big(\mathrm{Hol}_{R_i^{(s)}}(\cdot, A_c^\perp, B)\big)\Big)$$

since the function $\mathrm{Hol}_{R_i^{(s)}}(\cdot, A_c^\perp, B)$ was defined as a function on $\check{\mathcal{A}}^\perp$ and not as a function on all of $\overline{\check{\mathcal{A}}^\perp} = \mathcal{N}'$.

Let us now fix $j \leq m$ and $s > 0$ temporarily (until Proposition 1) and set $R := R_j^{(s)}$. Using[r] $\check{A}^\perp((R_u)'(t)) = \check{A}^\perp((\pi_\Sigma \circ R_u)'(t))$ we obtain

$$\mathrm{Hol}_R(\check{A}^\perp, A_c^\perp, B)$$

$$= \lim_{n\to\infty} \prod_{j=1}^n \exp\big(\tfrac{1}{n}\int_0^1 \big[(\check{A}^\perp + A_c^\perp + Bdt)(R_u'(t))\big]du\big)_{|t=j/n}$$

$$= \lim_{n\to\infty} \prod_{j=1}^n \exp\big(\tfrac{1}{n}\int_0^1 \big[\check{A}^\perp((\pi_\Sigma \circ R_u)'(t)) + (A_c^\perp + Bdt)(R_u'(t))\big]du\big)_{|t=j/n}.$$

$$(62)$$

Clearly, for a general element $\check{A}^\perp$ of $\overline{\check{\mathcal{A}}^\perp} = \mathcal{N}'$ the expression $\check{A}^\perp((\pi_\Sigma \circ R_u)'(t))$ appearing in the last expression does not make sense.

In order to get round this complication we will now make use of "point smearing", i.e. replace points by suitable test functions.

In order to do so we choose, for each $\sigma \in \Sigma$ a Dirac family[s] $(\delta_\sigma^\epsilon)_{\epsilon>0}$ around σ w.r.t. $d\mu_{\mathbf{g}}$. Moreover, for each $t \in S^1$ we choose a Dirac family $(\delta_t^\epsilon)_{\epsilon>0}$ around t w.r.t. the measure dt on S^1.

For every $p = (\sigma, t) \in \Sigma \times S^1$ and $\epsilon > 0$ we define $\delta_p^\epsilon \in C^\infty(\Sigma \times S^1, \mathbb{R})$ by

$$\delta_p^\epsilon(\sigma', t') := \delta_\sigma^\epsilon(\sigma')\delta_t^\epsilon(t') \quad \text{for all } \sigma' \in \Sigma \text{ and } t' \in S^1.$$

For technical reasons we will assume[t] also that for each fixed $\epsilon > 0$ the family $(\delta_\sigma^\epsilon)_{\sigma\in\Sigma}$ was chosen such that the function $\Sigma \times \Sigma \ni (\sigma, \bar\sigma) \to \delta_\sigma^\epsilon(\bar\sigma) \in$

---

[r]Of course, we also have $A_c^\perp((R_u)'(t)) = A_c^\perp((\pi_\Sigma \circ R_u)'(t))$ but we will not need this.
[s]I.e. for each fixed $\sigma \in \Sigma$ we have the following: δ_σ^ϵ, $\epsilon > 0$, is a non-negative and smooth function $\Sigma \to \mathbb{R}$. Moreover, $\int \delta_\sigma^\epsilon d\mu_{\mathbf{g}} = 1$, and we have $\delta_\sigma^\epsilon \to \delta_\sigma$ weakly as $\epsilon \to 0$ where δ_σ is the Dirac distribution in the point σ.
[t]In view of Question 2 in Sec. 4 below we remark that if we want to treat the case of general ribbon links L (i.e. ribbon links for which Assumption 1 need not be fulfilled) we will probably have to make some additional technical assumptions on the family $(\delta_\sigma^\epsilon)_{\sigma\in\Sigma}$.

$\mathbb{R}$ is smooth and, moreover, to have the property that for each ϵ and $\sigma \in \Sigma$ the support of δ_σ^ϵ is contained in the ϵ-ball w.r.t. $d_{\mathbf{g}}$ around σ.

Recall that $R = R_j^{(s)}$, $j \leq m$, $s > 0$. Let us now also introduce the notation $\bar{R} := R_j$. Let $\epsilon_j(s)$ be the supremum of all $\epsilon > 0$ such that for all $t \in S^1$, and $u \in [0,1]$ we have $\mathrm{supp}(\delta_{R_u(t)}^\epsilon) \subset \bar{R}_\Sigma$. (Observe that $\delta_{R_u(t)}^\epsilon$ depends on s even though this is not reflected in the notation.) Let $X_{\bar{R}_\Sigma}$ be the vector field on $\mathrm{Image}(\bar{R}_\Sigma) \subset \Sigma$, which is induced by the collection of loops $S^1 \ni t \mapsto \bar{R}_\Sigma(t,u) \in \Sigma$, $u \in [0,1]$.

After these preparations we can now introduce "smeared" analogues for the expression $\check{A}^\perp((\pi_\Sigma \circ R_u)'(t))$ appearing in Eq. (62) above. More precisely, we now replace, for fixed $\epsilon \in (0, \epsilon_j(s))$ the expression $\check{A}^\perp((\pi_\Sigma \circ R_u)'(t))$ by the expression $\check{A}^\perp(X_{\bar{R}_\Sigma} \delta_{R_u(t)}^\epsilon)$. Here we made the identification $VF(\Sigma) \cong \mathcal{A}_{\Sigma,\mathbb{R}}$ using the Riemannian metric $\mathbf{g}$. On the other hand $\mathcal{A}_{\Sigma,\mathbb{R}} \subset C^\infty(S^1, \mathcal{A}_{\Sigma,\mathbb{R}}) \cong \mathcal{A}_\mathbb{R}^\perp$, so according to Convention 2 above the expression $\check{A}^\perp(X_{\bar{R}_\Sigma} \delta_{R_u(t)}^\epsilon)$ is well-defined for every $\check{A}^\perp \in \overline{\mathcal{A}^\perp} = \mathcal{N}'$. We can now set

$$\mathrm{Hol}_R^{(\epsilon)}(\check{A}^\perp, A_c^\perp, B) :=$$
$$\lim_{n \to \infty} \prod_{j=1}^n \exp\left(\frac{1}{n} \int_0^1 \left[\check{A}^\perp(X_{\bar{R}_\Sigma} \delta_{R_u(t)}^\epsilon) + (A_c^\perp + B\,dt)(R_u'(t)) \right] du \right)_{|t=j/n} \in G.$$

$$(63)$$

Using similar methods as in the proof of Proposition 6 in[7] it is not difficult to prove the following result:

Proposition 1. *For every $s \in (0,1)$ and every $\epsilon \in (0, \epsilon(s))$ where $\epsilon(s) := \min_{i \leq m} \epsilon_i(s)$ we have*

$$\prod_i \mathrm{Tr}_{\rho_i}\left(\mathrm{Hol}_{R_i^{(s)}}^{(\epsilon)}(\cdot, A_c^\perp, B)\right) \in (\mathcal{N})^0. \quad (64)$$

Consequently, the expression

$$I_{rig}^{(s,\epsilon)}(L)(A_c^\perp, B) := \Phi_B^\perp\left(\prod_i \mathrm{Tr}_{\rho_i}\left(\mathrm{Hol}_{R_i^{(s)}}^{(\epsilon)}(\cdot, A_c^\perp, B)\right)\right) \quad (65)$$

is well-defined.

Proposition 2. *For every L fulfilling Assumption 1 above we have*

$$I_{rig}^{(s,\epsilon)}(L)(A_c^\perp, B) = \prod_i \mathrm{Tr}_{\rho_i}\left(\exp\left(\int_0^1 \left(\int_{(R_i^{(s)})_u} (A_c^\perp + B\,dt)\right) du\right)\right) \quad (66)$$

where for a loop $l : S^1 \to M$ *and a 1-form* α *on* M *we use the notation* $\int_l \alpha = \int_{S^1} l^*(\alpha)$. *Recall that we also assume that Assumption 2 above is fulfilled. Observe that the RHS of Eq. (66) actually does not depend on* ϵ.

Remark 7. We remark that Eq. (66) implies that $I_{rig}^{(s,\epsilon)}(L)(A_{ex}+A_{ex}^*, B) = I_{rig}^{(s,\epsilon)}(L)(A_{ex}^*, B)$ (cf. the notation in Sec. 2.6 above) which can be seen as a rigorous justification of Eq. (41) in Sec. 2.6 above.

Remark 8.

(1) Here is an alternative way for defining a "smeared" analogue of the expression $\check{A}^\perp((\pi_\Sigma \circ R_u)'(t))$. Observe that since Σ is compact there is a $\epsilon_0 > 0$ such that for all $\sigma_0, \sigma_1 \in \Sigma$ with $d_{\mathbf{g}}(\sigma_0,\sigma_1) < \epsilon_0$ there is a unique (geodesic) segment starting in σ_0 and ending in σ_1. Using parallel transport along this geodesic segment w.r.t. the Levi-Civita connection of $(\Sigma, \mathbf{g})$ we can transport every tangent vector $v \in T_{\sigma_0}\Sigma$ to a tangent vector in $T_{\sigma_1}\Sigma$. Thus every $v \in T_{\sigma_0}\Sigma$ induces in a natural way a vector field X_v on the open ball $B_{\epsilon_0}(\sigma_0) \subset \Sigma$.
   For every $\epsilon < \epsilon_0$ we replace $\check{A}^\perp((\pi_\Sigma \circ R_u)'(t))$ by the expression $\check{A}^\perp(X_{(\pi_\Sigma \circ R_u)'(t)}\delta^\epsilon_{R_u(t)})$.
(2) The alternative method also works when L does not fulfill Assumption 1 while the original method must be modified (in a relatively straightforward way) if L does not fulfill Assumption 1.
(3) When both Assumption 1 and Assumption 2 are fulfilled then both methods described here are equivalent. Indeed, for each fixed t and u the vector field $X_{(\pi_\Sigma \circ R_u)'(t)}$ coincides with the vector field $X_{\bar{R}_\Sigma}$ on the subset $S \subset \Sigma$ where both vector fields are defined. For sufficiently small $\epsilon > 0$ we therefore have $X_{(\pi_\Sigma \circ R_u)'(t)}\delta^\epsilon_{R_u(t)} = X_{\bar{R}_\Sigma}\delta^\epsilon_{R_u(t)}$.
(4) If Assumption 2 is not fulfilled then the two methods described here will probably not be equivalent. In particular, it seems that non-trivial self-linking terms appear when using the alternative method while no such self-linking term will arise when the original method is used.

3.4. *Step 4*

Finally, let us make rigorous sense of the full heuristic expressions on the RHS of Eq. (46) above.

For similar reasons as in Step 3 above we will use again "point smearing". Recall that above we chose for each $\sigma \in \Sigma$ a "Dirac family" $(\delta^\epsilon_\sigma)_{\epsilon>0}$ such that for every $\epsilon > 0$ the function $\Sigma \times \Sigma \ni (\sigma, \bar{\sigma}) \to \delta^\epsilon_\sigma(\bar{\sigma}) \in \mathbb{R}$ is smooth. This implies[u] that for each fixed $\epsilon > 0$ and each $B \in \mathcal{B}'$ the function $B^{(\epsilon)} : \Sigma \to \mathfrak{t}$ given by $B^{(\epsilon)}(\sigma) = B(\delta_\sigma^\epsilon)$ for all $\sigma \in \Sigma$ (cf. Convention 3 above) is smooth. Consequently, the function $\check{B}^{(\epsilon)} : \Sigma \to \mathfrak{t}$ given by $\check{B}^{(\epsilon)} = B^{(\epsilon)} - \int B^{(\epsilon)} d\mu_\mathbf{g}$ is a well-defined element of $\check{\mathcal{B}}$. (For $\check{B} \in \check{\mathcal{B}}' \subset \mathcal{B}'$ we will simply write $\check{B}^{(\epsilon)}$ instead of $\check{B}^{(\epsilon)}$.)

For fixed $y \in I$, $b \in \mathfrak{t}$, $s \in (0,1)$, and $\epsilon \in (0, \epsilon(s))$ we could now introduce the function $J_{b,y}^{(s,\epsilon)}(L) : \mathcal{E}' \to \mathbb{R}$ by

$$J_{b,y}^{(s,\epsilon)}(L)(\check{B}_1, \check{B}_2) := I_{rig}^{(s,\epsilon)}(L)(\star d\check{B}_1^{(\epsilon)}, \check{B}_2^{(\epsilon)} + b) \cdot \exp(-2\pi i k \langle y, \check{B}_2^{(\epsilon)}(\sigma_0) + b\rangle)$$
$$\times \mathrm{Det}_{rig}(\check{B}_2^{(\epsilon)} + b) 1_{\mathcal{B}_{reg}}(\check{B}_2^{(\epsilon)} + b) \quad (67)$$

for all $(\check{B}_1, \check{B}_2) \in (\check{\mathcal{B}})' \times (\check{\mathcal{B}})' \cong \mathcal{E}'$.

However, the last two factors $\mathrm{Det}_{rig}(\check{B}_2^{(\epsilon)} + b)$ and $1_{\mathcal{B}_{reg}}(\check{B}_2^{(\epsilon)} + b)$ are problematic since neither of these two factors (considered as functions $\mathcal{E}' \to \mathbb{R}$) is an element of $(\mathcal{E})^0$.

This is why, in addition to "point smearing", we will use an additional regularization and introduce regularized versions $1_{\mathcal{B}_{reg}}^{(n)} : \mathcal{B} \to \mathbb{R}$ and $\mathrm{Det}_{rig}^{(n)} : \mathcal{B} \to \mathbb{R}$, $n \in \mathbb{N}$, of $1_{\mathcal{B}_{reg}}$ and Det_{rig}. There are several ways to do this. In Appendix B we explain one possible regularization. The following definitions, results, and conjectures refer to the choice of $1_{\mathcal{B}_{reg}}^{(n)} : \mathcal{B} \to \mathbb{R}$ and $\mathrm{Det}_{rig}^{(n)} : \mathcal{B} \to \mathbb{R}$ of Appendix B.

Let $y \in I$, $b \in \mathfrak{t}$, $s \in (0,1)$, $\epsilon \in (0, \epsilon(s))$, and $n \in \mathbb{N}$ be fixed. We introduce the function $J_{b,y}^{(s,\epsilon,n)}(L) : \mathcal{E}' \to \mathbb{R}$ by

$$J_{b,y}^{(s,\epsilon,n)}(L)(\check{B}_1, \check{B}_2) := I_{rig}^{(s,\epsilon)}(L)(\star d\check{B}_1^{(\epsilon)}, \check{B}_2^{(\epsilon)} + b) \cdot \exp(-2\pi i k \langle y, \check{B}_2^{(\epsilon)}(\sigma_0) + b\rangle)$$
$$\times \mathrm{Det}_{rig}^{(n)}(\check{B}_2^{(\epsilon)} + b) 1_{\mathcal{B}_{reg}}^{(n)}(\check{B}_2^{(\epsilon)} + b) \quad (68)$$

for all $(\check{B}_1, \check{B}_2) \in (\check{\mathcal{B}})' \times (\check{\mathcal{B}})' \cong \mathcal{E}'$.

Proposition 3. *For all $b \in \mathfrak{t}$, $y \in I$, $s \in (0,1)$, $\epsilon \in (0, \epsilon(s))$, and $n \in \mathbb{N}$ we have*

$$J_{b,y}^{(s,\epsilon,n)}(L) \in (\mathcal{E})^0.$$

Consequently, the expression $\Psi\big(J_{b,y}^{(s,\epsilon,n)}(L)\big)$ *is well-defined.*

---

[u]In the special case $B = f \cdot d\mu_\mathbf{g}$ where $f : \Sigma \to \mathfrak{t}$ is continuous the smoothness of $B^{(\epsilon)} : \Sigma \to \mathfrak{t}$ follows easily from the assumption that $\Sigma \times \Sigma \ni (\sigma, \bar{\sigma}) \mapsto \delta_\sigma^\epsilon(\bar{\sigma}) \in \mathbb{R}$ is smooth. Moreover, the smoothness of $B^{(\epsilon)}$ follows also if B is any derivative of a distribution of the form $f \cdot d\mu_\mathbf{g}$ with $f \in C^0(\Sigma, \mathfrak{t})$. This covers already the general situation since, by a well-known theorem, every distribution $D \in \mathcal{D}'(\Sigma)$ can be written as a linear combination of derivatives of distributions of the form $f \cdot d\mu_\mathbf{g}$ where f is a continuous function $\Sigma \to \mathbb{R}$ and this result can immediately be generalized to the case of $\mathfrak{t}$-valued functions and distributions.

After these preparations we can finally write down a rigorous version of the heuristic expression on the RHS of Eq. (46) above:

$$\mathrm{WLO}_{rig}(L) := \lim_{n\to\infty}\lim_{s\to 0}\lim_{\epsilon\to 0}\sum_{y\in I}\int_{\sim}\Psi\big(J^{(s,\epsilon,n)}_{b,y}(L)\big)db \qquad (69)$$

where^{v} $\int_{\sim}\cdots db$ is given by

$$\int_{\sim}f\,db = \lim_{T\to\infty}\tfrac{1}{(2T)^d}\int_{[-T,T]^d}f\,db$$

for any measurable bounded function $f : \mathfrak{t}\to\mathbb{R}$. Here $d = \dim(\mathfrak{t})$ and we have identified $\mathfrak{t}$ with $\mathbb{R}^d$ using any fixed orthonormal basis $(e_i)_{i\leq d}$ of $\mathfrak{t}$.

Conjecture 1. $\mathrm{WLO}_{rig}(L)$ *is well-defined. In particular, all limits involved exist.*

If Conjecture 1 above is correct then in view of the semi-rigorous computations in[6,10] (and the rigorous computations in[12,13]) one naturally arrives at the following conjecture:

Conjecture 2. *Assume that* $k > c_{\mathfrak{g}}$ *where* $c_{\mathfrak{g}}\in\mathbb{N}$ *is the dual Coxeter number of* $\mathfrak{g}$ *(cf. Appendix A). Then we have for every* L *fulfilling Assumption 1 above*

$$\mathrm{WLO}_{rig}(L) \sim |L| \qquad (70)$$

where $\sim$ *denotes equality up to a multiplicative constant* $C = C(G,k)$ *and where* $|\cdot|$ *is the shadow invariant for* $M = S^2\times S^1$ *associated to the pair* $(\mathfrak{g},k)$, *cf. Appendix A for the definitions and concrete formulas (cf., in particular, Eq. (A.10)).*

Remark 9.

(1) As the notation $C = C(G,k)$ suggests the constant C referred to above is allowed to depend on G and k but will be independent of L. It will also be independent of the particular choice of the orthonormal basis $(e_i)_i$ of $\mathfrak{t}$ and the Dirac families $\{\delta^{\epsilon}_{\sigma}\mid\epsilon > 0,\sigma\in\Sigma\}$ and $\{\delta^{\epsilon}_t\mid\epsilon > 0, t\in S^1\}$ above. Finally, it will be independent of the particular choice of the auxiliary Riemannian metric $\mathbf{g}$ (as long as $\mathbf{g}$ fulfills Assumption 2 above).

---

^{v}Recall that db is the normalized Lebesgue measure on $\mathcal{B}_c \cong \mathfrak{t}$. We expect that the function $\mathcal{B}_c \ni b \mapsto \Psi\big(J^{(s,\epsilon,n)}_{b,y}(L)\big) \in \mathbb{R}$ appearing in Eq. (69) is periodic. This is why instead of using the proper Lebesgue integral $\int\cdots db$ we use the "mean value" $\int_{\sim}\cdots db$.

(2) Obviously we cannot expect Eq. (70) to hold with "∼" replaced by "=" since in Sec. 2 we have omitted several multiplicative constants. Moreover, Eq. (46) contains "∼" as well.

(3) In the standard literature the shadow invariant $| \cdot |$ associated to the pair $(\mathfrak{g}, k)$ is only defined when $k > c_{\mathfrak{g}}$. It can easily be generalized in a natural way so that it includes the situation $k \leq c_{\mathfrak{g}}$ but it turns out that the so defined generalization of $| \cdot |$ vanishes for $k < c_{\mathfrak{g}}$ and is essentially trivial for $k = c_{\mathfrak{g}}$. We expect that the same applies to $\mathrm{WLO}_{rig}(L)$.

Remark 10. Recall from Remark 2 above that I expect that both Conjecture 1 and Conjecture 2 can be generalized to the situation where L is a certain type of torus ribbon knot in $S^2 \times S^1$. In particular, it is very likely that using the approach above and one can obtain a rigorous continuum analogue of Theorem 5.7 in[14].

4. Open Questions

Question 1. *Are Conjectures 1–2 above indeed true?*

If the answer to Question 1 is "yes", then one arrives naturally at the following question:

Question 2. *Can Assumption 1 be dropped? In other words: will the more or less straightforward[w] generalizations of Conjectures 1–2 to the case of generic[x] ribbon links also be true?*

Before one studies Question 2 on a rigorous level it is reasonable to consider this issue first on an informal level.

Apart from Questions 1 and 2, which are obviously the main questions, also the following two questions are of interest:

---

[w]Recall that when Assumption 1 is dropped we need to modify Assumption 2 (cf. Remark 6 in Sec. 2.5) and some of the constructions & definitions in Sec. 3.3. Moreover, we need to give a (heuristic) derivation/justification for formula (41) in Sec. 2.6 also in the case of general ribbon links L.

[x]In fact, we expect that the class of ribbon links for which our approach is applicable cannot be the class of general ribbon links $L = (R_1, R_2, \ldots, R_m)$. All "singular" twists of the ribbons R_i, $i \leq m$, must probably be excluded. One sufficient condition on L which excludes such singular twists is that each R_{Σ}^i is a local diffeomorphism. Of course, crossings and self-crossing of the projected ribbons R_{Σ}^i, $i \leq m$ (and certain "regular" twists) are nor excluded by this condition.

Question 3. *Can Assumption 2 be dropped? If not, then is there a deeper reason why we have to make such an assumption?*

Question 4. *Is it possible to find regularized versions* $1_{\mathcal{B}_{reg}}^{(n)}$ *and* $\mathrm{Det}_{rig}^{(n)}$, $n \in \mathbb{N}$, *of* $1_{\mathcal{B}_{reg}} : \mathcal{B} \to \mathbb{R}$ *and* $\mathrm{Det}_{rig} : \mathcal{B} \to \mathbb{R}$ *which are more natural than the ones given in Appendix B?*

Appendix A. Turaev's shadow invariant

Let us briefly recall the definition of Turaev's shadow invariant in the situation relevant for us, i.e. for the base manifold $M = \Sigma \times S^1$ with $\Sigma = S^2$.

A.1. Lie theoretic notation

Let G, T, $\mathfrak{g}$, $\mathfrak{t}$, $\langle \cdot, \cdot \rangle$, and $\mathfrak{k}$ be as in Sec. 2 above. Using the scalar product $\langle \cdot, \cdot \rangle$ we can make the identification $\mathfrak{t} \cong \mathfrak{t}^*$. Let us now fix a Weyl chamber $\mathcal{C} \subset \mathfrak{t}$ and introduce the following notation:

- $\mathcal{R} \subset \mathfrak{t}^*$: the set of real roots associated to $(\mathfrak{g}, \mathfrak{t})$
- $\mathcal{R}_+ \subset \mathcal{R}$: the set of positive (real) roots corresponding to $\mathcal{C}$
- ρ: half sum of positive roots ("Weyl vector")
- θ: unique long root in the Weyl chamber $\mathcal{C}$.
- $c_{\mathfrak{g}} = 1 + \langle \theta, \rho \rangle$: the dual Coxeter number of $\mathfrak{g}$.
- $I \subset \mathfrak{t}$: the kernel of $\exp_{|\mathfrak{t}} : \mathfrak{t} \to T$. We remark that from the assumption that G is simply-connected it follows that I coincides with the lattice Γ which is generated by the set of real coroots associated to $(\mathfrak{g}, \mathfrak{t})$.
- $\Lambda \subset \mathfrak{t}^*(\cong \mathfrak{t})$: the *real* weight lattice associated to $(\mathfrak{g}, \mathfrak{t})$, i.e. Λ is the lattice which is dual to $\Gamma = I$.
- $\Lambda_+ \subset \Lambda$: the set of dominant weights corresponding to $\mathcal{C}$, i.e. $\Lambda_+ := \bar{\mathcal{C}} \cap \Lambda$.
- $\Lambda_+^k \subset \Lambda$, $k \in \mathbb{N}$: the subset of Λ_+ given by $\Lambda_+^k := \{\lambda \in \Lambda_+ \mid \langle \lambda, \theta \rangle \leq k - c_{\mathfrak{g}}\}$ (the "set of dominant weights which are integrable at level $l := k - c_{\mathfrak{g}}$").
- $\mathcal{W} \subset \mathrm{GL}(\mathfrak{t})$: the Weyl group of the pair $(\mathfrak{g}, \mathfrak{t})$.
- $\mathcal{W}_{\mathrm{aff}} \subset \mathrm{Aff}(\mathfrak{t})$: the "affine Weyl group of $(\mathfrak{g}, \mathfrak{t})$", i.e. the subgroup of $\mathrm{Aff}(\mathfrak{t})$ generated by $\mathcal{W}$ and the set of translations $\{\tau_x \mid x \in \Gamma\}$ where $\tau_x : \mathfrak{t} \ni b \mapsto b + x \in \mathfrak{t}$.
- $\mathcal{W}_k \subset \mathrm{Aff}(\mathfrak{t})$, $k \in \mathbb{N}$: the subgroup of $\mathrm{Aff}(\mathfrak{t})$ given by $\{\psi_k \circ \sigma \circ \psi_k^{-1} \mid \sigma \in \mathcal{W}_{\mathrm{aff}}\}$ where $\psi_k : \mathfrak{t} \ni b \mapsto b \cdot k - \rho \in \mathfrak{t}$ (the "quantum Weyl group corresponding to the level $l := k - c_{\mathfrak{g}}$").

The following formulas are used in Sec. 2.5 above and Appendix B below. For $b \in \mathfrak{t}$ we have

$$b \in \mathfrak{t}_{reg} \quad \Leftrightarrow \quad [\forall \alpha \in \mathcal{R}_+ : \alpha(b) \notin \mathbb{Z}] \tag{A.1}$$

and

$$\det(\mathrm{id}_\mathfrak{e} - \exp(\mathrm{ad}(b))_{|\mathfrak{e}}) = \prod_{\alpha \in \mathcal{R}} (1 - e^{2\pi i \alpha(b)})$$

$$= \prod_{\alpha \in \mathcal{R}+} (1 - e^{2\pi i \alpha(b)})(1 - e^{-2\pi i \alpha(b)}) = \prod_{\alpha \in \mathcal{R}+} 4\sin^2(\pi\alpha(b)). \tag{A.2}$$

A.2. The shadow invariant

Let $L = (l_1, l_2, \ldots, l_m)$, $m \in \mathbb{N}$, be a framed link in $M = \Sigma \times S^1$. For simplicity we will assume that each l_i, $i \leq m$ is equipped with a "horizontal" framing[y] Let $V(L)$ denote the set of points $p \in \Sigma$ where the loops l_Σ^i, $i \leq m$, cross themselves or each other (the "crossing points") and $E(L)$ the set of curves in Σ into which the loops $l_\Sigma^1, l_\Sigma^2, \ldots, l_\Sigma^m$ are decomposed when being "cut" in the points of $V(L)$. We assume that there are only finitely many connected components $Y_0, Y_1, Y_2, \ldots, Y_{m'}$, $m' \in \mathbb{N}$ ("faces") of $\Sigma \backslash (\bigcup_i \mathrm{arc}(l_\Sigma^i))$ and set

$$F(L) := \{Y_0, Y_1, Y_2, \ldots, Y_{m'}\}.$$

As explained in[26] one can associate in a natural way a half integer $\mathrm{gleam}(Y) \in \frac{1}{2}\mathbb{Z}$, called "gleam" of Y, to each face $Y \in F(L)$. In the special case where the link L is a framed link which is induced by a ribbon link $L = L_{ribb}$ fulfilling Assumption 1 in Sec. 2.4 we have the explicit formula

$$\mathrm{gleam}(Y) = \sum_{i \text{ with } \mathrm{arc}(l_\Sigma^i) \subset \partial Y} \mathrm{wind}(l_{S^1}^i) \cdot \mathrm{sgn}(Y; l_\Sigma^i) \in \mathbb{Z} \tag{A.3}$$

where $\mathrm{wind}(l_{S^1}^i)$ is the winding number of the loop $l_{S^1}^i$ and where $\mathrm{sgn}(Y; l_\Sigma^i)$ is given by

$$\mathrm{sgn}(Y; l_\Sigma^i) := \begin{cases} 1 & \text{if } Y \subset Z_i^+ \\ -1 & \text{if } Y \subset Z_i^- \end{cases} \tag{A.4}$$

---

[y]Here we use the terminology of Remark 4.5 in[12]. We remark that in the special case when L is the framed link which is induced by a ribbon link $L = L_{ribb}$ fulfilling Assumption 1 in Sec. 2.4 then each l_i will automatically be equipped with a horizontal framing.

Here Z_i^+ (resp. Z_i^-) is the unique connected component Z of $\Sigma\backslash\mathrm{arc}(l_\Sigma^i)$ such that l_Σ^i runs around Z in the "positive" (resp. "negative") direction.

Assume that each loop l_i in the link L is equipped with a "color" ρ_i, i.e. a finite-dimensional complex representation of G. By $\gamma_i \in \Lambda_+$ we denote the highest weight of ρ_i and set $\gamma(e) := \gamma_i$ for each $e \in E(L)$ where $i \leq n$ denotes the unique index such that $\mathrm{arc}(e) \subset \mathrm{arc}(l_i)$. Finally, let $col(L)$ be the set of all mappings $\varphi : \{Y_0, Y_1, Y_2, \ldots, Y_{m'}\} \to \Lambda_+^k$ ("area colorings").

We can now define the "shadow invariant" $|L|$ of the (colored and "horizontally framed") link L associated to the pair $(\mathfrak{g}, k)$ by

$$|L| := \sum_{\varphi \in col(L)} |L|_1^\varphi\, |L|_2^\varphi\, |L|_3^\varphi\, |L|_4^\varphi \tag{A.5}$$

with

$$|L|_1^\varphi = \prod_{Y \in F(L)} \dim(\varphi(Y))^{\chi(Y)} \tag{A.6a}$$

$$|L|_2^\varphi = \prod_{Y \in F(L)} \exp(\tfrac{\pi i}{k}\langle \varphi(Y), \varphi(Y) + 2\rho\rangle)^{\mathrm{gleam}(Y)} \tag{A.6b}$$

$$|L|_3^\varphi = \prod_{e \in E_*(L)} N_{\gamma(e)\varphi(Y_e^+)}^{\varphi(Y_e^-)} \tag{A.6c}$$

$$|L|_4^\varphi = \Big(\prod_{e \in E(L)\backslash E_*(L)} S(e, \varphi) \Big) \times \Big(\prod_{x \in V(L)} T(x, \varphi) \Big). \tag{A.6d}$$

Here Y_e^+ (resp. Y_e^-) denotes the unique face Y such that $\mathrm{arc}(e) \subset \partial Y$ and, additionally, the orientation on $\mathrm{arc}(e)$ described above coincides with (resp. is opposite to) the orientation which is obtained by restricting the orientation on ∂Y to e. Moreover, we have set (for $\lambda, \mu, \nu \in \Lambda_+^k$)

$$\dim(\lambda) := \prod_{\alpha \in \mathcal{R}_+} \frac{\sin \frac{\pi\langle\lambda+\rho,\alpha\rangle}{k}}{\sin \frac{\pi\langle\rho,\alpha\rangle}{k}} \tag{A.7}$$

$$N_{\mu\nu}^\lambda := \sum_{\tau \in \mathcal{W}_k} \mathrm{sgn}(\tau) m_\mu(\nu - \tau(\lambda)) \tag{A.8}$$

where $m_\mu(\beta)$ is the multiplicity of the weight β in the unique (up to equivalence) irreducible representation ρ_μ with highest weight μ and $\mathcal{W}_k$ is as above. $E_*(L)$ is a suitable subset of $E(L)$ (cf. the notion of "circle-1-strata" in Chap. X, Sec. 1.2 in[27]).

The explicit expression for the factors $T(x, \varphi)$ appearing in $|L|_4^\varphi$ above involves the so-called "quantum 6j-symbols" (cf. Chap. X, Sec. 1.2 in[27]) associated to the quantum group $U_q(\mathfrak{g}_\mathbb{C})$ where q is the root of unity

$$q := \exp(\tfrac{2\pi i}{k}). \tag{A.9}$$

We omit the explicit formulae for $T(x, \varphi)$ and $S(e, \varphi)$ since they are irrelevant for our purposes. Indeed, if L is the framed link which is induced by a ribbon link $L = L_{ribb}$ fulfilling Assumption 1 in Sec. 2.4 the set $V(L)$ is empty and the set $E_*(L)$ coincides with $E(L)$, so Eq. (A.5) then reduces to

$$|L| = \sum_{\varphi \in col(L)} \left(\prod_{i=1}^{m} N_{\gamma(l_i)\varphi(Y_i^+)}^{\varphi(Y_i^-)} \right) \left(\prod_{Y \in F(L)} \dim(\varphi(Y))^{\chi(Y)} \right.$$

$$\left. \times \exp(\tfrac{\pi i}{k} \langle \varphi(Y), \varphi(Y) + 2\rho \rangle)^{\mathrm{gleam}(Y)} \right) \quad \text{(A.10)}$$

where we have set $Y_i^{\pm} := Y_{l_{\Sigma}^i}^{\pm}$.

Appendix B. Construction of $1_{\mathcal{B}_{reg}}^{(n)}$ and $\mathrm{Det}_{rig}^{(n)}$

We will now describe the regularized versions $1_{\mathcal{B}_{reg}}^{(n)} : \mathcal{B} \to \mathbb{R}$ and $\mathrm{Det}_{rig}^{(n)} : \mathcal{B} \to \mathbb{R}$, $n \in \mathbb{N}$, of the functions $1_{\mathcal{B}_{reg}}$ and Det_{rig} which we used in Sec. 3.4 above.

- Let $\triangle_0$ be a fixed finite triangulation of Σ which is "compatible" with L in the sense that $\mathrm{arc}(L_{\Sigma})$ is contained in the 1-skeleton of $\triangle_0$. For each $n \in \mathbb{N}$ let $\triangle_n$ be the barycentric sub division of $\triangle_{n-1}$. We denote by $\mathcal{F}_2(\triangle_n)$ the set of 2-faces of $\triangle_n$. For $B \in \mathcal{B}$ and $F \in \mathcal{F}_2(\triangle_n)$ let $B(F)$ be the "mean value" of B on F, i.e.

$$B(F) := \frac{\int_F B d\mu_{\mathbf{g}}}{\int_F 1 d\mu_{\mathbf{g}}} \in \mathfrak{t} \quad \text{(B.1)}$$

- We approximate the indicator function $1_{\mathfrak{t}_{reg}} : \mathfrak{t} \to [0, 1]$ by a suitable sequence of trigonometric polynomials[z] $(1_{\mathfrak{t}_{reg}}^{(n)})_{n \in \mathbb{N}}$ such that for all $n \in \mathbb{N}$ we have

  $$1_{\mathfrak{t}_{reg}}^{(n)}(b) = 0 \quad \text{for all } b \in \mathfrak{t}_{sing} := \mathfrak{t} \backslash \mathfrak{t}_{reg}, \text{ and}$$

  $$|1_{\mathfrak{t}_{reg}}^{(n)}(b) - 1| \leq 1/N_n^2 \quad \text{for all } b \in \mathfrak{t} \text{ outside the } \tfrac{1}{n}\text{-neighborhood of } \mathfrak{t}_{sing}$$

---

[z] In order to see that such a sequence always exist we first choose, for each fixed $n \in \mathbb{N}$, a smooth 1-periodic function $\psi^{(n)} : \mathbb{R} \to \mathbb{R}$ such that $\psi^{(n)}(x) = 0$ for all $x \in \mathbb{Z}$ and $\psi^{(n)}(x) = 1$ for all x outside the $\tfrac{C}{n}$-neighborhood of $\mathbb{Z} \subset \mathbb{R}$ for some fixed $C > 0$. Since $\psi^{(n)}$ is a smooth periodic function its Fourier series converges uniformly. Accordingly, for every fixed $\epsilon = \epsilon_n > 0$ we can find a (1-periodic) trigonometric polynomial $p^{(n)}$ such that $\|\psi^{(n)} - p^{(n)}\|_{\infty} < \epsilon$ and, consequently, $\|\psi^{(n)} - \bar{p}^{(n)}\|_{\infty} < 2\epsilon$ where $\bar{p}^{(n)} := p^{(n)} - p^{(n)}(0)$. (Clearly, $\bar{p}^{(n)}(x) = 0$ for all $x \in \mathbb{Z}$.) In view of the identity $1_{\mathfrak{t}_{reg}}(b) = \prod_{\alpha \in \mathcal{R}_+} 1_{\mathbb{R} \backslash \mathbb{Z}}(\alpha(b))$, cf. Eq. (A.1) above, we now define $1_{\mathfrak{t}_{reg}}^{(n)}$ by $1_{\mathfrak{t}_{reg}}^{(n)}(b) := \prod_{\alpha \in \mathcal{R}_+} \bar{p}^{(n)}(\alpha(b))$ for all $b \in \mathfrak{t}$. Clearly, if $C > 0$ was chosen small enough and for every $n \in \mathbb{N}$ the number $\epsilon = \epsilon_n$ was chosen small enough we obtain a family $(1_{\mathfrak{t}_{reg}}^{(n)})_{n \in \mathbb{N}}$ with the desired properties.

where we have set $N_n := \#\mathcal{F}_2(\triangle_n)$.

For each $n \in \mathbb{N}$ we then introduce $1^{(n)}_{\mathcal{B}_{reg}} : \mathcal{B} \to \mathbb{R}$ by

$$1^{(n)}_{\mathcal{B}_{reg}}(B) = \prod_{F \in \mathcal{F}_2(\triangle_n)} 1^{(n)}_{t_{reg}}(B(F)) \quad \forall B \in \mathcal{B}$$

- Recall from Sec. 2.5 that for $B \in \mathcal{B}_{reg}$ we have

$$\text{Det}_{rig,\alpha}(B) = \exp\left(\int_\Sigma \left[\log(2\sin(\pi\alpha(B(\sigma)))) \tfrac{1}{4\pi} R_{\mathbf{g}}(\sigma)\right] d\mu_{\mathbf{g}}(\sigma)\right)$$

where $\log : \mathbb{R}\backslash\{0\} \to \mathbb{C}$ is the restriction to $\mathbb{R}\backslash\{0\}$ of the principal branch of the complex logarithm, i.e. is given by

$$\log(x) = \ln(|x|) + \pi i H(-x) \quad \forall x \in \mathbb{R}\backslash\{0\}$$

where $H(x) = (1 + \text{sgn}(x))/2$ is the Heaviside function.

For $B \in \mathcal{B}$ and $n \in \mathbb{N}$ we now define the "step function" $B^{(n)} : \Sigma \to \mathfrak{t}$ by

$$B^{(n)} = \sum_{F \in \mathcal{F}_2(\triangle_n)} B(F) \cdot 1_F \tag{B.2}$$

where $B(F)$ is as above. Moreover, for fixed $n \in \mathbb{N}$ we now replace $\exp(x)$, $x \in \mathbb{R}$, by the nth Taylor polynomial $\exp^{(n)}(x) = \sum_{k=0}^n \frac{x^k}{k!}$ and $\log(x)$ by $\log^{(n)}(x)$ where $(\log^{(n)})_{n \in \mathbb{N}}$ is any fixed sequence of polynomial functions $\log^{(n)}(x) : I \to \mathbb{C}$ which converges uniformly to log on every compact subinterval of $I := [-2, 2]\backslash\{0\}$.

After these preparations we set for each $B \in \mathcal{B}$.

$$\text{Det}^{(n)}_{rig,\alpha}(B) := \exp^{(n)}\left(\int_\Sigma \left[\log^{(n)}(2\sin(\pi\alpha(B^{(n)}(\sigma)))) \tfrac{1}{4\pi} R_{\mathbf{g}}(\sigma)\right] d\mu_{\mathbf{g}}(\sigma)\right) \tag{B.3a}$$

and define $\text{Det}^{(n)}_{rig} : \mathcal{B} \to \mathbb{R}$, $n \in \mathbb{N}$, by

$$\text{Det}^{(n)}_{rig}(B) := \prod_{\alpha \in \mathcal{R}_+} \text{Det}^{(n)}_{rig,\alpha}(B) \quad \forall B \in \mathcal{B}. \tag{B.3b}$$

Remark 11.

(1) After carrying out the $\epsilon \to 0$ and the $s \to 0$-limits on the RHS of Eq. (69) step functions of the type $B = \sum_{i=1}^r b_i 1_{Y_i}$ appear in the calculations, cf. the paragraph after Eq. (30) in Sec. 2.5. It is useful (and straightforward) to generalize the definitions of $1^{(n)}_{\mathcal{B}_{reg}}(B)$ and $\text{Det}^{(n)}_{rig}(B)$ to such functions B.

(2) In order to justify that $(1_{\mathcal{B}_{reg}}^{(n)})_{n\in\mathbb{N}}$ is indeed a regularization of $1_{\mathcal{B}_{reg}}$ observe that if $B \in \mathcal{B}_{reg}$ then

$$\lim_{n\to\infty} 1_{\mathcal{B}_{reg}}^{(n)}(B) = 1.$$

On the other hand, we do not necessarily have $\lim_{n\to\infty} 1_{\mathcal{B}_{reg}}^{(n)}(B) = 0$ if $B \notin \mathcal{B}_{reg}$. However, if B is a step function of the type $B = \sum_{i=1}^{r} b_i 1_{Y_i}$ as in Eq. (30) in Sec. 2.5 (which is the only case which will play a role in our computations after the $\epsilon \to 0$ and the $s \to 0$-limits on the RHS of Eq. (69) have been taken) then we do have

$$\lim_{n\to\infty} 1_{\mathcal{B}_{reg}}^{(n)}(B) = \begin{cases} 1 & \text{if } B \text{ takes only values in } t_{reg} \\ 0 & \text{if } B \text{ takes at least one value in } t_{sing} \end{cases}$$

where $1_{\mathcal{B}_{reg}}^{(n)}(B)$ is the aforementioned generalization (and where we use the condition mentioned above that the triangulation $\triangle_0$ of Σ and therefore also all barycentric subdivisions $\triangle_n$ of $\triangle_0$ are compatible with L in the sense above).

(3) In order to justify that $(\mathrm{Det}_{rig}^{(n)})_{n\in\mathbb{N}}$ is a regularization of $\mathrm{Det}_{rig}(B)$ we will now verify that for all $B \in \mathcal{B}_{reg}$ we have indeed

$$\lim_{n\to\infty} \mathrm{Det}_{rig}^{(n)}(B) = \mathrm{Det}_{rig}(B). \tag{B.4}$$

Observe first that since $B \in \mathcal{B}_{reg}$ we have $B(\sigma) \in t_{reg}$ and therefore (cf. Eq. (A.1)) $\sin(\pi\alpha(B(\sigma))) \neq 0$ for all $\sigma \in \Sigma$. Now let $\mathcal{S}_n$ be the 1-skeleton of $\triangle_n$ and let $\mathcal{N} = \bigcup_{n\in\mathbb{N}} \mathcal{S}_n$. Then for all $\sigma \in \Sigma$ which do not lie in $\mathcal{N}$ we have $\lim_{n\to\infty} B^{(n)}(\sigma) = B(\sigma)$ and therefore also $\sin(\pi\alpha(B^{(n)}(\sigma))) \neq 0$ if $n \in \mathbb{N}$ is sufficiently large. According to the choice of $\log^{(n)}$ we therefore obtain

$$\lim_{n\to\infty} \log^{(n)}(2\sin(\pi\alpha(B^{(n)}(\sigma)))) = \log(2\sin(\pi\alpha(B(\sigma)))). \tag{B.5}$$

Since $\mathcal{N}$ is a $\mu_{\mathbf{g}}$-zero set Eq. (B.5) holds for $\mu_{\mathbf{g}}$-almost all $\sigma \in \Sigma$. From Eqs. (B.3) it now easily follows that Eq. (B.4) is indeed fulfilled.

Finally, observe that for step functions $B = \sum_{i=1}^{r} b_i 1_{Y_i}$ of the type mentioned above which satisfy $1_{\mathcal{B}_{reg}}^{(n)}(B) \neq 0$ for sufficiently large n we have again Eq. (B.4) (and, in fact, even $\mathrm{Det}_{rig}^{(n)}(B) = \mathrm{Det}_{rig}(B)$ for sufficiently large n).

I want to emphasize once more that the regularized versions $1_{\mathcal{B}_{reg}}^{(n)} : \mathcal{B} \to \mathbb{R}$ and $\mathrm{Det}_{rig}^{(n)} : \mathcal{B} \to \mathbb{R}$ of the functions $1_{\mathcal{B}_{reg}}$ and Det_{rig} introduced above are probably not the best ones. It would be desirable to find a more elegant and more natural regularization (cf. Question 4 in Sec. 4 above).

References

1. S. Albeverio and A.N. Sengupta. A Mathematical Construction of the Non-Abelian Chern-Simons Functional Integral. *Commun. Math. Phys.*, 186:563–579, 1997.
2. M. Blau and G. Thompson. Derivation of the Verlinde Formula from Chern-Simons Theory and the G/G model. *Nucl. Phys.*, B408(1):345–390, 1993.
3. M. Blau and G. Thompson. Lectures on 2d Gauge Theories: Topological Aspects and Path Integral Techniques. In E. Gava et al., editor, *Proceedings of the 1993 Trieste Summer School on High Energy Physics and Cosmology*, pages 175–244. World Scientific, Singapore, 1994.
4. M. Blau and G. Thompson. On Diagonalization in $Map(M, G)$. *Commun. Math. Phys.*, 171:639–660, 1995.
5. M. de Faria, J. Potthoff, and L. Streit. The Feynman integrand as a Hida distribution. *J. Math. Phys.*, 32(8):2123–2127, 1991.
6. S. de Haro and A. Hahn. Chern-Simons theory and the quantum Racah formula. *Rev. Math. Phys.*, 25:1350004, 41 pp., 2013 [arXiv:math-ph/0611084]
7. A. Hahn. The Wilson loop observables of Chern-Simons theory on $\mathbb{R}^3$ in axial gauge. *Commun. Math. Phys.*, 248(3):467–499, 2004.
8. A. Hahn. Chern-Simons models on $S^2 \times S^1$, torus gauge fixing, and link invariants I. *J. Geom. Phys.*, 53(3):275–314, 2005.
9. A. Hahn. Chern-Simons models on $S^2 \times S^1$, torus gauge fixing, and link invariants II. *J. Geom. Phys.*, 58:1124–1136, 2008.
10. A. Hahn. An analytic Approach to Turaev's Shadow Invariant. *J. Knot Th. Ram.*, 17(11): 1327–1385, 2008 [see arXiv:math-ph/0507040v7 (2011) for the most recent version]
11. A. Hahn. White noise analysis in the theory of three-manifold quantum invariants. In A.N. Sengupta and P. Sundar, editors, *Infinite Dimensional Stochastic Analysis*, volume XXII of *Quantum Probability and White Noise Analysis*, pages 201–225. World Scientific, 2008.
12. A. Hahn. From simplicial Chern-Simons theory to the shadow invariant I, J. Math. Phys. 56: 032301, 52 pp., 2015 [arXiv:1206.0439v6]
13. A. Hahn. From simplicial Chern-Simons theory to the shadow invariant II, J. Math. Phys. 56: 032302, 46 pp., 2015 [arXiv:1206.0441v5].
14. A. Hahn. Torus Knots and the Chern-Simons path integral: a rigorous treatment. Preprint, 2015 [arXiv:1508.03804].
15. T. Hida, H.-H. Kuo, J. Potthoff, and L. Streit. *White Noise. An infinite dimensional Calculus*. Dordrecht: Kluwer, 1993.
16. Y. Kondratiev, P. Leukert, J. Potthoff, L. Streit, W. Westerkamp. Generalized Functionals in Gaussian Spaces – the Characterization Theorem Revisited. *J. Funct. Anal.*, 141(2), 301–318, 1996
17. H.-H. Kuo, J. Potthoff, and L. Streit. A characterization of white noise test functionals. *Nagoya Math. J.*, 121, 185–194, 1991
18. G. Kuperberg. Quantum invariants of knots and 3-manifolds (book review). *Bull. Amer. Math. Soc.*, 33(1):107–110, 1996.
19. P. Leukert and J. Schäfer. A Rigorous Construction of Abelian Chern-Simons

Path Integrals using White Noise Analysis. *Rev. Math. Phys.*, 8(3):445–456, 1996.

20. H. P. McKean and I. M. Singer. Curvature and the eigenvalues of the Laplacian. *J. Differential Geometry*, 1(1):43–69, 1967.

21. V. K. Patodi. Curvature and the eigenforms of the Laplace Operator. *J. Differential Geometry*, 5(1):233–249, 1971.

22. N.Y. Reshetikhin and V.G. Turaev. Ribbon graphs and their invariants derived from quantum groups. *Commun. Math. Phys.*, 127:1–26, 1990.

23. N.Y. Reshetikhin and V.G. Turaev. Invariants of three manifolds via link polynomials and quantum groups. *Invent. Math.*, 103:547–597, 1991.

24. S. Sawin. Jones-Witten invariants for non-simply connected Lie groups and the geometry of the Weyl alcove. arXiv: math.QA/9905010, 1999.

25. L. Streit and T. Hida. Generalized Brownian functionals and the Feynman integral. *Stochastic Process. Appl.*, 16(1):55–69, 1984.

26. V. G. Turaev, "Shadow Links and Face Models of Statistical Mechanics", J. Differential Geometry 36:35-74, 1992

27. V. G. Turaev, "Quantum Invariants of Knots and 3-Manifolds", De Gruyter, 1994

28. E. Witten. Quantum Field Theory and the Jones Polynomial. *Commun. Math. Phys.*, 121:351–399, 1989.

Chapter 6

A White Noise Approach to Insider Trading

Bernt Øksendal[1,2] and Elin Engen Røse[1]

[1]*Department of Mathematics, University of Oslo, P.O. Box 1053
Blindern, N–0316 Oslo, Norway*
and
[2] *Norwegian School of Economics (NHH), Helleveien 30, N–5045 Bergen,
Norway.*
E-Mail:oksendal@math.uio.no and elinero@math.uio.no

We present a new approach to the optimal portfolio problem for an insider with logarithmic utility. Our method is based on white noise theory, stochastic forward integrals, Hida-Malliavin calculus and the Donsker delta function.

1. Introduction

The purpose of this paper is to use concepts and methods from anticipating stochastic calculus, particularly from white noise theory and Hida-Malliavin calculus, to study optimal portfolio problems for an insider in a financial market driven by Brownian motion $B(t)$. Our basic problem setup is related to the setup in[28] :

We assume that the insider at any time $t \in [0, T]$ has access to the information (σ-algebra) $\mathcal{F}_t$ generated by the driving Brownian motion up to time t, and in addition knows the value of some $\mathcal{F}_{T_0}$-measurable random variable Y, where $T_0 > T$ is some given future time. With this information flow $\mathbb{H} = \{\mathcal{H}_t\}_{t \in [0,T]}$ with $\mathcal{H}_t = \mathcal{F}_t \vee \sigma(Y)$ to her disposal, she tries to find the $\mathbb{H}$-adapted portfolio π^* that maximises the expected logarithmic utility of the corresponding wealth at a given terminal time $T < T_0$.

In[28] it is assumed that the insider filtration $\mathbb{H}$ allows an *enlargement of filtration*, i.e. that there exists an $\mathbb{H}$-adapted process $\alpha(s)$ such that

$$\tilde{B}(t) := B(t) - \int_0^t \alpha(s)ds \tag{1}$$

is a Brownian motion with respect to $\mathbb{H}$. If this holds, the original problem, which was a priori a problem with anticipating stochastic calculus, can be transformed back to a semimartingale setting and (in some cases) solved using classical solution methods. In terms of the process α,[28] prove that the optimal insider portfolio can be written

$$\pi^*(t) = \frac{b(t) - r(t)}{\sigma^2(t)} + \frac{\alpha(t)}{\sigma(t)} \tag{2}$$

where $r(t)$ is the interest rate of the risk free asset, and $b(t), \sigma(t)$ are the drift term and the volatility of the risky asset, respectively.

In this chapter, we do not assume (1), but instead we follow the approach in[6] and in the recent paper[14] and work with *anticipative stochastic calculus*. This means that the stochastic integrals involved in the anticipating insider portfolio are represented by *forward integrals*. The forward integral, originally introduced in[30], is an extension of the Itô integral, in the sense that it coincides with the Itô integral if the integrand is adapted. (See below.) It was first applied to insider trading in[6], where it is pointed out why this integral appears naturally in the modelling of portfolio generated wealth processes in insider trading. In[6] a kind of converse to the result in[28] is proved, namely that if an optimal insider portfolio exists, then the underlying Brownian motion $B(t)$ is indeed a semimartingale with respect to the insider filtration $\mathbb{H}$, and hence (1) holds.

This chapter differs also fundamentally from[6], because we use white noise theory and Hida-Malliavin calculus to solve the anticipative optimal portfolio problem directly. The paper closest to ours is[14]. Indeed, this chapter might be regarded as a discussion of a special case in[14], although the method used here is different and specially adapted to the logarithmic utility case. One of our main results is that if the *Donsker delta function* $\delta_Y(y)$ of Y exists in the Hida space $(\mathcal{S})^*$ of stochastic distributions, and the conditional expectations $E[\delta_Y(y)|\mathcal{F}_t]$ and $E[D_t\delta_Y(y)|\mathcal{F}_t]$ both belong to $L^2(\lambda \times P)$, where λ is Lebesgue measure on $[0, T]$ and P is the probability law of $B(\cdot)$, then the optimal insider portfolio is

$$\pi^*(t) = \frac{b(t) - r(t)}{\sigma^2(t)} + \frac{E[D_t\delta_Y(y) \mid \mathcal{F}_t]_{y=Y}}{\sigma(t)E[\delta_Y(y) \mid \mathcal{F}_t]_{y=Y}} \tag{3}$$

where D_t denotes the Hida-Malliavin derivative at t. See Theorem 2.

Comparing (2) and (3) we get the following *enlargement of filtration formula*, which is of independent interest (see Theorem 3):

$$\alpha(t) = \frac{E[D_t\delta_Y(y) \mid \mathcal{F}_t]_{y=Y}}{E[\delta_Y(y) \mid \mathcal{F}_t]_{y=Y}}. \tag{4}$$

For more general results in this direction, see[15] .

For simplicity we only discuss the Brownian motion case in this paper. For more information about Hida-Malliavin calculus in a white noise setting and extensions to Lévy processes and more general insider control problems, see[14] .

2. Background in white noise theory and Hida-Malliavin calculus

In this section we summarise the basic notation and results we will need from white noise theory and the associated Hida-Malliavin calculus. For more details see e.g.[5,8,10,13,14,29] and the references therein. For a general introduction to white noise theory see[16,17] and Ch.1.

2.1. *List of notation*

- $F \diamond G =$ the Wick product of random variables F and G.
- $F^{\diamond n} = F \diamond F \diamond F... \diamond F$ (n times). (The nth Wick power of F).
- $\exp^{\diamond}(F) = \Sigma_{n=0}^{\infty} \frac{1}{n!} F^{\diamond n}$. (The Wick exponential of F.)
- $\varphi^{\diamond}(F) =$ the Wick version of the random variable $\varphi(F)$.
- $D_t F =$ the Hida-Malliavin derivative of F at t with respect to $B(\cdot)$. This is denoted by $\partial_t F$ in[23] , see page 30 there.
- $D_t(\varphi(F)) = ((\varphi)')(F) \diamond D_t F$. (The Hida-Malliavin chain rule.)
- $D_t(\varphi^{\diamond}(F)) = ((\varphi)')^{\diamond}(F) \diamond D_t F$. (The Wick chain rule.)
- $(\mathcal{S}), (\mathcal{S})' =$ the Hida stochastic test function space and stochastic distribution space, respectively.
- $(\mathcal{S}) \subset L^2(P) \subset (\mathcal{S})'$.
  Here, as usual, $L^2(P)$ is the set of random variables F with $E[F^2] < \infty$, where $E[\cdot]$ denotes expectation with respect to the probability measure P. If we choose P to be the white noise probability measure μ with $d = 1$ (Ch.1, Definition 2), then $L^2(P)$ coincides with (L^2) defined in Ch.1, Definition 3.

For more information about the Wick calculus we refer to Section 7 of Ch. 1.

2.2. *The forward integral with respect to Brownian motion*

The forward integral with respect to Brownian motion was first defined in the seminal paper[30] and further studied in[31,32]. This integral was introduced in the modelling of insider trading in[6] and then applied by several authors in questions related to insider trading and stochastic control with advanced information (see, e.g.,[11]). The forward integral was later extended to Poisson random measure integrals in[10].

Definition 1. We say that a stochastic process $\phi = \phi(t), t \in [0, T]$, is *forward integrable* (in the weak sense) over the interval $[0, T]$ with respect to B if there exists a process $I = I(t), t \in [0, T]$, such that

$$\sup_{t \in [0,T]} |\int_0^t \phi(s) \frac{B(s + \epsilon) - B(s)}{\epsilon} ds - I(t)| \to 0, \quad \epsilon \to 0^+ \qquad (5)$$

in probability. In this case we write

$$I(t) := \int_0^t \phi(s) d^- B(s), t \in [0, T], \qquad (6)$$

and call $I(t)$ the *forward integral* of ϕ with respect to B on $[0, t]$.

The following results give a more intuitive interpretation of the forward integral as a limit of Riemann sums:

Lemma 1. *Suppose ϕ is càglàd and forward integrable. Then*

$$\int_0^T \phi(s) d^- B(s) = \lim_{\triangle t \to 0} \sum_{j=1}^{J_n} \phi(t_{j-1})(B(t_j) - B(t_{j-1})) \qquad (7)$$

with convergence in probability. Here the limit is taken over the partitions $0 = t_0 < t_1 < ... < t_{J_n} = T$ of $t \in [0, T]$ with $\triangle t := \max_{j=1,...,J_n}(t_j - t_{j-1}) \to 0, n \to \infty$.

Remark 1. From the previous lemma we can see that, if the integrand ϕ is $\mathcal{F}$-adapted, then the Riemann sums are also an approximation to the Itô integral of ϕ with respect to the Brownian motion. Hence in this case the forward integral and the Itô integral coincide. In this sense we can regard the forward integral as an extension of the Itô integral to a nonanticipating setting.

We now give some useful properties of the forward integral. The following result is an immediate consequence of the definition.

Lemma 2. *Suppose ϕ is a forward integrable stochastic process and G a random variable. Then the product $G\phi$ is forward integrable stochastic process and*

$$\int_0^T G\phi(t)d^-B(t) = G\int_0^T \phi(t)d^-B(t). \tag{8}$$

The next result shows that the forward integral is an extension of the integral with respect to a semimartingale:

Lemma 3. *Let $\mathbb{G} := \{\mathcal{G}_t, t \in [0,T]\}(T > 0)$ be a given filtration. Suppose that*

(1) B is a semimartingale with respect to the filtration $\mathbb{G}$.
(2) ϕ is $\mathbb{G}$-predictable and the integral

$$\int_0^T \phi(t)dB(t), \tag{9}$$

with respect to B, exists.
Then ϕ is forward integrable and

$$\int_0^T \phi(t)d^-B(t) = \int_0^T \phi(t)dB(t). \tag{10}$$

We now turn to the Itô formula for forward integrals. In this connection it is convenient to introduce a notation that is analogous to the classical notation for Itô processes.

Definition 2. A *forward process* (with respect to B) is a stochastic process of the form

$$X(t) = x + \int_0^t u(s)ds + \int_0^t v(s)d^-B(s), \quad t \in [0,T], \tag{11}$$

(x constant), where $\int_0^T |u(s)|ds < \infty$, **P**-a.s. and v is a forward integrable stochastic process. A shorthand notation for (11) is that

$$d^-X(t) = u(t)dt + v(t)d^-B(t). \tag{12}$$

Theorem 1. The one-dimensional Itô formula for forward integrals.
Let

$$d^-X(t) = u(t)dt + v(t)d^-B(t) \tag{13}$$

be a forward process. Let $f \in \mathbf{C}^{1,2}([0,T] \times \mathbb{R})$ and define

$$Y(t) = f(t, X(t)), \quad t \in [0,T]. \tag{14}$$

Then $Y(t), t \in [0,T]$, is also a forward process and

$$d^-Y(t) = \frac{\partial f}{\partial t}(t, X(t))dt + \frac{\partial f}{\partial x}(t, X(t))d^-X(t) + \frac{1}{2}\frac{\partial^2 f}{\partial x^2}(t, X(t))v^2(t)dt. \tag{15}$$

We also need the following forward integral result, which is obtained by an adaptation of the proof of Theorem 8.18 in[8] :

Proposition 1. *Let φ be a càglàd and forward integrable process in $L^2(\lambda \times P)$. Then*

$$E[D_{s^+}\varphi(s)|\mathcal{F}_s] := \lim_{\epsilon \to 0^+} \frac{1}{\epsilon} \int_{s-\epsilon}^{s} E[D_s\varphi(t)|\mathcal{F}_s]dt$$

exists in $L^2(\lambda \times P)$ and

$$E[\int_0^T \varphi(s)d^-B(s)] = E[\int_0^T E[D_{s^+}\varphi(s)|\mathcal{F}_s]ds]. \tag{16}$$

Similar definitions and results can be obtained in the Poisson random measure case. See[10] and[8].

2.3. The Donsker delta function

As in[27], Chapter VI, we define the *regular conditional distribution* with respect to $\mathcal{F}_t$ of a given real random variable Y, denoted by $Q_t(dy) = Q_t(\omega, dy)$, by the following properties:

- For any Borel set $\Lambda \subseteq \mathbb{R}$, $Q_t(\cdot, \Lambda)$ is a version of $E[\chi_{Y \in dy}|\mathcal{F}_t]$
- For each fixed ω, $Q_t((\omega), dy)$ is a probability measure on the Borel subsets of $\mathbb{R}$

It is well-known that such a regular conditional distribution always exists. See e.g.[4], page 79.

From the required properties of $Q_t(\omega, dy)$ we get the following formula

$$\int_{\mathbb{R}} f(y)Q_t(\omega, dy) = E[f(Y)|\mathcal{F}_t]. \tag{17}$$

Comparing with the definition of the Donsker delta function, we obtain the following representation of the regular conditional distribution:

Proposition 2. *Suppose $Q_t(\omega, dy)$ is absolutely continuous with respect to Lebesgue measure on $\mathbb{R}$. Then*

$$\frac{Q_t(\omega, dy)}{dy} = E[\delta_Y(y)|\mathcal{F}_t]. \tag{18}$$

Explicit formulas for the Donsker delta function are known in many cases. For the Gaussian case, see Section 3.2. For details and more general cases, see[2,12,13,20–22] and[14]. See also Example 3 in Ch. 1.

3. The market model and the optimal portfolio problem for the insider

Suppose we have a market with the following two investment possibilities:

- A risk free investment (e.g. a bond or a (safe) bank account), whose unit price $S_0(t)$ at time t is described by

$$\begin{cases} dS_0(t) = r(t)S_0(t)dt; & 0 \le t \le T \\ S_0(0) = 1 \end{cases} \tag{19}$$

- A risky investment, whose unit price $S(t)$ at time t is described by a stochastic differential equation (SDE) of the form

$$\begin{cases} dS(t) = S(t)[b(t)dt + \sigma(t)dB(t)]; & 0 \le t \le T \\ S(0) > 0. \end{cases} \tag{20}$$

Here T is a fixed, given constant terminal time, $r(t) = r(t, \omega)$, $b(t) = b(t, \omega)$ and $\sigma(t) = \sigma(t, \omega)$ are given $\mathbb{F}$-adapted processes, and $B(t)$ is a Brownian motion on a filtered probability space $(\Omega, \mathbb{F} = \{\mathcal{F}_t\}_{t \ge 0}, P)$. We assume that $\sigma(t) > 0$ is bounded away from 0, and that

$$E[\int_0^T \{|b(t)| + |r(t)| + \sigma^2(t)\}dt] < \infty. \tag{21}$$

3.1. *The optimal portfolio problem*

We consider an optimal portfolio problem for a trader with inside information. Thus we assume that a filtration $\mathbb{H} = \{\mathcal{H}_t\}_{t \ge 0}$ is given, which is an *insider filtration*, in the sense that

$$\mathcal{F}_t \subseteq \mathcal{H}_t$$

for all t.

Suppose a trader in this market has the inside information represented by $\mathbb{H}$ to her disposal. Thus at any time t she is free to choose the *fraction* $\pi(t)$ of her current portfolio wealth $X(t) = X^\pi(t)$ to be invested in the risky asset, and this fraction is allowed to depend on $\mathcal{H}_t$, not just $\mathcal{F}_t$. If the portfolio is *self-financing* (which we assume), then the corresponding wealth process $X(t) = X^\pi(t)$ will satisfy the SDE

$$dX(t) = (1 - \pi(t))X(t)\rho(t)(t)dt + \pi(t)X(t^-)[\mu(t)dt + \sigma(t)dB^-(t)]. \tag{22}$$

For simplicity we put $X(0) = 1$. Since we do not assume that π is $\mathbb{F}$-adapted, the stochastic integrals in (22) are *anticipating*. Following the argument in[6] we choose to interpret the stochastic integrals as *forward integrals*, indicated by $dB^-(t)$.

By the Itô formula for forward integrals the solution of this SDE (22) is

$$X(t) = \exp[\int_0^t \{\rho(s) + [\mu(s) - \rho(s)]\pi(s) - \frac{1}{2}\sigma^2(s)\pi^2(s)\}ds$$

$$+ \int_0^t \pi(s)\sigma(s)d^-B(s)]. \tag{23}$$

Let $U : [0, \infty) \mapsto [-\infty, \infty)$ be a given *utility function*, i.e. a concave function on $[0, \infty)$, smooth on $(0, \infty)$, and let $\mathcal{A}_{\mathbb{H}}$ be a given family of $\mathbb{H}$-adapted portfolios. The *insider optimal portfolio problem* we consider, is the following:

PROBLEM Find $\pi^* \in \mathcal{A}_{\mathcal{H}}$ such that

$$sup_{\pi \in \mathcal{A}_{\mathcal{H}}} E[U(X_\pi(T))] = E[U(X_{\pi^*}(T))]. \tag{24}$$

In this paper we will restrict ourselves to consider the *logarithmic utility* U_0, defined by

$$U(x) = U_0(x) := \ln(x). \tag{25}$$

We will also assume that the inside filtration is of *initial enlargement* type, i.e.

$$\mathcal{H}_t = \mathcal{F}_t \vee Y \tag{26}$$

for all t, where Y is a given $\mathcal{F}_{T_0}$ -measurable random variable, for some $T_0 > T$.

Thus we assume that the trader at any time t knows all the value of a given $\mathcal{F}_{T_0}$ -measurable random variable Y, together with the values of the underlying noise process $B(s)$ for all $s \leq t$. Thus $\pi(t)$ is assumed to be measurable with respect to the σ-algebra $\mathcal{H}_t$ generated by Y and $B(s)$ for all $s \leq t$. In particular, the trader knows at time t the exact values of all the coefficients of the system and the values of the price processes at time t.

From (23) and (16) we get

$$E[\ln(X^\pi(T)] = E[\int_0^T \{r(s) + [b(s) - r(s)]\pi(s) - \frac{1}{2}\sigma^2(s)\pi^2(s)\}ds$$

$$+ \int_0^T \pi(s)\sigma(s)d^-B(s)]$$

$$= E[\int_0^T \{r(s) + [b(s) - r(s)]\pi(s) - \frac{1}{2}\sigma^2(s)\pi^2(s) + \sigma(s)D_s\pi(s)\}ds]$$

$$= E[\int_0^T E[r(s) + [b(s) - r(s)]\pi(s) - \frac{1}{2}\sigma^2(s)\pi^2(s) + \sigma(s)D_s\pi(s) \mid \mathcal{F}_s]ds].$$

$$(27)$$

Here and in the following we use the notation

$$D_s\pi(s) := D_{s+}\pi(s) := \lim_{t \to s^+} D_t\pi(s)$$

where as before D_t denotes the Hida-Malliavin derivative at t. Since π is assumed to be $\mathbb{H}$-adapted, it has the form

$$\pi(t, \omega) = f(t, Y, \omega) \qquad (28)$$

for some function $f : [0, T] \times \mathbb{R} \times \Omega \to \mathbb{R}$ such that $f(\cdot, y)$ is $\mathbb{F}$-adapted for each $y \in \mathbb{R}$.

Thus we can maximize (27) over all $\pi \in \mathcal{A}_{\mathbb{H}}$ by maximizing over all functions $f(t, Y)$ the integrand

$$J(f) := E[(b(s) - r(s))f(s, Y) - \frac{1}{2}\sigma^2(s)f^2(s, Y) + \sigma(s)D_sf(s, Y) \mid \mathcal{F}_s] \quad (29)$$

for each s. To this end, suppose the random variable Y has a Hida-Malliavin differentiable Donsker delta function $\delta_Y(y) \in (\mathcal{S})'$, so that

$$g(Y) = \int_{\mathbb{R}} g(y)\delta_Y(y)dy$$

for all functions g such that the integral converges. Then

$$f(s, Y) = \int_{\mathbb{R}} f(s, y)\delta_Y(y)dy, \qquad (30)$$

$$f^2(s, Y) = \int_{\mathbb{R}} f^2(s, y)\delta_Y(y)dy \qquad (31)$$

and

$$D_sf(s, Y) = \int_{\mathbb{R}} f(s, y)D_s\delta_Y(y)dy. \qquad (32)$$

Substituting this into (29) we get

$$
\begin{aligned}
J(f) &:= E[\int_{\mathbb{R}} \{(b(s) - r(s))f(s,y)\delta_Y(y) - \frac{1}{2}\sigma^2(s)f^2(s,Y)\delta_Y(y) \\
&\quad + \sigma(s)f(s,y)D_s\delta_Y(y)\}dy \mid \mathcal{F}_s] \\
&= \int_{\mathbb{R}} \{(b(s) - r(s))f(s,y)E[\delta_Y(y) \mid \mathcal{F}_s] - \frac{1}{2}\sigma^2(s)f^2(s,Y)E[\delta_Y(y) \mid \mathcal{F}_s] \\
&\quad + \sigma(s)f(s,y)E[D_s\delta_Y(y) \mid \mathcal{F}_s]\}dy \mid \mathcal{F}_s].
\end{aligned}
\tag{33}
$$

We can maximize this over $f(s,y)$ for each s, y. If we assume that

$$
0 < E[\delta_Y(y) \mid \mathcal{F}_s] \in L^2(\lambda \times P) \text{ and } E[D_s\delta_Y(y) \mid \mathcal{F}_s] \in L^2(\lambda \times P) \tag{34}
$$

for all s, y, then we see that the unique maximizing value of $f(s,y)$ is

$$
f^*(s,y) = \frac{b(s) - r(s)}{\sigma^2(s)} + \frac{E[D_s\delta_Y(y) \mid \mathcal{F}_s]}{\sigma(t)E[\delta_Y(y) \mid \mathcal{F}_s]}. \tag{35}
$$

We have proved the following, which extends a result in[28] (and is a special case of results in[14]):

Theorem 2. [Optimal insider portfolio]
Suppose Y has a Donsker delta function $\delta_Y(y)$ satisfying (34). Then the optimal insider portfolio is given by

$$
\pi^*(s) = \frac{b(s) - r(s)}{\sigma^2(s)} + \frac{E[D_s\delta_Y(y) \mid \mathcal{F}_s]_{y=Y}}{\sigma(t)E[\delta_Y(y) \mid \mathcal{F}_s]_{y=Y}}. \tag{36}
$$

Combining this result with the results of[28] and[6] given in the Introduction, we get the following result of independent interest. It is a special case of results in[15]:

Theorem 3. [Enlargement of filtration and semimartingale decomposition]
Suppose Y has a Donsker delta function $\delta_Y(y)$ satisfying (34). Then the $\mathbb{F}$-Brownian motion B is a semimartingale with respect to the inside filtration $\mathbb{H}$, and it has the semimartingale decomposition

$$
B(t) = \tilde{B}(t) + \int_0^t \alpha(s)ds, \tag{37}
$$

where $\tilde{B}(s)$ is an $\mathbb{H}$-Brownian motion, and $\alpha(s)$ (called the information drift) is given by

$$
\alpha(s) = \frac{E[D_s\delta_Y(y) \mid \mathcal{F}_s]_{y=Y}}{\sigma(t)E[\delta_Y(y) \mid \mathcal{F}_s]_{y=Y}}. \tag{38}
$$

This result is a special case of semimartingale decomposition results for Lévy processes in[15]. For information about enlargement of filtration in general, see[19] and[18] and the references therein.

3.2. Examples

Example 1. Consider the special case when Y is a Gaussian random variable of the form

$$Y = Y(T_0), \text{ where } Y(t) = \int_0^t \psi(s)dB(s); \text{ for } t \in [0, T_0] \qquad (39)$$

for some deterministic function $\psi \in L^2[0, T_0]$ with

$$\|\psi\|_{[t,T]}^2 := \int_t^T \psi(s)^2 ds > 0 \text{ for all } t \in [0, T].$$

In this case it is well known that the Donsker delta function exists in $(\mathcal{S})'$ and is given by

$$\delta_Y(y) = (2\pi v)^{-\frac{1}{2}} \exp^\diamond[-\frac{(Y-y)^{2\diamond}}{2v}] \qquad (40)$$

where we have put $v := \|\psi\|_{[0,T_0]}^2$. See e.g.[2], Proposition 3.2.
Using the Wick rule when taking conditional expectation, using the martingale property of the process $Y(t)$ and applying Lemma 3.7 in[2] we get

$$E[\delta_Y(y)|\mathcal{F}_t] = (2\pi v)^{-\frac{1}{2}} \exp^\diamond[-E[\frac{(Y(T_0)-y)^{2\diamond}}{2v}|\mathcal{F}_t]]$$

$$= (2\pi\|\psi\|_{[0,T_0]}^2)^{-\frac{1}{2}} \exp^\diamond[-\frac{(Y(t)-y)^{2\diamond}}{2\|\psi\|_{[0,T_0]}^2}]$$

$$= (2\pi\|\psi\|_{[t,T_0]}^2)^{-\frac{1}{2}} \exp[-\frac{(Y(t)-y)^2}{2\|\psi\|_{[t,T_0]}^2}]. \qquad (41)$$

Similarly, by the Wick chain rule and Lemma 3.8 in[2] we get

$$E[D_t\delta_Y(y)|\mathcal{F}_t] = -E[(2\pi v)^{-\frac{1}{2}} \exp^\diamond[-\frac{(Y(T_0)-y)^{2\diamond}}{2v}] \diamond \frac{Y(T_0)-y}{v}\psi(t)|\mathcal{F}_t]$$

$$= -(2\pi v)^{-\frac{1}{2}} \exp^\diamond[-\frac{(Y(t)-y)^{2\diamond}}{2v}] \diamond \frac{Y(t)-y}{v}\psi(t)$$

$$= -(2\pi\|\psi\|_{[t,T_0]}^2)^{-\frac{1}{2}} \exp[-\frac{(Y(t)-y)^2}{2\|\psi\|_{[t,T_0]}^2}]\frac{Y(t)-y}{\|\psi\|_{[t,T_0]}^2}\psi(t) \qquad (42)$$

Substituting (41) and (42) in (36) we obtain:

Corollary 1. *Suppose that Y is Gaussian of the form* (39). *Then the optimal insider portfolio is given by*

$$\pi^*(t) = \frac{b(t) - r(t)}{\sigma^2(t)} + \frac{(Y(T_0) - Y(t))\psi(t)}{\sigma(t)\|\psi\|_{[t,T_0]}^2}. \qquad (43)$$

202 *B. Øksendal and E. E. Røse*

In particular, if $Y = B(T_0)$ we get the following result, which was first proved in[28] (by a different method):

Corollary 2. *Suppose that* $Y = B(T_0)$. *Then the optimal insider portfolio is given by*

$$\pi^*(t) = \frac{b(t) - r(t)}{\sigma^2(t)} + \frac{B(T_0) - B(t)}{\sigma(t)(T_0 - t)}. \tag{44}$$

References

1. K. Aase, B. Øksendal, N. Privault and J. Ubøe: White noise generalizations of the Clark-Haussmann-Ocone theorem with application to mathematical finance. Finance Stoch. 4 (2000), 465-496.
2. K. Aase, B. Øksendal and J. Ubøe: Using the Donsker delta function to compute hedging strategies. Potential Analysis 14 (2001), 351-374.
3. N. Agram and B. Øksendal: Malliavin calculus and optimal control of stochastic Volterra equations. arXiv 1406.0325, June 2014.
4. L. Breiman: Probability. Addison-Wesley 1968.
5. O. E. Barndorff-Nielsen, F.E. Benth and B. Szozda: On stochastic integration for volatility modulated Brownian-driven Volterra processes via white noise analysis. arXiv:1303.4625v1, 19 March 2013.
6. F. Biagini and B. Øksendal: A general stochastic calculus approach to insider trading. Appl. Math. & Optim. 52 (2005), 167-181.
7. G. Di Nunno and B. ksendal: The Donsker delta function, a representation formula for functionals of a Lvy process and application to hedging in incomplete markets. Sminaires et Congrès, Societ Mathmatique de France, Vol. 16 (2007), 71-82.
8. G. Di Nunno, B. ksendal and F. Proske: Malliavin Calculus for Lvy Processes with Applications to Finance. Universitext, Springer 2009.
9. K.R. Dahl, S.-E. A. Mohammed, B. Øksendal and E. R. Røse; Optimal control with noisy memory and BSDEs with Malliavin derivatives. arXiv: 1403.4034 (2014).
10. G. Di Nunno, T. Meyer-Brandis, B. Øksendal and F. Proske: Malliavin calculus and anticipative Itô formulae for Lévy processes. Inf. Dim. Anal. Quantum Prob. Rel. Topics 8 (2005), 235-258.
11. G. Di Nunno, T. Meyer-Brandis, B. Øksendal and F. Proske: Optimal portfolio for an insider in a market driven by Lévy processes. Quant. Finance 6 (2006), 83-94.
12. G. Di Nunno and B. Øksendal: The Donsker delta function, a representation formula for functionals of a Lévy process and application to hedging in incomplete markets. Séminaires et Congrèes, Societé Mathématique de France, Vol. 16 (2007), 71-82.
13. G. Di Nunno and B. Øksendal: A representation theorem and a sensitivity result for functionals of jump diffusions. In A.B. Cruzeiro, H. Ouerdiane and

N. Obata (editors): Mathematical Analysis and Random Phenomena. World Scientific 2007, pp. 177 - 190.

14. O. Draouil and B. Øksendal: A Donsker delta functional approach to optimal insider control and applications to finance. Communications in Mathematics and Statistics (CIMS) 2015 (to appear). http://arxiv.org/abs/1504.02581.

15. O. Draouil and B. Øksendal: Optimal control and semimartingale decompositions. Preliminary version June 2015.

16. T. Hida, H.-H. Kuo, J. Potthoff and L. Streit: White Noise: An Infinite Dimensional Calculus. Springer 1993.

17. H. Holden, B. Øksendal,J. Ubøe and T. Zhang:Stochastic Partial Differential Equations. Universitext, Springer, Second Edition 2010.

18. M. Jeanblanc: Enlargements of Filtrations. In M. Jeanblanc, M.Yor and M. Chesney (editors): Mathematical Methods in Financial Markets, Springer 2010.

19. T. Jeulin and M. Yor (editors): Grossissements de filtrations, exemples et applications. Lecture Notes in Mathematics 1118, Springer 1985.

20. A. Lanconelli and F. Proske: On explicit strong solution of Itô-SDEs and the Donsker delta function of a diffusion. Inf. Dim. Anal. Quatum Prob Rel. Topics 7 (2004),437-447.

21. S. Mataramvura, B. Øksendal and F. Proske: The Donsker delta function of a Lévy process with application to chaos expansion of local time. Ann. Inst H. Poincaré Prob. Statist. 40 (2004), 553-567.

22. T. Meyer-Brandis and F. Proske: On the existence and explicit representability of strong solutions of Lévy noise driven SDEs with irregular coefficients. Commun. Math. Sci. 4 (2006), 129-154.

23. M. J. Oliveira: White Noise Analysis: An Introduction. This volume.

24. B. Øksendal and A. Sulem: Applied Stochastic Control of Jump Diffusions. Second Edition. Springer 2007

25. B. Øksendal and A. Sulem: Risk minimization in financial markets modeled by Itô-Lévy processes. Afrika Matematika (2014), DOI: 10.1007/s13370-014-02489-9.

26. I. Pikovsky and I. Karatzas: Anticipative portfolio optimization. Adv. Appl. Probab. 28 (1996), 1095-1122.

27. P. Protter: Stochastic Integration and Differential Equations. Second Edition. Springer 2005

28. I. Pikovsky and I. Karatzas: Anticipative portfolio optimization. Adv. Appl. Probab. 28 (1996), 1095-1122.

29. E. Røse: White noise extensions of Hida-Malliavin calculus. Manuscript University of Oslo, 5 February 2015

30. F. Russo and P. Vallois: Forward, backward and symmetric stochastic integration. Probab. Theor. Rel. Fields 93 (1993), 403-421.

31. F. Russo and P. Vallois. The generalized covariation process and Itô formula. Stoch. Proc. Appl., 59(4):81-104, 1995.

32. F. Russo and P. Vallois. Stochastic calculus with respect to continuous finite quadratic variation processes. Stoch. Stoch. Rep., 70(4):1-40, 2000.

Chapter 7

Outlook of White Noise Theory

Takeyuki Hida

Professor Emeritus at Nagoya University and at Meijo University

1. Introduction

This section is devoted to making a short visit to the original basic idea of white noise theory and proposals for some of future directions.

The white noise theory has extensively developed by the great contributions by many authors, although it has short history about four decades or so. We are however very proud of the beautiful dazzling results on this theory.

At present, it seems, however, fitting to revisit the original idea and for exploring new directions; some of them will be in line with the original idea and others are now coming from the fruitful interactions with other fields of science.

2. Revisiting the original ideas

What we have proposed is as follows: In order to analyze a random complex system which is represented by a family of random variables, we first find a system of *independent* random variables containing the same information as the given random system, then express the given random variables as functions of the independent random variables just constructed. We shall be ready to analyze those functions to investigate the given random system.

There is a road map:

Random system $\rightarrow$ Reduction (form independent random system) $\rightarrow$ Functions of independent variables $\rightarrow$ Analysis $\rightarrow$ Applications.

The results of the analysis would tell us the structure of the random system in question then follow controls, forecasting, applications and so on.

There are two characteristics; one is the Reductionism which means that the given random system should be expressed in terms of independent random variables, maybe i.i.d. each of which is atomic (or elemental), and the other we shall deal with the independent system consisting of continuously many variables, namely **analogue** system. Such a system is simply called a **noise**. The other characteristic is that for our analysis we deal with **generalized** functionals of a noise. The class of them is quite fitting for our purpose.

3. Three kinds of a noise

Now one may ask what kind of the noise can be available. We claim that the important, significant, and standard cases are discussed as follows.

1. The system, as was discussed, should be analogue and naturally those random variables are idealized (or generalized) random variables. We wish to explain why we emphasize the analogue case. Good applications depend on a parameter either time or space, which are continuum. We wish to associate an independent random variable with each point of the parameter set.

It may be claimed that we can approximate such a system by taking countably many variables. Unfortunately, in such a case, it is impossible to describe the propagation of random phenomenon as the parameter develops continuously. The same for the case where random event propagate on a surface or space.

2. The next question is how to obtain such a system, It may be formed by a measure theoretical way, that is the direct product of continuously many measure spaces, but the measure space cannot be separable, so that ordinary calculus cannot be proceeded. Separability is just a natural requirement, so that we have to pay a price in some way. To this end, we can propose an idea to overcome this difficulty, that is, the use of a so-to-speak, *underlying process*. Indeed, it is an additive process in the time parameter case. The time-derivative defines a system of independent variables. The derivative is no more an ordinary random variable, but an **idealized** random variable. Such a property can be quite acceptable.

We have another requirement on noise in order to realize our purpose called reductionism. Each derivative must be atomic, in a sense it cannot be decomposed into two non-trivial independent random variables without adding new information in addition to that contained in the given system of random variables.

There is a remark. We claim that we prefer noises depending on a continuous parameter. One might say the case in question may be approximated by independent variables indexed by discrete parameter. This is possible in some sense. But essential cases, like the case where the time causality is required, are impossible to be approximated by series with discrete parameter.

Parameters are usually supposed to be time or space or even general linearly ordered set. It is important to investigate how the given random phenomena would propagate as the parameter moves. That is the discussion related to the so-called *causality*. [Note] The following fact may be worth mentioning. In Gaussian case, if we use the approximation by the Lévy's method (see his 1948 book), this is taken into account. While, the method of using a base, say ξ_n in the Hilbert space $L^2(R^1)$ is hopeless so far as the causality is concerned. The system of smeared variables like $< \dot{B}, \xi_n >$'s will fail to describe the propagation of random phenomena in question as time goes by.

3. Classification of standard noises. They are parametrized, say by R^1 or its subinterval. Two cases are possible:

i) depending on the time, or

ii) depending on the space variable.

Probability distribution of a noise.

a) In the case i) above, there are two possibilities; namely Gaussian and Poisson type.

b) For the case ii) above, only Poisson type is possible.

The proofs and other details of what have been described above are omitted here, since they can be seen in many literatures on white noise analysis, and even we rush to state many further developments.

4. Again, we wish to emphasize the significance of the facts that the noise we have chosen involves continuously many independent random

variables that come from an additive process. Thus, the analysis of their functions (and functionals) is very much different from those with functions of countable number of variables. One of the differences is nonlinear functions, like polynomials, need renormalization that will be illustrated later. The other is the definition of the differential operators in those variables. Anyhow it comes from the infinitesimal difference of one variable; in our case, the difference depends on the underlying additive process, so that the definition of the differential operator used the Fréchet derivative, not the Gâteaux derivative. We have to remind this fact when we discuss differential calculus of functionals of the noise.

The third one to be mentioned for the analogue case is the existence of the **multiplicity**. *Roughly speaking as an analogue in the Hilbert space, the multiplicity of a one-parameter family of vectors is the minimal number of cyclic subspaces that generate the entire space. Some more details will be shown in the concrete examples.*

4. Functionals of a noise

We shall discuss functions, actually functionals of a noise. To fix the idea, we shall take a Gaussian noise, that is, the **white noise**, which is realized by taking the time derivatives of a Brownian motion $B(t)$. We therefore take $\dot{B}(t)$ as a basic variable, and we often take its sample function $x(t)$, where x is a member of E^* which is a space of generalized functions. Indeed, the probability distribution of the $\{\dot{B}(t), t \in R^1\}$ is supported by E^* which is the dual space of some nuclear space E dense in $L^2(R^1)$.

Formally writing a functional is expressed as $\varphi(\dot{B})$ or by $\varphi(x)$. Since the variable is no more ordinary random variables, but idealized variables, we have to introduce actual method on how to define a polynomial: one comes from intuitive method starting from a space of polynomials in the $\dot{B}(t)$'s. The latter is defined by analogy with the definition of (non-random) Schwartz distributions. Here we explain the first method.

As in the classical non-random calculus, we start with polynomials which is an extension of the primitive form

$$\sum a_{\mathbf{j}} : \dot{B}(t_1)^{p_1} : \cdots : \dot{B}(t_n)^{t_n} :, \ \mathbf{j} = 1, 2, \cdots, n$$

where : · : denotes the renormalization, which makes any polynomial in $\dot{B}(t)$'s a generalized white noise functional.

Now let such a primitive polynomial extend to general form of a polynomial expressed in the form

$$\varphi(\dot{B}) = \int \cdots \int F(t_1, \cdots, t_n) : \dot{B}(t_1)^{p_1} : \cdots : \dot{B}(t_n)^{t_n} : dt^n,$$

where F is a generalized function which is a member of the Sobolev space of order $-(\sum p_j + 1)/2$ over $R^{\sum p_j}$. Thus defined functions of $\dot{B}(t)$'s form the space $(L^2)^-$ of **generalized white noise functionals**.

There is another way to define generalized functionals of white noise. It is a random and infinite dimensional extension of the Schwartz distribution (generalized function). The white noise case given by the Gel'fant triple

$$(S) \subset (L^2) \subset (S)^*.$$

There is the Potthoff-Streit characterization of the $(S)^*$-functional. This is a big advantage to take $(S)^*$.

The S-transform: There is a transformation of a generalized white noise functionals to have their representations. It is defined by

$$(S\varphi)(\xi) = e^{-\|\xi\|^2} \int e^{<\dot{B},\xi>} \varphi(\dot{B})\mu(\dot{B}), \xi \in E.$$

The S-transform gives us a good expression of generalized white noise functionals. For example the above polynomial is transformed to be

$$\int \cdots \int F(t_1, \cdots, t_n)\xi(t_1)^{p_1} \cdots \xi(t_n)^{t_n} dt^n.$$

Such a representation is good to apply differential operators and some others, so that we have been quite happy. Later, we will know that the transformation is just coming from the inner product of $\varphi(\dot{B})$ and the generalization of the generating function of the Hermite polynomials. We now understand the reason why everything goes well by using the S-transform.

Differential calculus

1) The idea of the white noise derivative has been given before. Having applied the S-transform we apply the Fréchet derivative not Gâteaux derivative. Namely, for $\varphi(x)$ (x is a path of $\dot{B}$) we apply the S-transform to have $U(\xi) = (S\varphi)(\xi)$ and have the Fréchet derivative $U'(\xi, t)$ in the sense that

$$\Delta U(\xi) = \delta U(\xi, \delta\xi) + o(\delta\xi)$$

and

$$\delta U(\xi, \delta\xi) = \int U'(\xi, t)\delta\xi(t)dt.$$

If $S^{-1}U'(\xi, t)$ evaluated at t exists and is a generalized white noise functional, then $\varphi(x)$ is said to be $\dot{B}(t)$-differentiable and $S^{-1}U'(\xi, t)$ is denoted by $\partial_t\varphi(x)$ or by $\frac{\partial}{\partial\dot{B}(t)}\varphi(\dot{B}(t))$. The ∂_t is the **differential operator** and its adjoint operator $\partial = t^*$ is the **creation** operator.

Note: Behind our definition of the differential operator is a reason that we use a Brownian motion $B(t)$ that is continuous in t.

The Lévy Laplacian Δ_L is defined by

$$\Delta_L = \int \partial_t^2 (dt)^2.$$

The Lévy Laplacian and the infinite dimensional rotation group play dominant roles in white noise analysis. We omit the details in this chapter.

5. Some further developments to be proposed

This is indeed the most important part of this chapter.

Although we understand that it is extremely difficult to propose some future directions, we think that it is worth mentioning something about unsolved problems as well as future directions that would hopefully be discussed within white noise theory.

1) Some more calculus using the operators ∂_t's and ∂_t^*'s shall be done in the standard line of the analysis. We shall introduce operator algebras and discuss their applications, where we meet continuously many degree of freedom. Note that the theory cannot be done straightforwardly, but it is interesting to find good applications in quantum field theory and others.

2) Non-commutative calculus. We have been suggested by the Hamiltonian path integrals.

We can see in Chapter 3 Feynman Path integrals by W. Bock where the Hamiltonian path integrals are discussed in line with the white noise analysis. There we can see how naturally the Hamiltonian path integrals appear in quantum dynamics, where the operators are well describing the state and momentum which are not commutative.

Incidentally, we remind that one of the original aims to have started white noise analysis is Feynman's path integrals. We are happy to see that the works for the path integrals are directing us to an ideal situation.

We are also familiar with the quantum probability theory uses non-commutative operations which play the basic part of the theory. L. Accardi has proposed unification of white noise analysis and quantum probability theory at the international conference "Probability theory towards 2000" in October, 1995 at Columbia University in New York under the title "Probability towards 2000". We have been keen to study in this direction and obtained some by-products. We are still far from the goal.

There have been other approaches in this direction with different ideas and some are satisfactory.

3) The multiplicity theory. This theory has a long history even though restricted to Hilbert space theory and the theory of stochastic processes, in particular applied to the Gaussian process theory. There we may understand that the multiplicity is one of the numerical values that express a dependence of classes of stochastic processes. Some years ago, we (with Si Si) have revisited this notion from a new viewpoint. In fact, it is better to say that we have come when we recognized the roles of the new noise depending on the space.

Stable processes have appeared in socioinfornomics where we need further understandings of the decomposition of a Lévy process. One of the interpretations has come from the step of considerations on the background to have double multiplicity.

4) Stochastic differential equations.

If a stochastic differential equation is expressed in the form by using a linear operator

$$LX(t) = c\dot{B}(t),$$

then, we may come to a new viewpoint that L is a **whitening operator** acting on $X(t)$, In other words, the operator L tells us how to form a white noise. This expression is in agreement with our idea of reductionism. Some modifications are necessary when we meet the cases where the $\dot{B}(t)$ is replaced by a noise with space-time parameter, or c is random where $\dot{B}(t)$ is replaced by the creation operator so that there is no need to have renormalization. Thus, we have many examples where whitening idea is efficiently used.

5) Multi-dimensional noise.

We should note very important examples where a space-time noise plays dominant roles. For example, there is a well known equation called the Kardar-Parisi-Zhang (KPZ) equation expressed in the form

$$\frac{\partial}{\partial t}h = \frac{1}{2}\lambda(\frac{\partial}{\partial x}h)^2 + \nu\frac{\partial^2}{\partial_x^2}h + \sqrt{D}\eta,$$

where $h = h(x, t)$ is a height profile of some random matter at time $t \geq 0$ and at place $x \in R^1$. The η in the last term stands for a space-time Gaussian noise, and the constant λ is the strength of the nonlinearity. With a suitable initial condition, this equation actually describes surface growth.

There have been many approaches to this equation, for example by T. Sasamoto and H. Spohn and in particular a beautiful results by M. Hairer (2013) "Solving KPZ equation".

We are, however, interested in an investigation by using the white noise analysis. Note that the noise is now two-dimensional, so that we have to prepare the new space of generalized functionals on R^2 and to proceed the analysis as required.

It is known that there is the Cole-Hopf transformation of h. Set

$$Z(x, t) = \exp[(\lambda/2\nu)h(x, t)].$$

Then, the $Z(x, t)$ satisfies

$$\frac{\partial}{\partial t}Z = \nu\frac{\partial^2}{\partial_x^2}Z + (\lambda\sqrt{D}/2\nu)\eta \cdot Z.$$

How about applying the S-transform to have a good application of the white noise theory? The equation formed above can be projected down to each subspace spanned by polynomials in noise with fixed degree, where the multiplication in the last term may be considered suitably, maybe the multiplication by η is understood as the creation operator.

We hoped that such an approach is successful.

6) Polynomials in the noises.

We often meet a mathematical concept in probability theory called "multiple Wiener integral". We may understand them as a polynomial in $\dot{B}(t)$'s. We have so far taken this viewpoint, but now we revisit after having studied some examples.

The expression is like an integral, so we may say integral, however it is better to be viewed as a reasonable generalization (to continuously many variables) of a polynomial like $\sum_{t_j \, different} a(t_1, \cdots, t_n) \dot{B}(t_1) \cdots \dot{B}(t_n)$, assuming that the t_j's are different. In fact, it is, as it were, a multi-linear form, where the renormalization of the form is unnecessary. It cannot be like definite integral or even not indefinite integral in elementary analysis.

Having expressed in a somewhat general form

$$\int \cdots \int F(t_1, \cdots, t_n) : \dot{B}(t_1)^{p_1} \cdots \dot{B}(t_n)^{p_n} : dt^n$$

$(: \cdots :$ denotes the renormalization) we recognize that each $\dot{B}(t_j)$ in the summand does survive even under the inegral sign. Hence, we can differentiate it by the variable $\dot{B}(t_j)$. Such a property can never be seen in the usual integral in the ordinary (non-random) calculus.

One more matter to be mentioned is that the degree (like in elementary calculus on polynomials) is associated with such a monomial. We can, therefore, proceed to consider their algebraic structures arising from those polynomials. Certainly to the analysis, as well.

7) More harmonic analysis. The infinite dimensional rotation group characterizes the white noise measure. Our systematic approach to the white noise analysis is based on that measure space. Thus, a harmonic analysis has naturally arisen from the rotation group. One may expect that the group in question will be an inductive limit of a system of finite dimensional rotation groups G_n isomorphic to $SO(n)$. In reality, the limit does occupy only a very small part of our rotation group $O(E)$. Indeed, such a limit may characterize the part that can be approximated by finite dimensional calculus. Those members are said to be in Class I.

Essential part of the present rotation group $O(E)$ consists of those linear transformations that act on infinite dimensional function space with particular topological structure. In other words, those transformations change the type of a member ξ in E. Namely, they change the variable of a function ξ; namely they come from, as it were, the internal structure of the test functions ξ's in E, while the finite dimensional case uses simply a system of mutually orthogonal vectors. With such a profound remark in mind, we come to the class II which defines, as it were, essentially continuously many dimensional subgroup.

We may focus our attention on the subgroups in the class II. A member in this class is defined in such a way that

$$g; \ (g\xi)(u) = \xi(\psi(u))\sqrt{|\psi'(u)|}$$

with a suitable choice of a smooth function ψ. A particular interest can be found in a one-parameter subgroup $\{g_t, t \in R^1\}$, often called a "whisker" of $O(E)$, involving such transformations. We have so far obtained some whiskers that play their own significant roles, like the special conformal group or the affine group, however we are sure that there will be more interesting whiskers; thus we are given various research problems worth to be investigated.

8) Other applications to wider fields of sciences be well explored. There is an example: Socioinfornomics: where we can see their approaches to random structures of non-visualized phenomena.

Revisiting Heisenberg uncertainty principle where we see that some randomness is involved.

We are interested in everything that is represented as functions of a noise on a probability space that we can actually construct. It is noted that they always suggest us to be back to the original idea on how to deal with complex phenomena that can be formulated as a random system mathematically. There we have valuable problems to be studied.

Addenda. So far we have discussed topics related only to Gaussian noise (white noise). Similar significant properties can be found in Poisson noise. For example, two dimensional parameter noise has appeared in the KPZ equation. Similar fact may be considered in Poisson noise case. It is known that a Poisson process has very profound, complex structure unlike its sample functions. For instance, a Poisson process has two parameters; time $t \in R^1$ and intensity (space parameter) $\lambda > 0$. It is better to write $P(t, \lambda)$. We can form it so as to be an additive process in t and in λ. The density will be denoted by $p(t, \lambda)$ that leads us to interesting stochastic calculus.

Subject Index